Genetics and Plant Breeding

Genetics and Plant Breeding

(SECOND REVISED EDITION)

L.D. Vijendra Das

Ph.D. (Australia), Post-doctorate (Germany)
Agro-biology (Israel)
Professor of Genetics and Plant Breeding
Centre for Plant Breeding and Genetics
Tamil Nadu Agricultural University, Tamil Nadu

PUBLISHING FOR ONE WORLD

NEW AGE INTERNATIONAL (P) LIMITED, PUBLISHERS

New Delhi • Bangalore • Chennai • Cochin • Guwahati
Hyderabad • Kolkata • Lucknow • Mumbai
Visit us at www.newagepublishers.com

Published by New Age International (P) Ltd., Publishers
First Edition: 1998
Second Revised Edition: 2005
Reprint: 2017

GLOBAL OFFICES

- **New Delhi** **NEW AGE INTERNATIONAL (P) LIMITED, PUBLISHERS**
  7/30 A, Daryaganj, New Delhi-110002, (INDIA)
  Tel.: (011) 23253771, 23253472, **Telefax:** 23267437, 43551305
  E-mail: contactus@newagepublishers.com • Visit us at www.newagepublishers.com
- **London** **NEW AGE INTERNATIONAL (UK) LTD.**
  27 Old Gloucester Street, London, WC1N 3AX, UK
  E-mail: info@newacademicscience.co.uk • Visit us at www.newacademicscience.co.uk
- **Nairobi** **NEW AGE GOLDEN (EAST AFRICA) LTD.**
  Ground Floor, Westlands Arcade, Chiromo Road (Next to Naivas Supermarket)
  Westlands, Nairobi, KENYA, **Tel.:** 00-254-713848772, 00-254-725700286
  E-mail: kenya@newagepublishers.com

BRANCHES

- **Bangalore** 37/10, 8th Cross (Near Hanuman Temple), Azad Nagar, Chamarajpet, Bangalore- 560 018
  Tel.: (080) 26756823, **Telefax:** 26756820, **E-mail: bangalore@newagepublishers.com**
- **Chennai** 26, Damodaran Street, T. Nagar, Chennai-600 017, **Tel.:** (044) 24353401, **Telefax:** 24351463
  E-mail: chennai@newagepublishers.com
- **Cochin** CC-39/1016, Carrier Station Road, Ernakulam South, Cochin-682 016, **Tel.:** (0484) 2377303, **Telefax:** 4051304
  E-mail: cochin@newagepublishers.com
- **Guwahati** Hemsen Complex, Mohd. Shah Road, Paltan Bazar, Near Starline Hotel, Guwahati-781 008,
  Tel.: (0361) 2513881, **Telefax:** 2543669, **E-mail: guwahati@newagepublishers.com**
- **Hyderabad** 105, 1st Floor, Madhiray Kaveri Tower, 3-2-19, Azam Jahi Road, Near Kumar Theater, Nimboliadda Kachiguda, Hyderabad-500 027, **Tel.:** (040) 24652456, **Telefax:** 24652457
  E-mail: hyderabad@newagepublishers.com
- **Kolkata** RDB Chambers (Formerly Lotus Cinema) 106A, 1st Floor, S N Banerjee Road, Kolkata-700 014
  Tel.: (033) 22273773, **Telefax:** 22275247, **E-mail: kolkata@newagepublishers.com**
- **Lucknow** 16-A, Jopling Road, Lucknow-226 001, **Tel.:** (0522) 2209578, 4045297, **Telefax:** 2204098
  E-mail: lucknow@newagepublishers.com
- **Mumbai** 142C, Victor House, Ground Floor, N.M. Joshi Marg, Lower Parel, Mumbai-400 013
  Tel.: (022) 24927869, **Telefax:** 24915415, **E-mail: mumbai@newagepublishers.com**
- **New Delhi** 22, Golden House, Daryaganj, New Delhi-110 002, **Tel.:** (011) 23262368, 23262370, **Telefax:** 43551305
  E-mail: sales@newagepublishers.com

ISBN: 978-81-224-1575-9
C-17-02-10147

Printed in India at Rashtriya Printers, Delhi.
Typeset at Goswami Associates, Delhi.

NEW AGE INTERNATIONAL (P) LIMITED, PUBLISHERS
7/30 A, Daryaganj, New Delhi-110002
Visit us at **www.newagepublishers.com**
(CIN: U74899DL1966PTC004618)

"In memory of late Dr. N. Thurling, my guide and Plant Breeder, Institute of Agriculture, The University of Western Australia, Nedlands, W.A., who dedicated his life for the cause of Plant Breeding."

FOREWORD

I have great pleasure to recommend this book entitled **"GENETICS AND PLANT BREEDING"** authored by Dr. L.D. Vijendra Das, Former Director of Publications, Tamil Nadu Agricultural University, Coimbatore and also Professor of Plant Breeding and Genetics. The book covers a vast area from Mendel and his discoveries and to the latest technology of RFLP in Plant Breeding, further the author has added **"Glossary"** in this 2nd edition which will be useful for those who appear for ICAR examinations.

I congratulate the author for his strenuous effort to write and publish this book incorporating UGC syllabus also.

C. RAMASAMY
Vice-Chancellor
Tamil Nadu Agricultural
University

Coimbatore–641 003

PREFACE TO THE SECOND EDITION

The first edition of the book "Plant Breeding" was published during 1998. The book was very useful among the students of botany, agricultural botany and bio-technology throughout India. Further the students felt that if the book covers University Grants Commission (UGC) syllabus also it will be much useful for them. Hence great care was taken to peruse the UGC syllabus and the book was revised as "Genetics and Plant Breeding".

The science of "genetics" is based on the fact that the hereditary material of an organism, be it virus, bacterium or mammal is ordered into one or more chromosomes. By means of wide variety of physical, chemical and biological technique, it has been possible to examine the structure and function of chromosomes and correlate with genetic functions and phenotypic inheritance and distribution. Such an ordered arrangement of heritable material possess greater degrees of complexity of structure and behaviour were introduced into the biological world. The Universal occurrence of chromosomes suggests that they made their appearance very early in the history of life. The adaptive value of chromosomes can probably be stated in a variety of terms: precision of replication, coupled with sufficient built–in copy–error or mutation to provide for a supply of diversity through segregation and recombination; efficient informational packaging of the genetic code and its decoding through the process of transcription and translation and selective control of gene activity. Genetics focuses its major attention on the chromosome. It is concerned with its chemical and genetic organization and with the processes through which this organization determines the transmission of heredity characteristics from one generation to the next. Genetics is also concerned with both the structure and processes which deal with the conservative aspects of heredity and introduction of diversity into the systems of inheritance.

The Darwinian concept of evolution through natural selection postulated that some of the variations observed among individuals in a population were heritable. He visualized each of these variations as having only a slight effect on an individual, or on a population, but a significant effect when viewed cumulatively through time. Being heritable and potentially at least, having an influence on the survival and reproductive capabilities of the individual, these variations were acted upon by the forces of natural selection. The Mendelian laws of inheritance, formulated in 1865, still form the basis of our understanding of the transmission of heritable variation and the inheritance is still the basic technique for following phenotypic characters through several generations.

Mendel's genetic discoveries were ignored or given no credence, from 1865 to 1900, when several scientists recognized that their breeding result had already been anticipated and previously published. By this time, however, the scientific world was in a more receptive mood and accepted Mendel's principles. The egg and sperm of animals despite obvious differences in size and shape have physical continuity between generations. Fertilization was shown to be the union of gametes, with the fusion of parental nuclei in the cytoplasm of the egg being the crucial event, and with each nucleus providing equal, or nearly equal, numbers of chromosomes to the newly formed zygote and hence to the developing embryo. As the embryo grew by adding to its cell numbers, the critical acts of cell division were the longitudinal replication of each chromosome, and the segregation of these longitudinal halves to the two daughter cells thus providing both a physical basis for the qualitative and quantitative chromosomal equality of each daughter nucleus and a functional basis for the conservation of the genotype. It was becoming more and more obvious that the pattern of inheritance were linked to, or mirrored in, the behaviour of chromosomes.

In retrospect, therefore, the discoveries made during the last quarter century of the 1900s and the first few years of the 20th century made it increasingly clear that the chromosomes were key elements whose behaviour in division and fertilization for the transmission of heredity factors from one generation to another. Sutton visualized the chromosomes as the physical carriers of Mendelian factors and the segregation of a pair of homologous chromosomes and the independent assortment of non-homologous chromosomes as the physical basis for the qualitative and quantitative aspects of Mendelian segregation.

This second edition provides detailed information on genetics covering UGC syllabus and the students will be certainly benefited. This book also provides "Laboratory exercises" in the last chapter, so that the students will be familiarized with cytological and breeding techniques.

Further a full list of "Glossary" has been added at the end of the book so that the students can refer quickly the definitions and will be useful for examinations.

Dr. L.D. VIJENDRADAS
Ph. D., (Australia)
Post-doctorate (Germany)
Agro-biology (Israel)
Centre for Plant Breeding and Genetics
Tamil Nadu Agricultural University
Coimbatore–641 003
Tamil Nadu, India

PREFACE TO THE FIRST EDITION

Adaptation of plants to extreme environments can best be defined by applying Levitt's terminology of 1980 which provides a coherent frame of reference for the analysis of the response of plants to environmental stress. Extreme environments include those environments that potentially can cause stress to the organism exposed to them. Any environmental factor capable of eliciting from a plant a harmful chemical or causing a physical strain (change), which may be either reversible (elastic) or permanent (plastic) will result in poor yield.

In the year 1980, Sinmena and Ziv putforth an agronomic definition for plant response as these including individual physiological processes and integrated plant management leading to economic yield. Accordingly, an environmental factor capable of reducing the yield below the potential level, for a given set of conditions will have to be looked into from the breeding point of view.

Sources of Environmental problems for plants

Physical	*Chemical*	*Biotic*
Drought	Air pollution	Competition
Temperature	(organic) chemicals	Allelopathy
Radiation	Nutrients (inorganic)	Lack of symbiosis
Flooding	Pesticides	Human activities
Mechanical	Toxins	Diseases
Electrical	Salts	Insects
Magnetic	pH of soil	
Wind		

Of the world's available land area (14 billion ha), only around 3.2 billion hectare arable or potentially arable with minimal environmental modification. Approximately 25 percent of this arable land is subject to excess salinity, a problem both in arid regions and in some areas near the ocean. Another 25 percent of arable land suffers from excess rainfall, and therefore, subject to soil acidity. Rain water contains dissolved carbon dioxide in the form of carbonic acid. As hydrogen ion concentration build up in soils, other more useful cations are leached from the soil particles; in addition, normally insoluble salts, such as aluminium, come into the soil solution.

Of the 1.4 billion hectare currently cultivated, 14 percent irrigated and produce half of the world's food. Prospects for markedly increasing the extent of irrigation are not bright, since 80 per cent of the world's readily available fresh water is already used in agriculture. Also all irrigation systems, no matter how carefully managed, gradually add dissolved salts to the soil, thereby limiting their useful lifetimes.

Energy intensive modern agriculture utilizes considerable economic resources needed either to further increase agricultural production or to augment other sectors of national economics. It is currently estimated that energy inputs per hectare of maize production are equivalent to 757 litres gasoline. Agriculture can be viewed either as an energy-intensive enterprise designed to modify the environment to suit the plant, or as an exercise in genetics that modifies the plant to suit available environments.

The study of genetic phenomena and the related resistance mechanisms in plant is important because large areas world wide cannot be utilized for food production because of the limitations imposed by natural or man-made environmental conditions. The increase of population pressure requires maximizing the efficiency of the use of cultivated land and increasing the utilization of marginal soils. Because of the economic impact of stresses and the large amount of energy required to alter the environment to suit the plant, it is becoming increasingly important to utilize the existing technologies and to develop new ones in order to develop plants that are better adapted to changing environments.

BIOLOGY DIVERSITY AND CONSERVATION

India has a geographic area of 329 million hectares. It is bound on the north by the youngest, the most fragile and the highest mountain systems (the Himalayas) in the world. Climate in the country ranges from perpetual snow cover to near-equatorial conditions, from mangroves to humid tropics and to hot and cold deserts. The 45,000 species of plants constitute roughly 12 per cent of the global plant wealth. There are about 15,000 species of flowering plants and 33 per cent of them are endemic located in 26 endemic centres.

The crop plant genetic resources can more or less be assigned to specific centres of diversity, identified by a Russian agrobotanist, Vavilov, in the thirties on the basis of varietal diversity, homologous variation, endemism, dominant allele frequencies and disease resistance. The centres, in different continents, are referred to as germplasm treasuries. It is ironical that these areas, where the ancestors of the crops were first domesticated, are regions of low agricultural productivity and constitute the "hunger belt".

Of the 2503 cultivated plant species, 320 (12.8 percent) have come from the Indo-Chinese-Indonesian region; 297 (11.9 per cent) from the Chinese-Japan region; 292 species (11.9 per cent) each from the African and the South American regions. The Indian region is placed seventh with 167 species (6.7 per cent). The contribution includes important crops such as rice, sugarcane, coir, minor millets, brassicas, Asiatic vignas, cucumber, eggplant, banana, citrus, mango, jackfruit, cardamom, tree cotton, jute, edible dioscorea, vegetable amaranthus, alocasia, amorpho-phalus, colocasia, black pepper, turmeric, ginger, several cucurbits and umbellifers, rauvolfia and several herbal drugs, rhododendrons, jasmines, some bamboos, some ornamental orchids, betel-vine and betelnut. This region is also a secondary centre of diversity for grain amaranthus, maize, red pepper, soyabean, potatoes and rubber plant.

These are several examples of plant germplasm from India making significant contributions to plant improvements. Two cases are distinctive. Nearly 20 cultivar of rice contain useful genes from wild rices of Kerala. The genes are responsible for resistance in cultivated rices. Equally important is the contribution to the cantaloup (musk melon) industry (14,000 ha) of California by powdery mildew-resistant genes from the wild musk melon of India. In economic terms, the turnover of these crops would run to billions of dollars. Similarly the hardy livestock breeds are in constant demand abroad.

MINERAL DEFICIENCY AND TOXICITY

Mineral deficiencies and toxicities are frequent problems in strongly acidic or strongly alkaline soils. Aluminium toxicity at low pH is one of the best known examples, and a genetic basis for intraspecific variation in tolerance to high aluminium has been well established in wheat and in sorghum amongst others. Selection in culture is technically difficult, since other minerals become unavailable or toxic at pH extremes. In addition, the known mechanism for tolerance appears to reside in the root, so a whole-plant screen may be the best approach. An effective, rapid, non-destructive seedling test for response to aluminium had been developed as early as 1978 by Poole, Konzak and Kittrick. Despite problems of selection in culture, aluminium-resistant cultures have been selected in tomato tissue culture in the early studies during 1978 by Meredith, put an epigenetic basis for the resistance could not be discounted.

USE OF GENETIC ENGINEERING

In the last decade tissue culture techniques have been recognised as a powerful tool for breeding work. One approach to genetic modification by using tissue culture is the selection of mutants. Another possible approach to genetic modification by using tissue culture may be the production of somatic hybrids by the fusion of naked protoplasts derived from plant species that cannot hybridize sexually. One preliminary accomplishment was reported by Smith as early as 1979. It was suggested that somatic hybridization may be used for transferring chilling resistance from potato (*Solanum tuberosum* L.) to tomato (Lycopersicon esculentum Mill.), a chilling-sensitive species.

Recent advances in biotechnology used to explore and manipulate the genome of plants hold promise for increasing our understanding of mechanisms regulating stress responses. One method utilizes the bacterium—*Agrobacterium tumefaciens,* which causes crown gall on many dicotyledenous plants. It is indeed ironic that a disease producing organism is being used as a primary tool to understand both biotic and abiotic stress inducing mechanisms.

The infection of a plant with agrobacterium results in the formation of gall-like swellings at the transmission zone between the root and shoot. When the bacterium infects the tissues, circular segments of DNA called plasmids which encompass several genes are inserted into the host cell DNA. The bacterium alters the genome of the plant and reprogrammes its metabolism. The result is the synthesis of nutrients used by the bacterium for growth and the synthesis of hormones used by the plant causing tumor development at the site of the infection.

It is now possible by using the techniques of genetic engineering to remove plasmids from agrobacterium, modify them by inserting new DNA obtained from another organism, and transfer the inserted code for "desirable traits to a target plant. To accomplish the transfers the new plasmid is inserted into an agrobacterium cell which is cultured in a manner that produces thousands of cells containing the new genetic information. A host plant is inoculated with the altered bacteria and the new genetic information is incorporated into the genome of the host plant where it may cause beneficial effects.

Other bacteria also are known to produce plasmids and have been used in a similar manner in recombinant DNA research. For example to improve water stress tolerance in some legumes, genes responsible for proline synthesis have been introduced into rhizobia species that have been allowed to colonize and form nodules on the roots of plants. The gene modification prevents the feedback inhibition of bacterial proline synthesis and high concentrations of proline are produced. In this relationship the genetics of the host plant are not altered but when the plant is subjected to drought stress the thousands of rhizobia associated with the root nodules are induced to synthesize proline that is available to the plant as an osmoprotectant.

As the author, I am indebted to the Tamil Nadu Agricultural University and I hope this book will be useful to post-graduate students and scientists in the field. I also thank the individuals and publishers for the source and use of ideas, illustrations, tables, citations and concepts. I am very much thankful to Dr. Gary Kochert, Professor of Botany, University of Georgia, USA for providing details of RFLP in plant Breeding.

School of Genetics
Tamil nadu Agricultural University
Coimbatore-641 003
Tamil Nadu, India

Dr. L.D. VIJENDRA DAS, Ph.D.
(Australia)
Post-doctorate (Germany)

CONTENTS

GENETICS

PLANT BREEDING

GENETICS

CHAPTER 1

Mendel's Experiments and Principles of Inheritance

The contribution of Mendel to Genetics is called Mendelism. Gregor Johann Mendel from 1822–1884, the father of genetics who was an Austrian monk made crosses in garden pea (*Pisum sativum*) and conducted experiments during 1856-1863. He presented his results of the experiments before the Natural History of Society at Brunn in 1865. His paper was published in 1866 in the Annual Proceedings of the Society and distributed to libraries in Europe and America but no one appreciated and it was neglected until 1900. In 1900 the principles of genetics worked by Mendel were rediscovered by three botanists, namely Correns (Germany), De Vries (Holland) and Tshermark (Austria). Bateson confirmed Mendel's work by a series of hybridization experiments.

MENDEL'S EXPERIMENTS

Mendel crossed varieties of edible peas (*Pisum sativum*). For example, he crossed a red-flowered variety with a white flowered variety. He did this by dusting the pollen of one variety (the red) on the pistils of other (the white). Of course, he prevented the white plant from pollinating itself. This he did by removing the stamens of the white flowers before the flowers had opened and shed their pollen. After pollinating the emasculated white flowers with the red pollen, he enclosed them in bags in order to prevent insects from getting to them with pollen from unknown sources. Thus he crossed the red variety and the white. The offspring of the cross was red. Mendel then self-fertilized the off-spring and he found that they produced off spring of their own in the ratio of 3 reds : 1 white.

The pea plant contains a number of contrasting characters. Out of these contrasting characters, Mendel selected only seven characters. Mendel in his first experiment crossed two plants differing in one character (height) only. A plant having a tall stem was crossed with another plant having dwarf stem. Tall and dwarf are the two varieties of a single character, height. Such crosses, where parents differ in one pair of alternative characters are known as monohybrid crosses. The resulting hybrids are known as monohybrids. When the behaviour of each single character was established, Mendel crossed two plants differing in two characters, such as flower position and height of the stem. A plant having axial flower and a tall stem was crossed with a plant having terminal flower and a short stem. Such crosses, where parents differ in two pairs of alternative characters are known as dihybrid crosses. The resulting hybrids are known dihybrids. The plants involved in the above crosses are called parent plants. They form the parental generation which is marked by P. The first hybrid generation resulting from a cross between parental plants is called the first filial generation and is marked as F_1. The second generation of hybrids arising from the self or cross fertilization of F_1 hybrid generation is called the second filial generation and is marked as F_2.

1. Law of Dominance

Each organism is formed of many characters and each character is controlled by a pair of factors or genes (T or *t*). Mendel's law of dominance states that one factor in a pair may mask or prevent the expression of the other. He called the character appeared in the F_1 generation or his monohybrid cross as dominant and those which did not appear in the F_1 generation as recessive. A recessive factor freely expresses itself in the absence of its dominant allele. This law is formulated based on the monohybrid experiment.

2. Law of Segregation (Segregation of Genes)

From his experiments Mendel concluded that each parent contributed one factor for a character to the F_1 hybrid. In this way the F_1 hybrid has two factors for each character. When the F_1 hybrid forms gametes the two factors separate from each other. There is no mixing up of factors thus emphasizing the purity of gametes. The phenomenon of separation became Mendel's second law of principle and was later termed as the Law of Segregation. This is explained diagrammatically as follows:

P	*Tall plant (true breeding)*	*X*	*Dwarf plant (true breeding)*
Factors Gametes F_1 hybrid F_1 gametes F_2 segregation	TT (T)	Tall Tt (Selfing) i.e. Tall Tt (T) (t) X (T) (t) 3 Tall : 1 Dwarf TT, Tt, Tt, tt	tt (t)

3. Law of Independent Assortment

This law is based on dihybrid experiment. According to this law, the genes for each pair of characters separate independently from those of other characters during gamete formation.

Example :

P	*Round Yellow RRYY*	*X*	*Wrinkled Green rryy*
Gametes : F_1 F_1 gametes	(RY)	Round Yellow RrYy (RY) (Ry) (rY) (ry) 1 : 1 : 1 : 1	(ry) X Self (Round Yellow) RrYy (RY) (Ry) (rY) (ry)

F_2 Checker board

	RY	**Ry**	**rY**	**ry**	
RY	RRYY	RRYy	RrYY	RrYy	
Ry	RRYy	RRyy	RrYy	Rryy	9:3:3:1 ratio
rY	RrYY	RrYy	rrYY	rrYy	
ry	RrYy	Rryy	rrYy	rryy	

Mendel applied the principles of a monohybrid cross in the dihybrid cross, the true breeding round yellow parent must be homozygous RRYY, and the wrinkled green parent rryy. Since each character is determined by two factors, in a dihybrid cross there must be four factors present in each parent. Likewise the F_1 hybrid must be RrYy. Mendel found that the pair of factors for roundness will be behaving independently of the pair of factors for yellow colour of seeds. In other words, one factor for a character must be passing independently of a factor for another character. Thus in the F_1 hybrids, R and r pass into different gametes. Now the probability of an R gamete formed is one-half, and of r gamete also one-half. Similar probabilities exist for Y and y gametes. It follows that the probability that R and Y will go to the same gamete is one fourth, as also of R and y, r and Y, and r and y. Therefore, gametes containing factors RY, Ry, rY and ry will form in equal proportions (1:1:1:1).

The F_1 hybrid producing the four types of gametes mentioned above was selfed. The results expected in the F_2 progeny can be predicted by making a checker board or a Punnett Square. Gametes produced by one parent are plotted on top of the checker board, and the gametes of the other parent on the vertical side. The 16 square of the checker board are filled up by making various possible combinations of male and female gametes during fertilization. The phenotypes read out from the checker board indicate a 9:3:3:1 ratio exactly as observed by Mendel.

Back Cross and Test Cross

"Back cross is the cross of F_1 hybrid with any one of its parents". "Test cross is the cross of F_1 hybrid with recessive parent". Mendel verified his results by performing the test cross. He crossed the F_1 hybrid heterozygous for both characters with a double recessive parent (rryy) which will produce only one type of gamete ry. The uniformity in the gametes of the recessive parent determines the differences in the types of gametes produced by the heterozygous parent. Now the hybrid RrYy produces gametes carrying RY, Ry, rY and ry with equal frequency (1:1:1:1). It follows that during fertilization if all these four types of gametes unite with ry gamete of the recessive parent, the resulting progeny will show all the four combinations of characters also in equal proportions. Indeed, Mendel observed the test cross progeny to consist of Round yellow, Round green, Wrinkled yellow and Wrinkled green plants in the ratio of 1:1:1:1.

F_1 :	Round yellow	X	Wrinkled green
	RrYy		rryy
Gametes :	(RY) (Ry) (rY) (ry)		(ry)
	1 : 1 : 1 : 1		
F_2	RY Ry rY ry		
	ry RrYy Rryy rrYy rryy 1:1:1:1		

From the results of his dihybrid crosses, Mendel realized the following facts. At the time of gamete formation the segregation of alleles R and r into separate gametes occurs independently of the segregation of alleles Y and y. That is why the resulting gametes contain all possible combinations of these alleles. i.e., RY, Ry, rY, ry. In this way Mendel proved that when two characters are considered in a cross, there is independent assortment of genes for each character, and this became the Law of Independent Assortment.

The Chromosomal Theory of Heredity

The Mendelian laws of inheritance, formulated in 1865, still form the basis of our understanding of the transmission of heritable variation, and the inheritance test is still the basic technique for following phenotypic characters through the several generations. It is the genius of Mendel that a few, but quantitatively predictable, results enabled him to postulate that an abstract entity—the element or factor, as he called it, the gene as we now designate it—was responsible for the phenotypic character he was following in his breeding experiments, that this entity was singly represented in gametes and doubly so in zygotes and somatic tissues, and that factors, when together in pairs, could exhibit dominance or recessiveness in expression.

The egg and sperm of animals, despite obvious differences in size and shape, had earlier been shown to be cells; the physical continuity between generations was, therefore, a slender cellular bridge. Fertilization was shown to be the union of gametes, with the fusion of parental nuclei in the cytoplasm of the egg being the crucial event, and with each nucleus providing equal, or nearly equal, numbers of chromosomes to the newly formed zygote and hence to the developing embryo. As the embryo grew by adding to its cell numbers, the critical acts of cell division were the longitudinal replication of each chromosome, and the segregation of these longitudinal halves to the two daughter cells, thus providing both a physical basis for the qualitative and quantitative chromosomal equality of each daughter nucleus and a functional basis for the conservation of the genotype. Meiosis, a kind of cell division leading to the formation of eggs or sperm in animals and spores in most plant species, was revealed as a mechanism for halving the chromosome number; thus providing a means for counteracting the chromosome doubling that took place earlier in the fusion of parental nuclei in fertilization and making clear the patterns of haploidy and diploidy in the life cycle. Chromosomes had been demonstrated to have a physical continuity from one cell to another, and from one generation to another, and to be qualitatively different from each other in so far as they affected developmental process. By 1902, Garrod, through his study of the metabolic diseases of humans—"in born errors of metabolism", as he called them—would forge the first link that would couple the biochemistry of disease syndromes to patterns of inheritance. With the further observation that species were characterized by a constant number of chromosomes, although the two sexes of a species may differ slightly from each other, it was becoming more and more obvious that the patterns of heredity or inheritance were somehow linked to, or mirrored in, the behaviour of chromosomes.

In retrospect, therefore, the discoveries made during the last quarter century of 1900s and the first few years of the 20th century made it increasingly clear that the chromosomes were key elements whose behaviour in division and fertilization revealed a regularity that could, on the one hand, account for the transmission of heredity factors from one generation to another, whether the cells or of individual organisms, and on the other for the conservation of species reproduction, i.e., the preservation of type, generation after generation. The science of cytogenetics was thus launched with a brilliant correlation of factor (gene) and chromosome behaviour.

The basis of the theory, as stated by Sutton, is as follows:

1. In somatic cells, arising from a fertilized egg, or zygote, the chromosomes consist of similar groups, one of maternal origin inherited through the egg, the other of paternal origin and inherited through the sperm. Each somatic nucleus, therefore, contains pairs of like chromosomes, or homologues, the number of pairs being the same as the haploid set of chromosomes in a gamete. Hence, the chromosomes, like Mendel's factors, are doubly represented in the somatic cells of an organism, and singly, or pure, in the gametes.
2. The chromosomes retain their structural individuality and their continuity throughout the life cycle of an organism. Again, the factors of Mendel retain their individuality and continuity even though the character they determine might not be expressed. The basis of genetic homogeneity and heterogeneity, dominance and recessiveness, was suggested by chromosomal behaviour.
3. In meiosis, synapsis brings together pairs of homologous chromosomes and leads to their subsequent segregation into different cells, establishing thereby a quantitative basis for both the segregation of factors and the independent assortment found when two pairs of contrasting factors are considered together.
4. Each chromosome, or chromosome pair, plays a definite role in the development of the individual. This conclusion derived not from any specific facet of Mendel's investigations but rather from studies of Boveri of abnormal larvae of Sea Urchins which lacked certain normal chromosomes, as well as Sutton's own studies of the size differences among the chromosomes of the insect *Brachystola.*

Sutton, consequently, visualized the chromosomes as the physical carriers of Mendelian factors and the segregation of a pair of homologous chromosomes and the independent assortment of non-homologous chromosomes as the physical basis for the qualitative and quantitative aspects of Mendelian segregation. Sutton, together with Roux and Boveri, also anticipated the phenomenon of linkage when he stated that all the factors in any one chromosome must be inherited together. As the theory of gene and the concept of Mendelian factors, the heredity constitution of an organism, and eventually of a species, the notions would eventually lead to a further refinement, namely, the molecular basis of hereditary phenomena.

Gene Interactions and Modified Dihybrid Ratios

According to Mendel each character is controlled by a pair of factors or genes. But later discoveries proved that in many cases the expression of a single character is controlled by the interaction of more than one pair of genes. This is called "Interaction of genes" or "Factor hypothesis". This was proposed by Bateson and Punnet. This hypothesis states that certain characters are controlled by the interaction of two or more genes. The genic interaction may occur in between genes located in the same chromosome or different chromosomes. This type of genic interaction is known as non-allelic genes interactions. The genic interaction may also occur between the two alleles of a single type of gene. This type of genic interaction is known as allelic gene interaction. However, Mendelian laws will hold good in all the type of segregation and interactions. As a matter of interaction the phenotypic ratio will alone be modified and not the genotypic ratio.

Complementary Genes (9:7)

Complementary genes may be defined as "two or more dominant genes occurring in different loci of the same chromosome or different chromosomes interact with one another to produce a character but neither of them produces that character in the absence of the other". The action of these independent genes are complementary.

Bateson and Punnet studied the inheritance of flower colour in sweet pea, *Lathyrus odoratus.* There are two varieties of pea plants, one producing red flower and the other white flower.

P:	White flower CCaa	X	White flower ccAA
Gametes :	(Ca)		(cA)
F_1	Red flower CcAa (Selfing)		
	Cc Aa	X	CcAa
Gametes :	(CA) (Ca) (cA) (ca)		(CA) (Ca) (cA) (ca)

	CA	*Ca*	*cA*	*ca*	
CA	CCAA Red	CCAa Red	CcAA Red	CcAa Red	
Ca	CCAa Red	CCaa White	CcAa Red	Ccaa White	9(Red) : 7 (White)
cA	CcAA Red	CcAa Red	ccAA White	ccAa White	
ca	CcAa Red	Ccaa White	ccAa White	ccaa White	

The cross between two white varieties can be explained by assuming two genes for red colour which must be present together, i.e., must act in a complementary way to each other. Thus each gene independently contributes something different but essential for synthesis of red pigment. If one of the two genes for red colour is absent, the result is a white flower.

The inheritance of the colour of aleurone layer in corn also demonstrates interaction of complementary genes. The outermost layers of endosperm in the maturing corn kernels become modified into a specialized aleurone tissue, so named because the cells have rich deposits of aleurone grains. In corn the aleurone layer is coloured due to anthocyanin pigments in the cells, and is controlled by complementary effect of two genes.

Supplementary Genes (9:3:4)

Supplementary genes may be defined as two independent pairs of dominant genes, which interact in such a way that one dominant gene produces its effect, when the second dominant gene is added to the first, a new character is expressed.

Coat Colour in Mice

Inheritance of coat colour in mice was studied by Castle. There are three different varieties of mice, they are agouti (grey), black and albino (white). Agouti colour is dominant to both black and albino. Black is dominant to albino but recessive to agouti. Albino is recessive to both agouti and black. Agouti is produced by dominant gene A in the presence of another dominant gene B. Dominant gene B alone produces black colour. Dominant gene A produces albino. The recessive condition of these genes causes albino.

Castle crossed a homozygous black mice (BBaa) with a homozygous albino (bbAA). The F_1 individuals are Agouti. When the F_1 Agouties are inbred their progeny consists of 9 agouti, 3 black and 4 albino.

Parents :	Black male BBaa (Ba)	X	Albino female bbAA (bA)
Gametes :			BbAa Agouti (two F_1 individuals crossed)
F_1 generation:	Bb Aa	X	Bb Aa
F_1 Gametes :	(BA) (Ba) (bA) (ba)		(BA) (Ba) (bA) (ba)

	BA	*Ba*	*bA*	*ba*	
BA	BBAA Agouti	BBAa Agouti	BbAA Agouti	BbAa Agouti	
Ba	BBAa Agouti	BBaa Black	BbAa Agouti	Bbaa Black	9 : 3 : 4 Agouti : Black : albino
bA	BbAA Agouti	BbAa Agouti	bbAA Albino	bbAa Albino	
ba	BbAa Agouti	Bbaa Black	bbAa Albino	bbaa Albino	

Duplicate Gene (15:1)

When two or more genes have the same effect on a given trait, they are referred to as duplicate genes. In maize the gene for yellow endosperm is dominant over white endosperm. A pure breeding yellow endosperm plant when crossed to a white endosperm plant yields yellow endosperm in F_1. On self pollination of F_1 hybrids in F_2 generation of 15 yellow and 1 white endosperm is obtained. The yellow endosperm results from two independent dominant genes Y_1 and Y_2. When anyone of these two dominant genes or both together are present, yellow endosperm is produced. When only recessive alleles are present in the homozygous condition ($y_1y_1y_2y_2$) it forms white endosperm. Thus the dominant genes Y_1 and Y_2 have an identical effect on endosperm colour and are consequently termed duplicate genes or isogenes.

Parent :	Yellow endosperm	X	Yellow endosperm
	Y1Y1y2y2		y1y1Y2Y2
Gametes :	(Y1y2)		(y1Y2)
F_1	Yellow endosperm		
	Y1y1Y2y2		
F_1 gametes	(Y1Y2) (Y1y2)		
	(y1Y2) (y1y2)	(Selfing)	

	Y1Y2	*Y1y2*	*y1Y2*	*y1y2*
Y1Y2	Y1Y1Y2Y2 Yellow	Y1Y1Y2y2 Yellow	Y1y1Y2Y2 Yellow	Y1y1Y2y2 Yellow
Y1y2	Y1Y1Y2y2 Yellow	Y1Y1y2y2 Yellow	Y1y1Y2y2 Yellow	Y1y1y2y2 Yellow
y1Y2	Y1y1Y2Y2 Yellow	Y1y1Y2y2 Yellow	y1y1Y2Y2 Yellow	y1y1Y2y2 Yellow
y1y2	Y1y1Y2y2 Yellow	Y1y1y2y2 Yellow	y1y1Y2y2 Yellow	y1y1y2y2 White
15 (Yellow) : 1 (White)				

George H. Shull (1914) reported a case of duplicate gene in the common weeds, shepherd purse *Bursa bursapastoris*. There are two varieties of seeds. Normal variety produces a triangular seed case and the other mutant variety produces oval seed case. When these two varieties are crossed, the resulting F_1 plants produce triangular seeds (which is thus dominant). When the F_1 hybrids are selfed, in F_2 plants with triangular seeds and oval seeds are produced in the ratio 15:1 instead of the normal Mendelian ratio.

Epistatic Genes (13:3)

Epistasis is a Greek word which means stopping or suppression. In some animals a gene at one locus on a chromosome suppresses or masks the expression of a gene at another locus. Such genes are known as inhibiting genes, since they inhibit the expression of other genes. A gene that inhibits or masks the expression of another gene (non-allelic gene) is said to be epistatic. The gene that is masked is said to be hypostatic.

In poultry white birds belong to two different varieties namely white leghorns or white wyandottes. Experiments reveal that the gene for white plumage of white leghorns is dominant over the gene for coloured plumage of coloured varieties. But the gene for white plumage of white wyandottes is recessive to the gene for coloured plumage of coloured varieties. Therefore, the gene which produces white plumage in white leghorns is different from the gene for white plumage in white wyanodottes.

A cross between a white leghorn and a white wyanodotte gives an F_1 of white birds. When such birds are inbred, the F_2 progeny segregates in the ratio of 13 white to 3 coloured birds. The experiment is explained below by postulating two gene C and I for the white leghorns.

Parents :	White Leghorn CCII	X	White Wyanodotte ccii
F_1	White CcIi		
F_1 gametes :	(CI) (Ci) (cI) (ci)	X Inbred	(CI) (Ci) (cI) (ci)

F_2

	CI	*Ci*	*cI*	*ci*
CI	CCII White	CCIi White	CcII White	CcIi White
Ci	CCIi White	CCii Colour	CcIi White	Ccii Colour
cI	CcII White	CcIi White	ccII White	ccIi White
ci	CcIi White	Ccii Colour	ccIi White	ccii White

F_2 ratio : 13 White : 3 colour

The F_2 ratio indicates that only three out of 16 genotypes, that is CCii, Ccii, Ccii produce coloured birds. The white leghorn obviously contain a gene I, which in the dominant state inhibits or suppresses the expression of the dominant colour gene C, resulting in white plumage. The recessive alleles of the inhibitor gene (ii) produce coloured birds due to expression of gene C. In other words gene I is epistatic to gene C. This is a case of dominant epistasis because even one dominant allele of gene I is able to express itself.

Lethal genes

A lethal gene produces an effect which differs from the normal condition that its possessor is unable to survive. There is a mutant recessive gene which causes internal adhesions of the lungs. A child homozygous for the lethal gene might survive upto embryonic development. But at birth, when it suddenly becomes dependent upon its lungs for its oxygen supply, it would die because its lungs could not expand properly. Being recessive, it could be carried by normal parents in heterozygous condition without any ill effects. In a mice cross heterozygous yellow (Yy) XYy produces offspring 1YY dies (homozygous yellow) : 2Yy (heterozygous yellow) : 1yy homozygous non-yellow. Hence a 2 : 1 ratio is realised. The lethal gene dominant in mouse, is thus lethal, causing death in homozygous condition YY.

CHAPTER 2

Multiple Allelism : Multiple Alleles

Mendel proposed that a character is controlled by a single pair of genes. The two genes of a character are located in the same locus of homologous chromosomes. These two genes are called alleles. Alleles are a pair of genes controlling the same character and located at the same locus of the homologous chromosomes. These alleles undergo mutation to give rise to three or more alleles located in the same locus of the homologous chromosomes. These mutant alleles express different alternatives of the same character. Such genes are called multiple alleles. Multiple alleles may be defined as a series of three or more genes which control the same character and occupy the same locus in the homologous chromosomes. The multiple alleles arise by the mutation of normal genes.

Nature of Wing in Drosophila

Normal wing in *Drosophila* is controlled by a dominant gene represented by *Vg*. The recessive allele of this gene produces vestigial wing. There is a third allele formed by the mutation of Vg allele. This allele produces a wing which is intermediate between normal and vestigial wing. This type of wing is called antlered wing and the allele controlling this wing is represented by vg^a. All these three games Vg, vg^a and vg are multiple alleles since they occupy the same locus in the homologous chromosomes. The inheritance of these genes follow the simple Mendelian ratios.

The gene Vg is dominant to vg^a and vg. The genes vg and vg^a are neither dominant or recessive to each other. Hence the *Drosophila* with the genotype vg vg^a has the wing which is intermediate between vestigial and antlered. Such flies are called vestigial antlered compound.

Parents :	Normal wing Vg Vg	X	Vestigial wing vg vg
F_1	Vg vg Normal	X	Vg vg Normal
F_2 Genotypic ratio :	1 Vg Vg : 2 Vg vg : 1 vg vg		
Phenotypic ratio :	3 normal : 1 vestigial		
Parents :	Vestigial vg vg	X	Antlered vg^a vg^a
F_1	vg vg^a	X	vg vg^a
F_2 genotypic ratio:	1 vg vg : 2 vg^a vg^a : 1 $vg^a vg^a$		
Phenotypic ratio :	1 vestigial : 2 vest-ant- compound : 1 antlered		

Eye Colour in Drosophila

An outstanding case is one concerned with eye colour is *Drosophila.* The normal eye colour of *Drosophila* is red and that a mutation in the X chromosome changed the normal red eye colour to white. Over a dozen other mutations have occurred all at the same locus and they have changed the eye colour to various lighter shades of red, known as cherry, blood, eosin, apricot, ivory, cream etc. The mutant genes have received corresponding names and symbols; for example eosin (w^e) where the letter w tells us that the mutation is at the white locus and the superscript e that the eye colour is eosin. When eosin and white are crossed, the daughters receive an X from each parent and are eosin-white compounds, w^e/w. These are intermediate between eosin and white in eye colour. A cross of eosin and cherry gives the compound w^e/w^{ch}, again intermediate in eye colour. In general, when any two of the mutants in the series are crossed, the F_1 are intermediate. The first case of multiple alleles of a gene controlling eye colour in *Drosophila* was made by T.H. Morgan in 1910 during his studies on mutation.

Multiple Allelism in Man

The red blood corpuscles of man contain a great many antigens (mostly proteins). Included among them are two antigens known as A and B. Some people's corpuscles contain antigen A, others contain B, still others contain both A and B, and finally some contain neither A nor B. On the basis of these antigens, people are classified into four blood groups; A, B, AB and O. These are sometimes referred to as the ABO blood groups. They are also sometimes called "Landsteiner" groups, after their discoverer.

The ABO blood groups are under the control of genes at a certain locus. There are other blood groups in man besides the ABO (such as the MN, the Rh, etc.). All of these are also under gene control, each series of blood groups being under the control of genes at a single locus. It is not known which of the four blood groups is the normal one. Ordinarily individuals with normal traits are the most numerous in any species. Both groups O and A are fairly numerous, in the European populations and either might be considered the normal. For convenience we shall regard class O as the normal. Then groups A and B arose from group O as the result of two dominant mutations (one for each group). The mutant genes can be given the symbols A and B, respectively. Both of these arose in the case locus from one of the normal genes in class O. We can designate the normal gene in question by the symbol +. Thus, the three genes +, A, and B occupy the same locus and are multiple alleles. Since + is a recessive, group O must be pure for +; that is, its formula is +/+. Since gene A is a dominant, group A might be either pure (A/A) or it might be hybrid (+/A). In like manner, group B might be either B/B or +/B. Group AB on the other hand is always hybrid and is of composition A/B.

It will be seen that, if both parents in a given family are of class O, they must both be of genotype +/+, and all of the children must, therefore, also be +/+; that is, of class O, like their parents. If on the other hand, both parents were of class A but both happened to be hybrid (A/+), then they might have some children of class O in addition to some of their own class, since A/+ X A/+ gives offspring in the ratio of 1A/A ; 2A/+ : 1 +/+. Hence, if both parents were of group A, then the children would have to belong to either group A or group O; we could affirm that a child belonging to any other group was not the offspring of a man of group A if the mother were also of group A. In cases where the paternity of a child is in question, therefore, blood tests could be used as evidence only under restrictions of the kind just indicated.

Group AB (hybrid for genes A and B) shows the full effect of both alleles (A and B). Often a hybrid that has two mutant alleles at the same locus is intermediate between the two mutants, as for example vestigial antlered or eosin white in Drosophila.

M and N Blood Groups

In 1927 Landsteiner and P. Levine described the MN blood group system. They discovered that the human blood contains another two types of antigens, namely M and N. Based on the presence or absence of antigens human beings are classified into three groups, namely M, N and MN. The M group person contains antigen M; N group persons contain antigen N and MN group person contains both antigens M and N. The MN blood group has no natural antibodies. These antigens have no relationship with the antigens of ABO blood group. The antigens are controlled by dominant genes represented by M and N. The genotypes of these MN groups are given below.

Blood group	Genotype
M	M N
N	N N
M N	M N

Rh Blood Group

Landsteiner and Wiener (1940)_discovered the existence of a special type of antigen in the RBC of Rhesus monkey. Since this factor was first discovered in Rhesus monkey, it was named rhesus factor or Rh factor. The Rh factor is also present in the RBC of some human beings and it is absent from others. Based on the presence or absence of Rh factor, the human beings are classified into two groups. They are Rh positive (Rh+) and Rh negative (Rh-). The Rh+ person has Rh antigen in the RBC; the Rh- person has no Rh antigen.

In the European country, 85% of the human beings are Rh+ and the remaining 15% are Rh-. In India 93% are Rh+ and 7% are Rh-. In China 99.5% are Rh+ and only 0.5% are Rh-.

The Rh antigen has no natural antibody. However, Rh antibody can be produced artificially. An Rh- person develops Rh antibody when he receives blood from an Rh+ person. Even a small amount of Rh+ blood (as small as 0.05 ml) can evoke the production of Rh antibody in the Rh- person. The antibody once formed remains throughout the life.

Genetics of Rh Factor

There are several varieties of Rh antigen and of antibody. The commonest Rh antigen is called antigen D and its antibody is called anti D. The production of antigen is controlled by multiple alleles. The antigen D is produced by a dominant gene represented by R. When this gene is recessive it cannot produce the antigen. Hence the Rh+ persons may be homozygous dominant (RR) or heterozygous (Rr). The Rh- persons are always homozygous recessive (rr). The Rh factor follows the Mendelian principle of inheritance. The blood type of the children can be easily visualized by knowing the blood types of parents. A few marriages and the resulting children are given below:

(1)	Male Rh+	X	Female Rh +
	Parents : RR		Rr
	Gametes: (R)		(R) (r)
	Children : RR (Rh+)	:	Rr (Rh+)
(2)	Male Rh+	X	Female Rh-
	Rr		rr
	(R) (r)		(r)
	Rr (Rh+)		rr (Rh-)

Weiner's Theory

A.S.Weiner, proposed that there are 8 main alleles. They are r, R^0, R^1, R^{11}, R_1, R_2, R_x, R_y at the locus for this particular character. This was possible because some of the alleles produce two or three of the Rh antigens. Thus A.S. Weiner proposed that the Rh blood groups are determined by a series of multiple alleles.

Sterility and Self Incompatibility in Plants

Sometimes bisexual plants produce functional male and female gametes, yet self pollination does not lead to fertilization and there is no seed setting. The pollen grains may fail to germinate on the stigma; or even if pollen tubes are formed their growth may be inhibited. Such a condition where pollen grains fail to fertilize ovules of the same plant is known as self-incompatibility or self sterility. *Nicotiana* (tobacco) was one of the first plants in which self-sterility was noticed. Breeding tests have established that self-incompatibility is controlled by multiple alleles of a single gene and may be of two types - gametophytic or sporophytic. The gametophytic self-incompatibility (GSI) is controlled by the genotype of the pollen grains. This type is quite common in families like Liliaceae, Gramineae and Solanaceae. Sporophytic self-incompatibility (SSI) is controlled by the somatic or sporophytic cells of the plant which produce pollen grains, as observed in families Crcuiferae and Compositae.

East and Mangelsdorf in 1925 explained the behaviour of sterility genes. The gene controlling incompatibility is designated S and its multiple alleles are represented by a series S_1, S_2, S_3, S_4, S_5 ..... and so on. It is known that two alleles are present in the diploid cells of a plant. In the GSI system pollen grains produced by an S_1 S_2 plant will have either S_1 or S_2 allele and are not able to pollinate a pistil with S_1S_2 genotype. But if S_1 or S_2 pollen grains land on an S_3 S_4 pistil, fertilization takes place. In plants belonging to the SSI system if even one of the alleles present in the sportphytic tissue of the male parent is present in the pistil of the female parent, then all the pollen grains produced by the male parent fail to effect fertilization of that pistil. Thus, if the male parent has the genotype S_1S_2 it would be incompatible with plants having S_1S_2; S_1S_4; S_1S_5; S_2S_3; S_2S_4; S_2S_5 genotypes, but would be compatible with S_3S_4; S_3S_5; S_4S_5 plants. *Oenothera* (evening primrose) has 37 different self-sterility alleles and red clover has more than 200. Since the incompatibility allele in the style opposes the penetration of pollen tubes with like genes, this explanation of incompatibility was called "Oppositional factor hypothesis" by East and Mangelsdorf.

CHAPTER 3

Multiple Factor Hypothesis

QUANTITATIVE GENETICS (QUANTITATIVE TRAITS) OR POLYGENIC INHERITANCE

According to Mendel, each character is controlled by two genes, located at the same locus. But certain characters are controlled by more than two genes located at different loci. These genes are called multiple genes or polygenes. Multiple genes or polygenes are defined as a group of genes located at different loci and involved in the expression of one character. The inheritance of multiple genes is called multiple gene (factor) inheritance. As a single character is expressed by many genes, each gene has a small effect on the expression of the character. The intensity of the character depends on the total number of genes involved. Hence these genes are also called cumulative genes. The characters controlled by cumulative genes are not sharply marked from one individual to other. But they are continuous with many grades of intermediates. Hence these characters are quantitative rather than qualitative. As these characters show continuous variation between two extremes, the inheritance is called quantitative inheritance.

Inheritance of Kernel Colour in Wheat

Nilsson-Ehle performed many crosses between varieties of wheat having red seeds and those having white seeds. In all crosses except one the F_2 generation gave a ratio of 3 red to 1 white. In one exceptional cross he used a very old red variety from the north of Sweden. The F_1 plants produced seeds intermediate in colour between the two parents, and the F_2 generation segregated in the ratio of 63 red : 1 white. The noteworthy feature of this experiment was the variation in the intensity of the red pigment in wheat grains produced by F_2 plants. This shows that 3 pairs of genes are involved for dark red colour and the F_2 ratio is 63:1 with different grades of intermediate types.

(1) When single pair of genes are involved

Dark Red X White

AA aa

F_1 Aa Intermediate red

F_2 AA Red : Aa Intermediate : aa White

1 : 2 : 1

Phenotypic ratio : same as genotypic ratio

If red alone grouped together the Phenotypic ratio : 3:1

(2) When two pairs of genes are involved

Dark Red X White

AA BB aa bb

F_1 Aa Bb Intermediate red

F_2

	AB	*aB*	*Ab*	*ab*
AB	AABB Dark red	AaBB Red	AABb Red	AaBb Intermediate red
aB	AaBB Red	aaBB Intermediate Red	AaBb Intermediate Red	aaBb Light red
Ab	AABb Red	AaBb Intermediate red	AAbb Intermediate red	Aabb Light red
ab	AaBb Intermediate red	aaBb Light red	Aabb Light red	aabb White

F_2 Phenotypic and Genotypic ratio :

Dark red	—	1
Red	—	4
Intermediate red	—	6
Light red	—	4
White	—	1
Total		**16**

If red alone grouped together, the Phenotypic ratio : 15:1

(3) Dark Red X White

AA BB CC aa bb cc

F_1 Intermediate red

Aa Bb Cc

1	6	15	20	15	6	1
Dark red	Red	Medium red	Intermediate red	Light red	Very light red	White

Phenotypic and Genotypic ratio are same

If red alone grouped together, the Phenotypic ratio : 63:1

Transgressive Variation

Punnet crossed a large variety of fowl Golden Hamburg with a small variety Sebright Bantam. The F_1 was intermediate between the two parents. In F_2 generation some birds were larger and some were smaller than the parental varieties. However, most of the F_2 individuals were intermediate between the original parents. Punnet explained that the genotype of the large parent variety was AA BB CC dd and small variety was aa bb cc DD. The F_1 birds would be Aa Bb Cc Dd having intermediate size. In the F_2 birds atleast one bird would have AA BB CC DD genotype and that would be larger than the large parent and at least one would have aa bb cc dd genotype and that would be smaller than the small parent variety. The extreme larger or smaller size was due to the presence of four pairs of dominant or recessive genes in an individual. This pattern, in which the extremes of the F_2 exceed those of parents is called "Transgressive variation" or "Transgressive segregation" (Figure A).

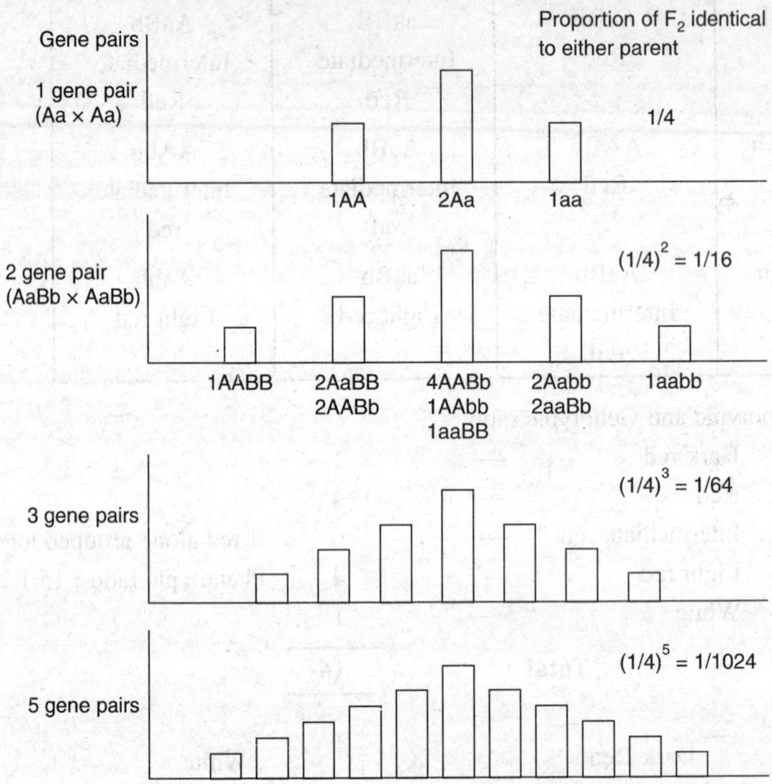

Fig. A. Graphs showing relation of polygenes with F_2 phenotypes.

Descriptive Statistics

In experimental work, we are always dealing with samples of limited size, and it is desirable to know how accurately the sample represents the population. In order to get the desired information, it is necessary, among other things, to know the amount of variability of the sample. We, therefore, need some sort of measure of the variability of a sample before we can tell how truly it represents the entire population. One way of measuring the variability of a sample would be simply to get the average amount of variation. For example, in a sample which consisted of three men whose weights were 154, 160 and 166 pounds, we might first get the mean or average weight of the sample (160) pounds, then we find out how much each man deviated from the mean (6, 0 and 6 pounds respectively), get the total deviation from the mean (12 pounds), and divide this total by the number of men (12/3 or 4 pounds).

However, statisticians prefer another measure of variation known as the Standard deviation. We can get this by first squaring the deviations of each individual from the mean (in the above example 6^2, 0^2, 6^2); getting the sum of these squared deviation (72); dividing the sum by the number of individuals (72/3 or 24); and then taking the square root of this sum, i.e., 24, or about $\pm$ 5 pounds. The average of the sum of the squared deviations (i.e., 24 in the above case) is known as the variance. Thus, in getting the standard deviation, we first get the variance, and then take the square root of the variance to get the standard deviation.

We might give another example of the way the variance and standard deviation are got, in connection with the corolla lengths of the F_2 tobacco plants given below.

Calculation of Variance and Standard Deviation of Corolla Lengths of F_2 Shown in Table

Class Frequency (f) X Class Size	*Mean (m)*	*Deviation (d) of Class from*	*d^2 Mean*	*fxd^2*
1 × 52 = 52		67 – 52 = 15	225	1 × 225 = 225
5 × 55 = 275		67 – 55 = 12	144	5 × 144 = 720
16 × 58 = 928		67 – 58 = 9	81	16 × 81 = 1296
23 × 61 = 1423		67 – 61 = 6	36	23 × 36 = 828
18 × 64 = 1152		67 – 64 = 3	9	18 × 9 = 162
62 × 67 = 4154		67 – 67 = 0	0	62 × 0 = 0
37 × 70 = 2590		70 – 67 = 3	9	37 × 9 = 333
25 × 73 = 1825		73 – 67 = 6	36	25 × 36 = 900
16 × 76 = 1216		76 – 67 = 9	81	16 × 81 = 1296
4 × 79 = 316		79 – 67 = 12	144	4 × 144 = 576
2 × 82 = 164		82 – 67 = 15	225	2 × 225 = 450
2 × 85 = 170		85 – 67 = 18	324	2 × 324 = 648
Total 211 14265	67 mm		1314	7434

Total number of f_2 = 211

$$\sum \frac{fd^2}{n} \text{(variance)} = \frac{7434}{211}, \text{ or } 35$$

$$\sqrt{\sum \frac{fd^2}{n}} \text{ (standard deviation)} = \sqrt{35} \quad \text{or} \quad = 6 \text{ (approx)}$$

The symbol Σ the Greek letter for S stand for sum.

However, statisticians prefer (for mathematically involved reasons) to use n – 1 instead of n as a divisor in the expression just given, so that in general, if we let σ (sigma, the Greek letter for s) stand for the standard deviation, then

$$\sigma \text{ (standard deviation)} - \sqrt{\Sigma fd^2} \ / \ n - 1$$

In the following figure certain area is included under the curve between the lines erected (on either side of the mean) about the values – σ (mean minus the standard deviation) and + σ (mean plus the standard deviation). This equals two-thirds of the area under the curve. Thus, suppose that we measured the weights of a number of men, say 200, and found that the mean weight of the sample was 160 pounds. Let us say that the standard deviation turned out to be ± 5 pounds. Then, about two-thirds of the men measured would fall in the classes included between 155 pounds and 165 pounds. That is to say, about two-thirds would be included within the limits of the mean minus the standard deviation on the one hand and the mean plus the standard deviation on the other hand (Figure B).

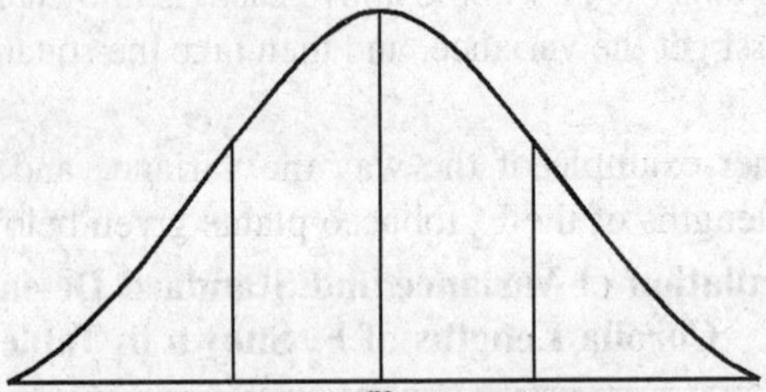

Fig. B. Area under the curve included between plus and minus standard deviation (±)

Coefficient of Variation

The natural variation or variability relative to the mean or average of a group is called Coefficient of variation. In general, the coefficient of variation if got by dividing the standard deviation by the mean and multiplying by 100 (to convert the product to per cent). Thus, if we let v stand for coefficient of variation, σ as before, the standard deviation, and m for mean, then

$$\text{V (coefficient of variation)} = \frac{\sigma\,(\text{standard deviation}) \times 100}{\text{m}}$$

The standard deviation gives us an idea of the variability of a sample, but it does not tell us just how good the sample is (in representing the general population). The probability that the mean of a sample represents the "true" mean, or the extent to which it probably does so, depends on,

(a) the variability of the sample

(b) the size of the sample (small samples of course, are not good as large).

The measure of this probability is known as the standard error of the mean. To get this quantity, we divide the standard deviation by the square root of the total number of individuals in the sample.

$$\text{S.E.m (Standard error of mean)} = \frac{\sigma\,(\text{standard deviation})}{\sqrt{\text{n}}}$$

The standard error of the mean of a sample really represents the standard deviation of the means of samples of the same size taken from the same population. The mean of a sample is often used to represent the sample, and usually the standard error of the mean is given along with the mean.

Standard Error of a Difference

This measures the standard error of the difference between two means.

$$\text{S.E.}_{\text{diff.}} = \sqrt{\text{S.E.}_1{}^2 + \text{S.E.}_2{}^2}$$

When the difference greater than twice its probable standard error, it is usually regarded as significant.

In summary then, some of the constants used in the statistical analysis of data are as follows.

1. The mode is the class of greatest frequency in a sample that is divided into classes.
2. The mean ($\overline{m}$) is the average size of the sample. If Σ is used for "sum of" f for the frequency of each class, V for the magnitude of the class (such as 60 pounds), and n for the total number of individuals, then :

$$\overline{m} = \frac{\Sigma fv}{n}$$

3. Variance is the mean of the squared deviations : i.e.,

$$\text{Variance} = \frac{\Sigma fd^2}{n}$$

4. The standard deviation (σ) is essentially the square root of the variance; i.e.,

$$\sigma \text{ (standard deviation)} = \sqrt{\frac{\Sigma fd^2}{n-1}}$$

The standard deviation includes within its plus and minus limits about two thirds of the individuals in a sample.

5. The coefficient of variation (v) is the magnitude of the standard deviation relative to the mean, expressed as a percent.

$$\text{v (coefficient of variation)} = \frac{\sigma \text{ (standard deviation)} \times 100}{\overline{m}}$$

6. The standard error of the mean (S.E.m) is the standard deviation of the means of a number of samples of the same size taken from the same population (but calculated from a single sample on the basis of its size and standard deviation).

$$\text{The standard error of the mean (S.E.m)} = \frac{\sigma \text{ (standard deviation)}}{\sqrt{n}}$$

The measure just given is important, for a sample is usually represented by its mean, along with the standard error of the mean.

7. If we let S.E.1 stand for the standard error of one mean, and S.E.2 for the standard error of another mean, then the standard error of the difference between the two means ($S.E._{diff}$) is got from the formula,

$$S.E._{diff} = \sqrt{(S.E._1)^2 + (S.E._2)^2}$$

In the above case, we are dealing with the mean of a sample or a difference in means of two samples, and our formulas tell us how well the mean represents the entire population.

CHAPTER 4

Linkage and Recombination

An organism possesses numerous genes for its various characters. The number of genes of an organism thus exceeds the number of its chromosomes because most organism contain limited number of chromosomes. Hence, a single chromosome carries a large number of genes.

All the genes on a chromosome are said to be linked to one another. Linkage is defined as the tendency of two or more genes to remain together in the original combination in the same chromosome. This theory of linkage was propounded by T.H. Morgan in 1911.

A Case of Linkage

In Drosophila, the ordinary normal fly has grey body and long wings. But a stock can be got having the recessive mutations, black body (b) and vestigial wings (v), both in the same

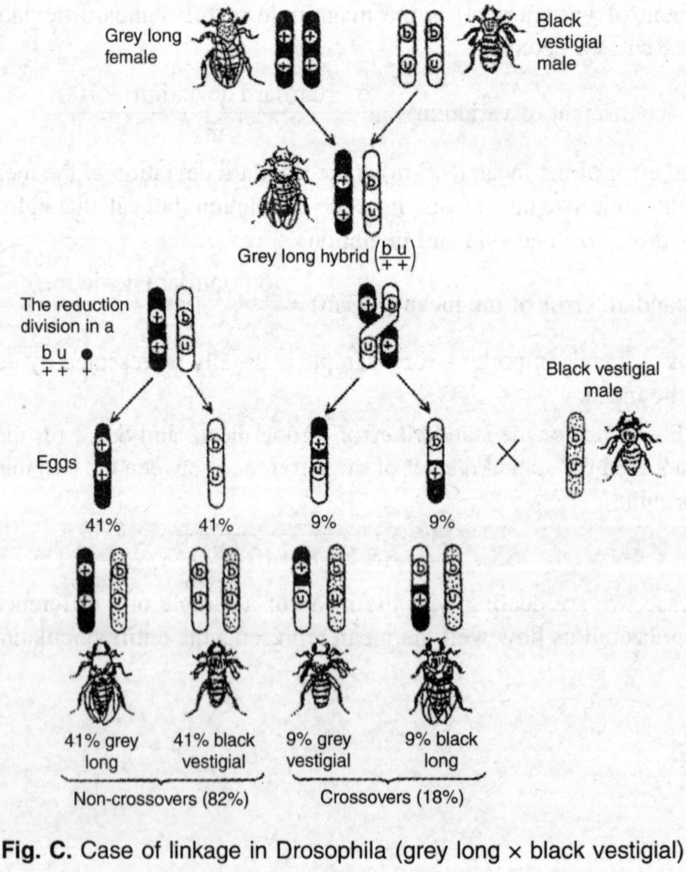

Fig. C. Case of linkage in Drosophila (grey long × black vestigial)

chromosome. The normal allele is indicated by + signs. When a pure gray long is crossed to a black vestigial (Figure C).

In the hybrid crossing over takes place and the two cross over classes +v and b+. It has been found that experimentally the above 4 classes of eggs are formed approximately in the following proportions. 41 ++ : 41 bv : 9 +v : 9 b+. When the F_1 was test crossed the offspring produced approximately, 41 grey long : 41 black vestigial : 9 grey vestigial : 9 black long. The last two classes are cross overs. They form 18 out of every 100 offspring. So there is 18 percent of crossing over between black and vestigial. In the linkage experiments the test cross is the simplest way in which to determine the percentage of crossing over.

Thus linkage as the tendency of two genes in the same chromosome to remain together in the process of inheritance. But crossing over sometimes separate genes which are linked. The further apart the two genes are, then the amount of crossing over is greater and vice versa. Usually the crossing over may be anything from 1 percent or less to almost 50 percent of the total.

Crossing Over

Linked genes being carried in the same chromosome. If however, the two genes located in the same chromosome should remain together in all cases. But this is not what actually happens and the linked genes sometimes separate. This as follows, the principle is interchange of parts or segments between homologous chromosomes which is called crossing over. Crossing over takes place in the segment of the chromosome between the location of two genes in some cells only. During meiotic prophase, homologous chromosomes come together and pair each other throughout their length. This is evidence that this pairing is brought about by mutual attraction of similar parts of the homologous chromosomes which contain allelic genes. Between the Pachytene and the diplotene stages, the paired chromosomes divide each into two chromatids so that the bivalent is now composed of four chromatids. At about the same time when the chromosomes are first seen to be divided, the chromatids establish one or more exchanges or chiasmata per bivalent. At each chiasma two of the four chromatids have become broken and then rejoined. This process first correctly understood by Janssens in 1904 has been studied later by Darlington.

In Drosophila crossing over rarely or never takes place in the male.

Double Crossing Over

Sometimes crossing over occurs at two points in the same chromosome pair as shown in the figure D. This is known as “double crossing over”. The gametes formed by double crossing over are called double crossovers. By contrast, crossing over just once is known as single crossing over and the resultant gametes are called single crossovers. The amount of double crossing over between two loci increases with the distance apart of the loci, but as a rule the doubles are fewer than the singles.

Fig. D

Crossing over might also occur at three points in the same chromosome pair (triple crossing over), but over short distances there are very few, if any, triples, quadruples would be still fewer.

Coupling and Repulsion Phases

The terms coupling and repulsion were coined by Bateson and Punnet in 1906. Coupling is the condition in linked inheritance in which an individual heterozygous for two pairs of factors received the two dominant member from one parent and the two recessive from the other parent. e.g., AB / ab.

Repulsion is the condition in linked inheritance in which an individual heterozygous for 2 pairs of linked factors received the dominant member of one pair and the recessive member of the other pair from one parent and the reverse condition from the other parent e.g. Ab / aB.

The cross AB / AB x ab / ab is said to be in the coupling phase and the cross Ab / Ab x aB / aB is said to be in the repulsion phase.

Chromosomal Map

The figure E shows the relative positions of y m and f in the chromosome. Such a figure is known as a chromosome map. It is based on two assumptions: (1) that the genes are arranged in a line and (2) that the per cent of crossing over between two genes is an index of their distance apart. A chromosome map, in fact, sometimes called a crossover map. It is constructed entirely from the results of breeding experiments (test crosses), not from microscopic examination of the chromosomes themselves. We might, then, define a chromosome map as a line on which the genes are represented by points separated by distances proportional to the amount of crossing over. Since a chromosome map is based on the percents of crossing over between genes, it is sometimes referred to as a crossover map. The first two chromosome maps were made in 1911 by Sturtevant and soon after additional maps were made by Bridges and others, *Drosophila* being the earliest material employed.

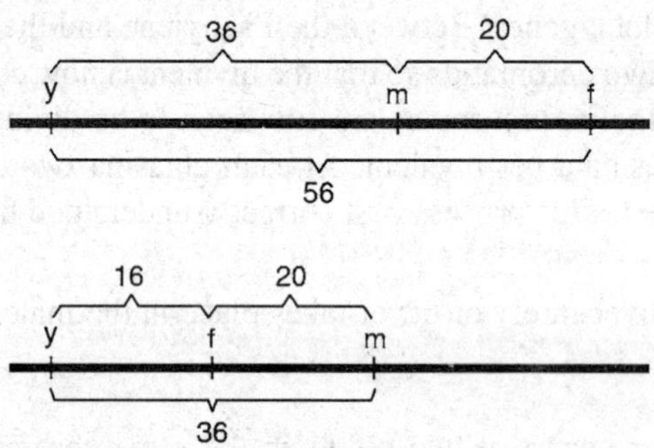

Fig. E. Diagram showing the possible arrangements of three genes when only two of the distances are known.

Two and Three Point Test Crosses with their Significance in Chromosome Mapping

When two genes are mapped by performing a cross, it is called a two point cross. The three point cross involves three pairs of linked genes, is a valuable method for determining positions of genes in relation to each other and for mapping distances between genes (Figure F). In 1926, Bridges and Olbryoht used this method for mapping three recessive sex-linked genes in *Drosophila*; scute sc (without bristles), echinus ec (rough eyes), and crossveinless wings cv (absence of transverse veins). The cross involved mating of a scute crossveinless fly with an echinus fly, and then testcrossing female F_1 heterozygotes with the recessive homozygous male showing all three recessive phenotypes.

In the data of Bridges and Olbrycht clearly the first group of progency $\frac{(+ + ec)}{(sc\ cv +)}$ represents the parental combinations, and the second two groups $\frac{(+ cv +)}{(sc + ec)}$ and $\frac{(+ cv\ ec)}{(sc + +)}$ are recombinations.

$$P : \frac{sc\ cv\ ec}{sc\ cv +} \times + + ec$$

$$\frac{+ + ec}{ec\ cv +} \times \begin{array}{c}\text{Test cross}\\ sc\ cv\ ev\end{array}$$

Progeny :	+ + ec	810	} Parentals	—	1638
	sc cv +	828			
	+ cv +	88	} Recombinations	—	150
	sc + ec	62			
	+ cv ec	89	} Recombinations	—	192
	sc + +	103			

Fig. F. The three point cross in *Drosophila*. Data from Bridges and Olbrycht (1926)

Since chiasma formation takes place between linked genes, in order to determine crossover percentage in a three point cross, the genes must be analysed two at a time, ignoring the third gene each time. Thus we can examine three combinations : cv – ec, cv-sc, and ec-sc in each group of progeny. Considering genes cv-ec, these are found to be present in progeny of the first group in the same way as in the parent. That is cv is present in its normal wild form (+) on one homologue along with ec; in the other homologue cv is present along with the wild form of ec. Thus in the first group of progeny cv and ec are present as cv + and ec+. Since this is the way they are present in the parent, we can infer that there is no recombination between cv and ec. The second group of progeny again shows cv + on one homologue and ec + on the other as in the parents. Therefore, the second group of progeny also does not represent recombination between genes cv and ec. In the third group of progeny both cv and ec are present on one homologous chromosome whereas the other homologue has wild forms (+ +) of both cv and ec. This arrangement of cv and ec i.e., + and + is different from that in the parental combination and has arisen due to crossing over between the loci for cv and ec.

The data in the above figure shows that there were 192 flies in the third group of progeny. We can calculate the percentage recombination between cv and ec by dividing 192 by the total progeny (1638 + 150 + 192 = 1980) and multiplying by 100. The amount of crossing over between cv and ec is thus found to be 9.7%.

Let us now consider genes sc and cv. In the first group of progeny they are present as in the parental combinations, that is + + on one chromosome and as sc cv on the other. In the second group of progeny their arrangements are different, sc + being together in one chromosome and cv + together on the other homologue. Their changed positions indicate recombination between sc and cv. In the third group of progeny also sc and cv are present differently than in parental combinations. Thus all the 150 members of second group of progeny and 192 of the third group represent recombination between sc and cv. The recombination percentagae between sc and cv is calculated as described earlier, i.e.,

$$\frac{150 + 192 \times 100}{1980} = 17.3\%$$

Analysing the genes ec–sc in the same way we find that in the first group of progeny they are present as in the parental combinations. In the second group they are present as + + on one chromosome, and as sc ec on the other i.e., they represent recombination. In the third group of progeny they are present as in the parental combinations. Therefore, percentage recombination between sc – ec is 150 × 100/1980 = 7.6%. We have thus found that percentage recombination between scute and crossveinless is 17.3, between scute and echinus 7.6 and between crossveinless and echinus 9.7. The recombination percentage also represents crossover percentage and map distance between the genes.

With the data available, it is now possible to map the genes. The chromosome is drawn as a line and the two genes showing lowest recombination frequency are marked first (figure G). In this case they are scute and echinus 7.6 map units apart. Next mark cv and ec which are 9.7 units apart by indicating cv either on the left or right of ec. If we mark cv 9.6 map units on the left of ec, then as seen in the figure (b), the distance between cv and sc would equal 9.7 minus 7.6 equal to 2.1 units. This does not agree with the data on recombination percentage found experimentally. But if we mark cv on the right of ec 9.7 units apart, then it indicates a map distance between sc and ev equal to 9.7 plus 7.6 i.e., 17.3 which is identical to the experimental data.

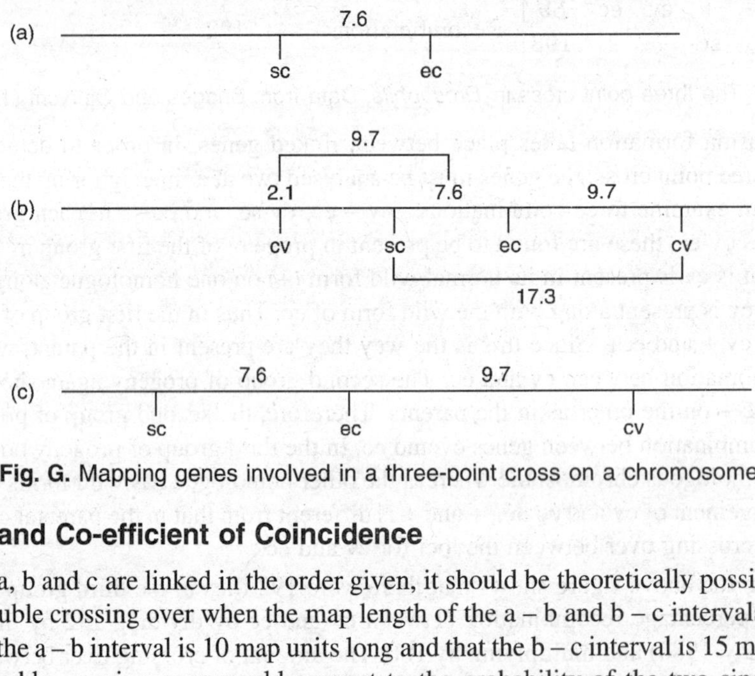

Fig. G. Mapping genes involved in a three-point cross on a chromosome.

Interference and Co-efficient of Coincidence

If genes a, b and c are linked in the order given, it should be theoretically possible to predict the frequency of double crossing over when the map length of the a – b and b – c intervals are known. If it is assumed that the a – b interval is 10 map units long and that the b – c interval is 15 map units, then the frequency of double crossing over would amount to the probability of the two single cross - overs occurring simultaneously. This would be 10% of 15%, or 1.5%, provided the two single crossovers are without influence on each other's occurrence, that is if they occur independently of each other.

Studies, however, have revealed that adjacent crossovers do not occur independently and that one occurring in, for example, the a - b interval would tend to suppress those occurring simultaneously in the b – c interval. The effect, therefore, would be to reduce the expected frequency of double crossing over. H.J. Muller, who first discovered the phenomenon, termed it crossover interference, and it may be defined as the tendency of one crossover to interfere with the occurrence of another in its vicinity. Cytologically, the term chiasma interference would be appropriate. Interestingly enough, interference does not generally extend across the centromere in a two-armed chromosome; each arm, therefore, acts independently in crossing over from each other.

Interference is more pronounced as the distance between successive genes becomes shorter; it decreases with increasing distance. An example will serve to illustrate its effect. The maize data illustrate partial interference. The map distances for the bm–pr and pr–v intervals are 22.3 and 43.4, respectively. The expected frequency of double crossing over, based on independent occurrence, would be 9.7%, that is, 22.3% of 43.4%. Only 7.8% doubles were recovered among the test cross progeny. To express the degree of interference, Muller coined the term coincidence, which is an inverse measure of interference. The coefficient of coincidence, which is equal to one minus interference when the latter is given in terms of a decimal fraction, is calculated by dividing the observed number of double crossovers by the expected number. In the above experiment this would be 0.078 / 0.097 = 0.804 = coefficient of coincidence. Therefore, only 80.4% of the expected double crossovers were recovered, indicating a partial interference of 19.6%.

Pleiotropism

Pleiotropism is defined as the effect of a single gene upon two or more characters which are not related. In *Drosophila*, the recessive gene for vestigial wings (vg) produces vestigial wings in homozygous condition. In addition to wing length, it also produces tiny wing like balancer behind the wings; certain bristles; the structure of spermatheca and low number of eggs. This phenomenon of multiple effect of a single gene is called pleiotropism.

In a F_1 of the cross between varieties virnar and rosi cotton belonging to the species *Gossypium arboreum* Linn., the leaf characters namely, narrow lobe, lanceolate shape and palmatipartite incision of the former were found dominant over broad lobe, ovate shape and palmatifid incision of the latter. In the F_2 population, out of the total of 277 plants all the 201 plants examined possessed narrow lobe, lanceolate shape and palmatipartite incision as against 76 plants which had broad lobe, ovate shape and palmatifid incision. The ratio between the above two phenotypic groups works out to 3 : 1. The gene symbols 'L' for narrow lobe and '1' for broad lobe was adopted as suggested by Hutchinson (1934) treating them as allelomorphs. As there are no recombinations in the F_2, it may be regarded that 'L' and '1' govern the expression of all the above three character pairs. It was proposed that a single gene pair 'L' and '1' is pleiotropic in effect, 'L' being responsible for the expression of narrow lobe, lanceolate shape and palmatipartite incision and '1' governing the expression of broad lobe, ovate shape and palmatifid incision. (Vijendra Das and Mensinkai, 1966*b*).

CHAPTER 5

Sex Chromosome in Man, Drosophila and Melandrium

Sex Determination in Man (Female XX Male XY)

The difference between the sexes is due to difference in their chromosomes. In the unreduced cells the chromosomes occur in pairs. This applies without exception to the female. But in the male there are two chromosomes which do not exactly match. One of these is known as the X chromosome, the other as the Y. In the female, there is a pair of X chromosomes instead of an XY pair. We can represent the difference between the sexes by simply showing an XX pair in the female, and XY pair in the male. When the reduction division takes place in the female, all the eggs receive an X. In the male, half the sperm cells receive an X and half receive a Y. An egg (X) might be fertilized by either an X or Y sperm. The fertilized eggs, therefore, might be either XX or XY. All the later cells of the body produced by mitosis are like the fertilized egg and are either XX or XY, and so they develop into either a girl or a boy in 1:1 ratio. The X and Y are referred to as the sex chromosomes. The remaining chromosomes, by contrast, are called autosomes. The process whereby offspring are produced to be either male or female is known as sex determination.

The 1:1 Sex ratio

Before the chromosome mechanism of sex determination was discovered, there was no known reason why the number of girls and boys born in a large population should be about equal. But the chromosome mechanism explains the approximate equality. We now know that half the sperm cells contain an X and half a Y. Hence, the number of fertilizations that produce girls (XX) and boys (XY) are about equal.

The X and Y Chromosomes in Man (Figure H)

In man, the X is larger than the Y. The X consists of two segments, one of which is no different from the corresponding (homologous) segment of the Y (lower segment) and another which is not represented in the Y. This latter part of the X, therefore, contains genes which are represented only once in the male but twice in the female. It is this which contains the genes for colour blindness, night blindness and hemophilia, previously referred to. The Y chromosome also contains a segment which is not represented in the X, and the genes in this segment are transmitted from father to son in each generation, just like the Y itself.

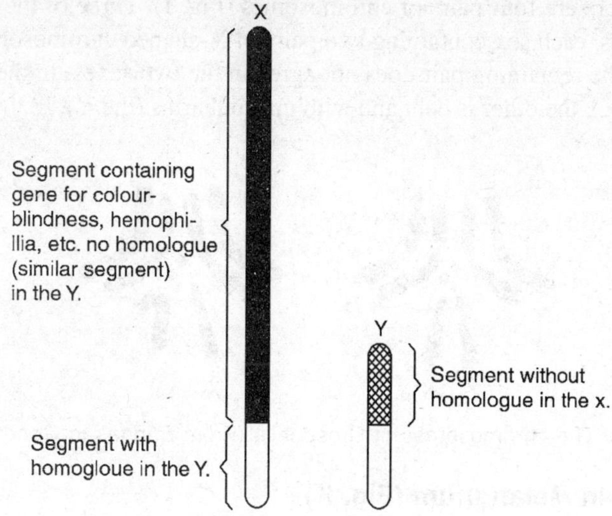

Fig. H. The X and Y chromosomes of man.

Sex Determination in Drosophila

The X chromosome was first clearly observed in 1891 in a male bug by Henking, but its true nature was not understood at that time (Fig. I). In 1901, McClung suggested that the X was concerned with sex determination, and in 1902 he clearly described in grasshoppers. McClung, however, thought that the female lacked an X ; that an X caused a male, and its absence, a female. The first complete account of sex determination was worked out in 1905 by Miss N.M. Stevens, using several insects as her material. Shortly after this E.B. Wilson traced a similar story in other insects. In describing sex determination in lower animals, we may take as our example the case of the fruit fly, *Drosophila,* as reported by Stevens in 1907 and 1908.

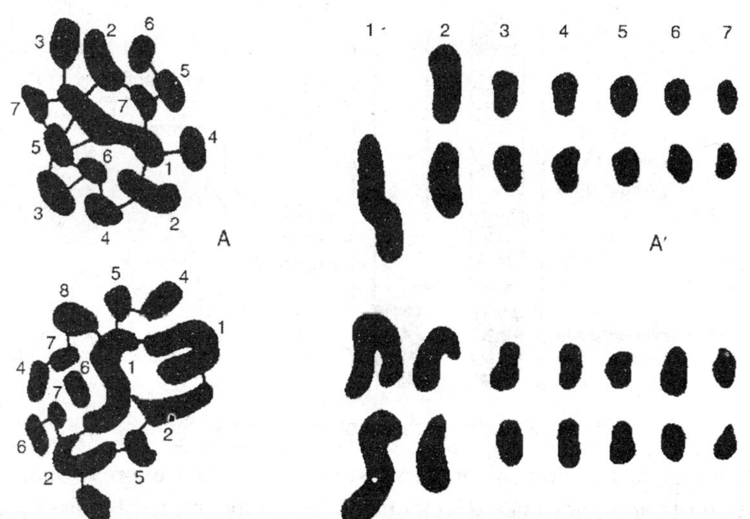

Fig. I. The choromosome of a bug (Protenor). Male, above (A, A′) ; female, below (B, B′). No. 1 is the X (unpaired in the male). (From Wilson, The Cell, by permission of the Macmillan Company, Publishers.)

In *Drosophila* there are four pairs of chromosomes (Fig. J). Three of these four pairs consist of exact mates in both sexes, each sex containing two pairs of V-shaped chromosomes and a small pair of dot like chromosome. The remaining pair does not agree in the two sexes. In the male, one member of this pair is straight (the X): the other is bent and with unequal arms (the Y). In the female, there are two X's instead of an X and a Y.

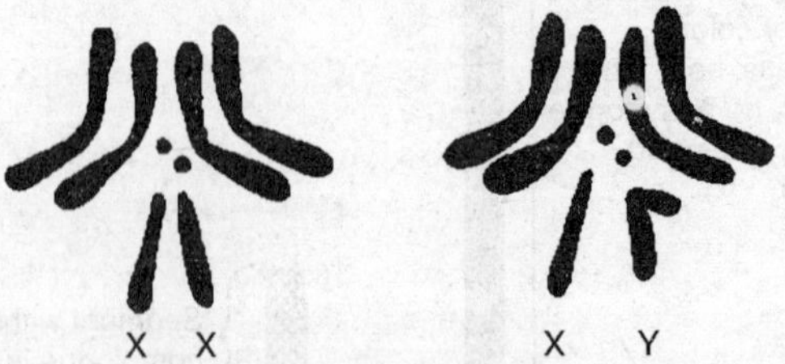

Fig. J. The chromosomes of Drosophila (From Bridges in Genetics).

Sex Chromosomes in Melandrium (Fig. K)

In *Melandrium album*, the diploid female plants have 22 autosomes plus 2 X chromosomes. The male plants have 22 autosomes plus X chromosomes. H.E. Warmke and others have shown that sex is determined by a balance between male-determining genes in the Y and female-determining genes in the X and in the autosomes. The Y chromosome is the largest and most conspicuous member of the complement. Only a small part of the X chromosome is homologous with the Y chromosome. But the major part is non-homologous, or differential. Three distinct regions influencing sex determinations and male fertility have been localized on the non-homologous regions of the Y chromosome. Region 1 suppresses femaleness and thus allows maleness to be expressed in a chromosome.

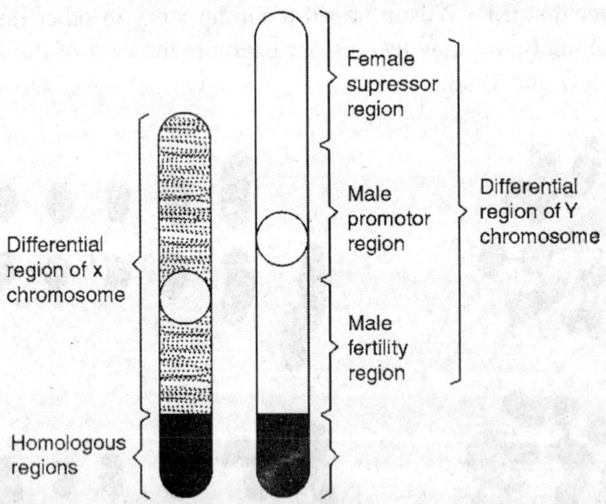

Fig. K. X and Y chromosomes of Melandrium.

In the absence of this region, plants are bisexual; i.e., they express both male and female characteristics. Region II promotes male development. When this region is missing a female plant is produced. Region III carries male fertility genes; loss of this region results in male sterility. The non-homologous region of X chromosome promotes femaleness in the absence of region of X chromosome promotes femaleness in the absence of region I of Y chromosome. The autosomes also carry female determining factor.

Balance Concept of Sex Determination in Drosophila; Mechanism of Sex Determination

This is also called the theory of genic balance as applied to sex determination. On general grounds it would seem likely that sex, like every other trait, depends upon a large number of genes and that these are distributed throughout all the chromosomes- autosomes as well as sex chromosomes. If this is so, then in animals in which the Y is inert or absent (and in which, therefore, we need not consider the Y), the sex genes in the X are of such a nature that one dose of them in conjunction with those in the autosomes causes one sex to develop; two doses-again in conjunction with those in the autosomes - the opposite sex.

If sex development depends on the interaction of genes in the X and the autosomes, then, obviously, it depends on the relative amounts of X and autosomes, not on the absolute number of X's (whether one or two). In sex determination one X produces a male if it interacts with two sets of autosomes, but if it interacted with one set, we should ordinarily expect it to produce a female, because one X and one set of autosomes is relatively the same as two X's and two sets of autosomes. In 1912, H.J. Muller proposed the theory that sex determination depends upon the relative number of X's and autosomes. This theory was later experimentally confirmed by Bridges and was then referred to as the theory of genic balance.

In the Drosophila type of sex determination, the X chromosome favours femaleness, in that a female has more X's than a male (the number of autosomes being the same in the two sexes). On the other hand, in the male there is relatively more autosomal material as compared with the X than in the female, and so the autosomes favour maleness (since in the male there are two sets of autosomes to one X, but in the female only one set per X). We might, therefore, think of the X chromosomes as containing the genes for femaleness and the autosomes the genes for maleness. However, it is not likely that all the female determiners predominant in the autosomes. Hence, the net effect of increasing the X's relative to the autosomes is to throw the balance of sex development in favour of femaleness, and the net effect of a relative increase in the autosomes is to throw the balance in favour of maleness.

The above discussion indicates that male and female do not differ in the kinds of genes they contain, but in their relative amount, each sex containing both male and female genes. It is the interaction of the two kinds in different proportions that produce the two sexes. In 1901, Goldschmidt suggested that every individual has the genes for both sexes, and that sex determination depends on the interaction of male and female genes in different proportions. This essentially is the theory of genic balance. Goldschmidt based his theory on the study of intersexes in the gypsy moth.

Sex Linked Inheritance in Man

Our knowledge of the X chromosome has shed light on the hereditary basis of certain traits associated with sex, as exemplified by colour blindness in man. Women are much less often colour blind than men. But if a woman does happen to be colour blind, and if she marries a normal man, all of her sons are colour blind but none of her daughters are.

We can readily understand this fact on a simple basis. The X chromosome contains genes not only for sex but also for other traits, such as normal colour vision. A recessive mutation in the X caused colour blindness. We can designate the mutated gene as c (for colour blindness), and its normal allele as +. A colour blind woman carries a c in each of her X's. Her formula is c/c. A normal man carries +, the normal allele of c, in his X. We may regard his Y chromosome as a partially empty sack that carries no gene at the locus under discussion. Thus, his formula is +/ (the blank space below the line indicating the "empty" Y). The sons in the mating under consideration receive their X from their mother. As this contains c, they are colour blind. The daughters, on the other hand, receive an X from their father (in addition to one that they receive from their mother) and, as the paternal X contains +, the daughters appear normal. Consider next the mating of a pure normal woman with a colour - blind man (Figure L).

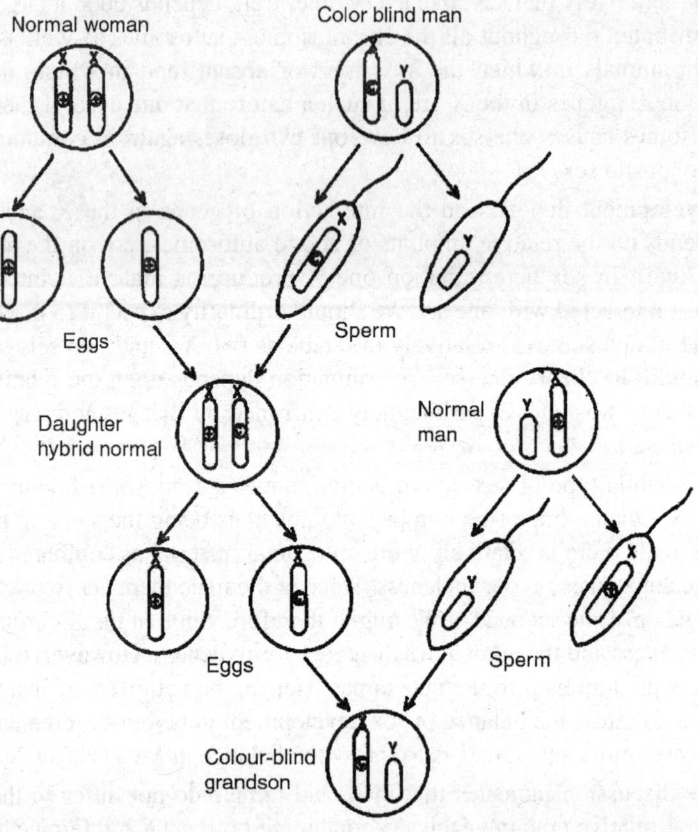

Fig. L. The transmission of the color-blind gene from a man to half his grandsons through his daughters.

The daughters from this mating are c/+ (hybrid normals) and the sons (not shown in the figure) are +/ (normal). If the daughters should marry normal man (c/+ x +/), then half of their sons would be colour blind (C/). The other half (not shown in the figure) would be normal (+/). All the second generation daughters (not shown in figure) would receive a + from their normal father, and therefore they would be normal, though they would also receive an X from their mother (with either + or c) and, therefore, half would be pure normals and half hybrids (or 1 + / + : 1 c / +).

It is now apparent why colour blindness is much scare among women than men. For, in order for a marriage to produce colour-blind sons, it is necessary that only one of the parents, the mother, carry the colour blind gene. But if the marriage is to produce colour-blind daughters both parents must carry the gene. Any such coincidence would be very unusual, unless the parents were related.

Mutations for a number of other abnormalities besides colour blindness are known to occur in the sex chromosomes of man; among these are the abnormalities known as hemophilia and night blindness. Persons affected with hemophilia have a tendence to profuse bleeding from slight wounds because their blood fails to clot upon being shed. In night blindness, unusual difficulty is experienced in seeing in subdued light. (The rods in the retina are affected in night blindness, the cones in colour blindness). Both the genes for hemophilia and night blindness are recessive to their normal alleles. Genes which are in the X chromosome are said to be sex linked. They are tied or linked to sex in inheritance and are said to show sex-linkage.

Sex Linked Inheritance in Drosophila

Numerous mutations have occurred in the X chromosome of *Drosophila* for example, one that changed thc eye colour from red to white. The mutant gene white (w) is recessive to its normal allele (+). When a white eyed female is crossed to a red eyed male (w/w x +/), the daughters are 2/+ (hybrid red) and the sons are w/(white). In all forms of life with sex chromosomes there are probably sex-linked genes. But before a sex linked gene can be discovered it is necessary to have a sex-linked mutation and to make the appropriate crosses of the mutant form to the normal. Sex linkage has been discovered in several forms of life besides Drosophila and man, including such diverse organisms as chickens, moths, and plants with separate sexes.

Sex Linked Lethals

In *Drosophila,* occasionally half the sons in a family are missing. The sex ratio in other words is 2 female ; 1 male. In a small family it might, of course, sometimes happen as a matter

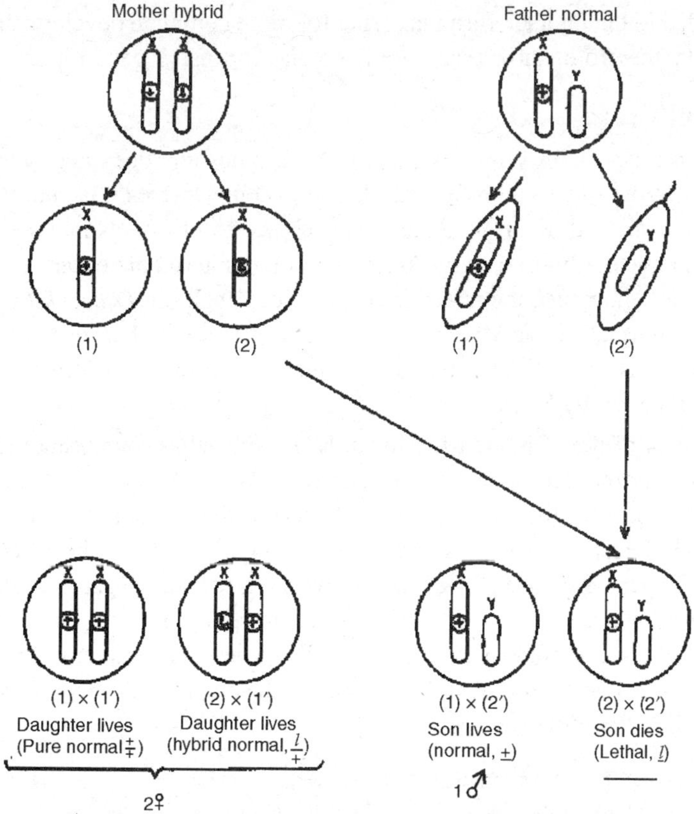

Fig. M. A sex-linked lethal

of chance that there would be only half as many sons as daughters. But the number of offspring in a *Drosophila* family might be very large (three or four hundred) and if such a family contained only half as many sons as daughters it would be safe to conclude that the 2:1 ratio was actually significant (i.e. not due simply to a chance departure from the 1:1 ratio).

In the families under discussion the mother contains a mutation known as a lethal (1) in one of her X's (figure upper left M). A lethal, as the term implies, expresses itself by causing the death of the developing young. The hybrid mother survives because she contains the dominant normal allele (+) in one of her X's, and this prevents the lethal from expressing itself. But half of their sons receive her lethal-containing X and die (figure lower right M). The remaining offspring receive at least one normal X and survive. Hence, a sex ratio of 2 female : 1 male.

The Distribution of the Sex Genes in the X Chromosome

It is possible to break up the X chromosome of *Drosophila* into segments by means of x-rays (applied to the entire animal), and it is then found that the sex-determining material is not limited to any one segment. This can be shown by means of sexually abnormal flies known as "intersexes". An inter sex, as the term implies, is intermediate between a normal male and female in sex development. It has a certain chromosomal composition, different from that of normal males and females. By the proper experimental procedure, it is possible to get flies which contain the usual chromosomes of an intersex and, in addition, a fragment of the X. It is then found that the fly develops more nearly into a female, and this is true regardless of from what part of the X the fragment is derived. Moreover, larger fragments of the X, if added to the intersex formula, make for more complete development than the smaller fragments. These facts, therefore, show that there is no one "primary" gene for sex in Drosophila.

The Y Chromosome

In some animals, especially some insects, the Y chromosome does not have very much to do with sex development. It often is smaller than the X and in some species it is entirely missing. In fact, it is possible, as Wilson showed, to arrange a number of species of insects in a series beginning with those that have a Y as large or almost as large as the X through intermediate steps to those that have no Y at all. In all these cases, however, the male has an X. Occasionally in *Drosophila* a male lacks the Y but not the X, yet it is normal in every respect except that it is sterile because a of immobility of its sperm cells. The Y contains about eight to ten genes which are necessary for the mobility of the sperm cells in *Drosophila,* as shown by Stern.

About 98 per cent of the Y is inert in *Drosophila*, Muller has found that the inert part is produced by about a half a dozen genes, each of which produces a segment or "block" of inert material about itself. The block genes (as they are called by Muller) have no auspicious effect on development proper but the removal of each one causes a certain reduction in the size of the Y (equal to the amount of inert material that the gene produces). There are, however, a few active genes of the ordinary kind in the Y; for instance, one that is concerned with bristle development. It must not be concluded from what has been said above that the Y chromosome is inert in all species. As a matter of fact, in fishes, moths, and some plants, the Y has a very decided influence on sex development. It may favour either maleness or femaleness (depending upon the species).

Sex Limited Characters : or Genes

Sex limited genes are those which express characters in only one sex. The sex limited genes may be located on any chromosome (sex linked genes are located only on the sex chromosome). Their expression in the vertebrates is governed by the sex hormones. Sex limited genes are responsible for the secondary sexual characteristics as well as primary sexual characters.

Examples

1. In man the beard is produced by sex-limited genes. A woman, normally, does not have a beard, yet she surely carries all the genes necessary to produce a beard. But the expression of that particular gene in ladies is prevented by the absence of a particular hormone. In rare cases abnormalities in hormone secretion may occur in a woman which allow these genes to express themselves and the result is a bearded lady.
2. Breast development is normally limited to woman but hormone unbalance may cause breast development in a man.
3. The genes for the deep masculine voice and masculine musculature in man will express themselves only when the male hormone is present. Genes for the feminine musculature, express themselves in the absence of male hormone. They do not require the presence of the female hormone.
4. In cattles the milk production is controlled by sex limited genes. He bulls also carry the gene for milk production. But it is not expressed in the male sex. In the cows the genes for milk production makes its expression in the presence of female hormones. The brilliant plumage of peacock is also due to some sex limited genes.
5. As excellent illustration of sex limited inheritance is provided by the plumage pattern in birds. In domestic fowls of Leghorn, males have long curved, fringed feathers on tail and neck, but feathers of female are shorter, straighter and with out fringe. Thus males are cock feathered and females are hen feathered. Results of various crosses show that hen feather is due to a dominant gene H and cock feathering is due it's recessive allele "h". It has been found that the expression of the genes H and h, depends upon the sex hormone. Any male that receives atleast one dominant gene will be hen feathered, but those males which are homozygous for the recessive gene (hh), will show the cock-feathered condition. Females are all hen-feathered regardless of genotype. It shows that particular type of feathering depends upon specific combination of genotype and sex hormones. The dominant gene H produces hen feathering in the presence of male or female sex hormone. The recessive gene, "h" produces cock feathering in the absence of female hormone and hen feathering in the presence of female hormone.

Genotypes	Phenotypes	
	Male	Female
HH	Hen-feathered	Hen feathered
Hh	Hen-feathered	Hen feathered
hh	Cock feathered	Hen feathered

Sex Influenced Genes

The sex influenced genes are those whose dominance is influenced by the sex of the bearer. They are located on the autosomes. The sex influenced genes express more frequently in one sex than in the other. The baldness in man is a very good example for sex influenced character. This particular character is dominant in men and recessive in women. This is because the gene for baldness, (B) in heterozygous condition (Bb) express itself in male but the heterozygous females are normal even though they carry the gene for baldness. It shows that only one gene is enough to produce a bald man, where as a woman must require two such genes to be bald. In man the single gene for baldness can operate only in the presence of male hormone.

Genotypes	Phenotypes	
	Male	Female
BB	Bald	Bald
Bb	Bald	Normal
bb	Normal	Normal

There is another interesting sex influenced gene in man which affects the length of the index finger. The short index finger is due to a gene which is dominant in the male and recessive in the female. The horned condition in sheep is also due to a sex influenced genes.

CHAPTER 6

Maternal Influence cn Inheritance

Most of the phenotypic characters are controlled by genes present in the chromosomes. But some characters are expressed by factors present in the cytoplasm. Very little cytoplasm is contained in the sperm cell of an animal; most of it is in the egg. Hence, if there are hereditary units in the cytoplasm, we should expect that they would more likely be transmitted to the offspring through the egg than through the sperm cell. The genes or factors lying in the cytoplasm are called plasmagenes. Plasmagenes also transmit characters from one generation to another. The transmission of characters controlled by plasmagenes is called cytoplasmic inheritance or extrachromosomal inheritance. This was first described by Correns in 1908.

Characteristics of Cytoplasmic Inheritance

1. Plasmagenes are situated outside the chromosomes.
2. Like chromosomal genes, plasmagenes are also self-replicating.
3. Plasmagenes are transmitted by means of female cytoplasm only.
4. Like chromosomal genes, plasmagenes are also capable of mutation.
5. The offspring receive cytoplasm only from female gamete and not from male gamete. Hence the plasmagenes of female parent alone are contributed to the offspring. As a result the cytoplasmic inheritance is said to be a maternal inheritance.
6. The results of reciprocal crosses are not the same.

   Examples : Kappa particles in Paramacium
   Shell coiling in snail
   Plastid inheritance in Mirabilis
   Inheritance of pigments in Ephestia

Shell Coiling in Snails (Delayed Gene Effect in the Snail (Limnaea)

An actual case in which the effects of a gene are delayed has been found in the fresh water snail, Limnaea (by Boycott, Diver, and Garstand). The shell of a snail normally turns to right (dextral), but there is a recessive mutant with a left-turning (or sinistral) shell. We can designate the mutant gene as s (or sinistral) and its normal allele an +. The coiling of the shell can be traced very far back in development, and the difference between a dextral and a sinistral snail can be seen as early as the second "cleavage" (or cell division) of the fertilized egg. In fact, the difference is potentially present in the unfertilized egg.

Now, the direction in which the developing egg coils is not determined by its own genes but by the genes of its mother. The egg, of course, is derived from an unreduced cell (the diploid cocyte), and this has the same genotype as any other unreduced cell of the mother. Suppose then that the mother was left-turning and of genotype s/s. This formula would apply to all her unreduced cells, including her oocytes. In these oocytes, the cytoplasm would become left-turning due to the s/s constitution of their nucleus. The eggs derived from the oocytes would, therefore, develop into left-turning snails, even though they were fertilized by + sperm (containing the dominant normal allele for dextral, (figure N). But in the hybrid, the unreduced eggs (oocytes) would have a nucleus of constitution + / s, and since + (dextral) is dominant, the cytoplasm of these unreduced eggs would become dextral. Hence, all the reduced eggs would have dextral cytoplasm and they would give rise to dextral offspring. Now, in the hybrid (s/+), half of the reduced eggs have a nucleus of genotype s. If the s eggs were fertilized by s sperm, the resulting offspring (s/s), though pure for s, would be dextral, because the egg cytoplasm was made dextral by the genotypic constitution of the hybrid mother (+/s).

The Problem of Cytoplasmic Inheritance

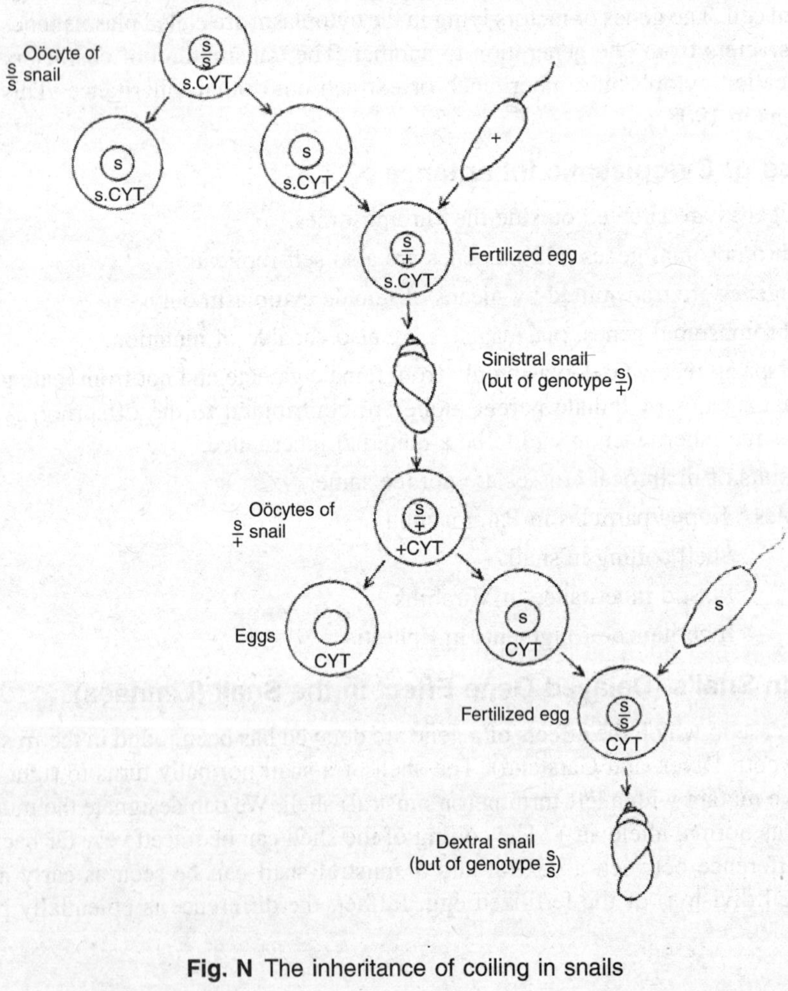

Fig. N The inheritance of coiling in snails

In brief, in the above hybrid (+/s) the + gene does not have any immediate effect on the developing embryo, but only on the eggs of the adult. That is to say, the effect of the gene is delayed until the hybrid becomes an adult. It then makes all her eggs dextral. Coiling in snails is often referred to as a case of "maternal inheritance" because the genotype of the mother determines the direction of coiling in the offspring. But the mother's genotype is determined by her father as well as by her mother, and so inheritance itself is really biparental, not exclusively maternal. It is, therefore, incorrect to refer to coiling as a case of maternal inheritance.

The Kappa Bodies in Paramecium

A certain race of Paramecium found by Sonneborn and called "killer" destroys other races of Paramecium by releasing a toxic substance into the water in which the Paramecia are swimming. The killer race owes its peculiar properties to certain bodies that are contained in its cytoplasm and that have been called "kappa" (the Greek letter k). The normal or "sensitive" race lacks these bodies. The kappa bodies pass from one generation of Paramecia to the next in the process of cell division, the kappa bodies multiplying by division in step with the Paramecia.

The killer race has a certain nuclear gene (K), allelic to k, found in sensitive strains. It is possible, by crossing, to substitute k for K in a line of killers. These then lose their kappa bodies and become sensitives. Thus, kappa is dependent for its continued existence on the genetic constitution of the nucleus (either K helps kappa, or k hinders it, possible by creating cytoplasmic conditions either favourable or unfavourable, as the case may be). The exact nature of kappa has not been determined. It was at first thought that it might be a gene, but later work showed that it behaved in some respects like an organism living in the cytoplasm. In particular, kappa was found to be somewhat independent of Paramecium in its division rate, less rapid at higher temperatures for kappa than for Paramecium. It is possible to reduce the number of kappa bodies in a line of killers by culturing the line at a high temperature. Each kappa particle does not then divide into two (on the average) every time the Paramecium does, and the number of kappa particles per Paramecium becomes fewer and fewer in the course of cell division. Eventually some Paramecia are left without any kappa bodies and become sensitives. This sort of thing would hardly be expected if kappa were a gene. However, if kappa were an organism, it might very well have a division rate of its own, somewhat different from that of Paramecium. Moreover, Preer has been able to stain the kappa particles and observe them under a microscope. Genes, as we usually know them, are too small to be visible under a microscope. Therefore, the relatively large size of the kappa bodies makes it improbable that they are genes and supports the theory that they are organisms. In summary, kappa is unlike a gene and more like an organism, both in respect to its independent division rate and its relatively large size. It is also unlike a gene in several other respects.

If kappa is an organism, it must be regarded as a symbiont, since it does not harm its host in the manner of parasite. On the contrary, it helps its host, by getting rid of competitors (other Paramecia) in the medium in which it lives. Symbionts are often green as in the case of the one-celled green algae (Zoochlorellae) found in the cells of the green Hydra. It happens that there is a species of Paramecium (*P. bursaria*) which harbours Zoochlorellae as symbionts, and it also happens that the number of Zoochlorellae per Paramecium in this species is about the same as the number of kappa bodies found by Preer in the killer race (about 1000 to 1500). This correspondence in numbers led E. Altenburg to suggest that kappa was derived from Zoochlorellae, which, in the course of evolution, had lost their chloroplaslts, become smaller, and assumed some additional new properties, including the killer property. In any event, until the true nature of kappa is definitely settled, kappa cannot be considered evidence for cytoplasmic genes.

Plastid Inheritanace in Mirabilis

The cytoplasm of plants contains minute bodies known as plastids. Certain of these (known as chloroplastids) contain chlorophyll, the colouring substance in green plants. Others, known as leucoplastids are colourless and store starch. Plastids resemble genes in two important respects: (1) they are capable of multiplying by division; (2) they are capable of mutation. Before a plant reproduces, a few plastids get into the eggs but none, as a rule, gets into the pollen. Thus, the offspring inherit their plastids almost solely from the female part of the parent plant.

A large part of a plastid consists of chlorophyll or of stored carbohydrate. The genetic element in a plastid is very small. It has not been definitely identified under the microscope. The egg contains its plastids in the form of "proplastids". These, at first are so small that they cannot be identified. If, indeed, they happen to be large enough be visible, they cannot be distinguished from the many other small granules in the cytoplasm. Possibly they are invisible when totally free of chlorophyll and other material not essential to their genic nature.

Sometimes a seedling is white because its plastids lack chlorophyll. As a rule, the white seedling is due to a recessive gene mutation which somehow prevents the chlorophyll from developing. The mutant gene in question is in a chromosome and is inherited like other genes; it might be transmitted to the offspring through either the pollen or the eggs. But sometimes the mutation is in the plastid itself, and in such cases it can be inherited, as a rule, only through the female line. In this way through mutant plastids, it is known that plastids in general are, as a rule, transmitted through the eggs.

An example of plastid inheritance is furnished by the four O' clock, *Mirabilis jalapa*, studied by Correns. In four o' clocks there is a race known as albomaculata in which the leaves are mottled with white and green patches. In the white patches, all the plastids are without chlorophyll but, in the green patches, atleast some of the plastids are green, though not necessarily all. In fact, some green areas contain only green plastids; other green areas contain both green and white plastids. Between the boundaries of a uniform green and a white area, a cell often contains both kinds of plastids, and this fact suggests that in the developing plant the cells with both kinds of plastids give rise to the cells with only one kind. In fact, at cell division, the plastids are distributed in a purely random manner, so that occasionally a cell gets only white or only green plastids. Such a cell might then give rise to a uniform patch by further cell division and growth. The size of a uniform area varies; it might include an entire branch. A seed from a uniform green branch gives rise to a seedling that is wholly green; and a seed from a white branch grows into a white seedling which dies before maturity. In a mottled branch, there are still large numbers of cells, including eggs, which contain both green and white plastids; and an egg cell with both kinds would, upon being fertilized, give rise to a mottled offspring. In any event, the source of the pollen would have not influenced on the colour of the offspring. Their colour is determined entirely by the egg, and what the egg produces depends on the character of its plastids whether green, white or both. Since the plastids are in the cytoplasm, we are clearly dealing with a case of cytoplasmic inheritance. Moreover, we are dealing with plasmagenes, since plastids (like genes) have the capacity of self-duplication and mutation.

Cytoplasmic Inheritance in Yeast (Mitochondria)

Mitochondria are cytoplasmic organelles containing their own DNA and some enzymes that catalyze the process of oxidative phosphorylation. Mitochondria are not autonomous bodies because they require their own genes and nuclear genes to function normally.The role of mitochondria in cytoplasmic inheritance was first understood when Ephrussi and his colleagues (1955) studied petite mutants (petite meaning small) in yeast (*Saccharomyces cerevisiae*). When a cell divides, approximately equal numbers of mitochondria pass into daughter cells. They can originate from pre-existing mitochondria, and can probably divide transversely.

Several mutant strains of yeast are referred to as petite mutants because they form smaller colonies on agar as compared to normal yeast. The growth rate of petites is slow due to absence of respiratory enzymes, cytochromes a, b and c, and due to deficiency in some dehydrogenases present in normal mitochondria.

The petite strains are of three types. In the first type called segregational petites when an individual is crossed with a normal strain, a 1:1 ratio of normal : petite results after segregation. This suggests Mendelian inheritance and the petite strain has originated due to a mutation in nuclear genes.

The second and third types of petite strains arise when normal yeast cells are treated with the acridine dye euflavin of with ethidium bromide dyes which are known to intercalate in double stranded DNA. The petites so formed do not segregate regularly when crossed with normal yeast, and fall under two types - neutral and suppressive. When a neutral petite is crossed with a normal type, all the progeny in the next and successive generations are normal and the petite trait never reappears. If a haploid cell of the neutral petite strain is mated to a haploid vegetative cell of a normal strain, a diploid zygote is formed. The diploid cells produced by this zygote reproduce asexually, sometimes may even divide meiotically to form spores, but the petite trait is never visible in the progeny. The genetic explanation of the trait is that there is a cytoplasmic factor (p+) which is present in the normal strains of yeast but absent in petites (p-).

The suppressive petites when crossed to normal cells of yeast show that petite trait in the progeny but in non-Mendelian ratios. Suppressive petite mutants are found to have mutant DNA in their mitochondria. The mutant mitochondria replicate and transmit the mutant phenotype to the progeny cells.

Mitochondrial mutants are also known in *Neurospora, Paramecium* and *Trypanosoma*.

The Genetics of Mitochondria and Chloroplasts

The presence of organelle—specific DNA in mitochondria and chloroplasts was identified with certainly over a decade ago. The occurrence of specific chloroplast DNA was first demonstrated by Chun *et al.* in 1963, and of mitochondrial DNA a year later by Luck and Reich (1964). It is now fairly well established that organelle DNA is distinct from nuclear DNA in several respects. That it can carry out synthesis of RNA and protein on its own synthesizing 'machinery' and replicate itself. There are mitochondrial and chloroplast ribosomes. The genomes of these organelles contain genes for the rRNA species present in both large and small subunits of their respective ribosomes.

Mitochondrial DNA (mtDNA)

Mitochondria contain an open or closed circular DNA molecule. DNA from different species shows a wide range in average base composition ranging between less than 20% to more than 50% of G + C is needed to code for an average protein. Thus regions of mtDNA containing very small amounts of G and C nucleotides (in cases where the average base composition is less than 30%) may not be able to code for a protein (G-Guanine; C-cytosine).

Recent evidences indicates that there is one copy of the gene for each of the r RNA species associated with the large and small subunits of mitochondrial ribosomes. The positions of these two genes have been mapped in some lower eukaryotes. In the mold *Neurospora crassa* the two genes for rRNA are adjacent to each other, whereas in yeast (*Saccharomyces cerevisiae)* they are separated, with the use of radioactive isotopes it has been shown that in *N. crassa* and *S. cerevisiae* some of the mitochondrial inner membrane proteins are synthesized by the mitochondrial protein synthesizing apparatus.

The existence of mutations have been recorded in the mitochondrial genome. A study of the oligomycine resistance mutations have led to the identification of 3 distinct linkage groups in mtDNA. Recently Tzagaloff *et al.* (1975) have been able to identify a number of mutants with defects in cytochrome oxidase and coenzyme Q-cytochrome c reductase.

mtDNA has also been studied in a limited number of higher plants. A closed circular DNA with molecular weight ranging between 60-80 x 106 daltons has been seen in spinach, lettuce and pea. However, higher plant mtDNA has not shown the presence of rRNA and tRNA gene. No protein synthesis is recorded for plant mtDNA.

Chloroplast DNA (ctDNA)

Circular DNA molecules have been observed in ctDNA of both higher plants and algae. It contains genes for both large and small rRNA species present in chloroplasat ribosomes. Hybridization studies have indicated one gene per chromosome for each rRNA in tobacco, whereas in bean, lettuce, pea, spinach and maize there are two genes per chromosome. Similar studies have demonstrated the presence of genes for about 25 tRNA species each of size about 25,000 daltons.

Radioactive isotopes studies in vivo have clearly shown that protein synthesis occurs in chloroplasts. One such protein is the large subunit of ribulose diphosphate carboxylase (termed fraction I protein) which is a major constituent of chloroplast as well as other membrane proteins. The large subunit of fraction I protein and some other membrane proteins are also synthesized by isolated chloroplasts. The small subunit of fraction I protein is coded for by nucleus DNA.

Replication of Mitochondrial and Chloroplast DNA

Both mtDNA and ctDNA replicate within their respective organelles. In vitro synthesis of DNA also occurs in the isolated organelles if the four deoxyribonucleotide triphosphates are supplied, and is blocked by inhibitors of DNA synthesis such as ethidium bromide, adriflavin and actinomycin-D. Whereas nuclear DNA synthesis is limited to S phase of the cellcycle, there is evidence for continuous synthesis of organelle DNA throughout the cell cycle. In vivo density label studies have shown that mtDNA in Hela cells and ctDNA in Chlamydomonas and Euglena replicates semiconservatively. As in bacteria, replication of organelle DNA appears to involve the membrane (called mesosome in bacteria).

Inheritance of Pigments in Ephestia (moth)

In the larvae of *Ephestia* moth pigments are produced by a hormone known as kyanurenine. The synthesis of this hormone is controlled by a pair of dominant nuclear genes AA. When these genes are present in the recessive condition (aa), the larvae are colourless. In a cross between recessive female aa and heterozygous male (Aa), 50% of the larvae of F_1 generation are pigmented and 50% are unpigmented. At the same time in a cross between heterozygous female and a recessive male all the larvae develop black pigments. But in the adults only 50% of the individuals have the pigementation. These pigmented adult animals carry the gene A which is responsible for the production of the pigments. Other 50% of the adult individuals lose their pigment in the adult because they carry recessive genes aa which have no power to produce the pigments. Their larval pigmentation is due to the kyanureine which is brought by the cytoplasm of the egg.

P :	Pigmented male	X	Unpigmented female	Pigmented female	X	Unpigmented male
	Aa		aa	Aa		aa
F_1 :	Aa		aa	Aa		aa
	Pigmented		Unpigmented	Pigmented		Pigmented in the larvae; but unpigmented in adult

CHAPTER 7

Alterations in Genetic Make Up-Changes at Genetic Level

The sudden heritable changes in genes, other than those due to Mendelian segregation and recombination constitute mutations. The idea of mutation first originated from observations of a Dutch botanist Hugo de Vries in 1880, on variations in plants of *Oenothera lamarckiana* (evening primrose) growing wild in Holland. This plant had been introduced from America and had grown wild in Europe. De Vries collected seeds from *Oenothera* plants, raised plants from them, and analysed the progeny for transmission of traits showing variation. He found that heritable variations were distinct from environmental variations. He gave the name mutation (Latin mutare meaning change) to heritable changes and in 1901 published a book entitled "The Mutation Theory". De Vries is credited with the discovery of the idea of mutations. Mutations have been broadly categorized as somatic and germinal mutations. When a mutation occurs in a somatic cell, it does not change the whole organism, but produces a phenotypic change in the organ to which the mutant cell belongs. The resulting individual is a mosaic for mutant and normal tissues. White sectors in the red eyes of *Drosophila* males is an example of somatic mutation. In plants somatic mutations are transmitted to the progeny by methods of vegetative propagation such as budding and grafting. Germinal mutations take place in cells of the germ line. A classic example of germinal mutations and perhaps also the one first recorded is that of short-legged sheep. In 1791 a farmer named Seth Wright in New England, U.S.A., noticed a short-legged lamb which could not jump over the fence and run away. Considering this to be an advantage, he started breeding work on the short-legged lamb. He had 15 eves (females) and one ram (male) out of which he established a line of short-legged sheep. The name ancon was given to this breed (Greek meaning elbow), because the crocked looking forelegs showed resemblance to the human elbow. The first short-legged lamb arose as the result of a recessive mutation in one of its recent ancestors. From the breeding work done in Wright's farm the crosses indicated that this lamb was homozygous for the mutant gene and several of the eves were heterozygous carriers. The same mutation is said to have occurred in a flock of sheep in Norway in 1925 and another breed of ancon sheep has been produced out of it.

Spontaneous Mutations

Normally genes are extremely stable structures, yet mutations have been occurring in nature all the time in a spontaneous manner. Although mutations are rare and occur at random, they have an advantage in providing a constant source of new variability which is required by organisms to adapt to the changing environment. There are numerous physical and chemical agents which can increase the frequency of mutations. If a mutation is dominant it is easily detected because it is always expressed. But dominant mutations are exceedingly rare. The mutant gene which causes aniridia (absence of iris in human eye) is dominant over the normal allele. Chondrodystrophic dwarfism is also due to a dominant gene. In this type of dwarfism, the head and trunk are of normal size, but arms and legs are extremely short. Mutations which cause lethal effects in adult individuals produce a clearly visible phenotype and are simple to detect. In *Drosophila* less than 20% of all mutations are lethal.

Mutations which are most frequent are usually recessive. They are most difficult to detect if the changed locus lies on an autosome. Since a single allele of the recessive mutant gene will not express itself, such mutations perpetuate themselves for generations in heterozygous carriers without being noticed. They are detected when both gametes involved in fertilization carry the mutation, thus bringing together two alleles of the recessive mutant gene.

Haploid organisms offer an advantage for the study of autosomal recessive mutations owing to the fact that a single allele expresses itself. If a recessive mutation is sex-linked it follows a specific pattern of transmission. Such a mutation is often detected when it expresses itself even in the single dose in homozygous males.

Various tests have been devised for detecting spontaneous mutation in a bacterial population.

Replica Plating

In 1946 Luria and Delbrueek noticed one of the first cases of spontaneous mutation in a bacterial population. When Escherischia *coli* cells were infected with the bacteriophage T_1 most of the cells lysed releasing progenies of newly synthesized viruses. However, a very small proportion of bacterial cells (about 1 in 10^8) did not lyse but could multiply and form colonies. These bacteria as well as their descendants obtained by subculturing the colonies were found to be resistant to T_I. It appeared that some of the originally infected cells had become mutant forms resistant to phage. The replica plating technique devised by Lederberg in 1952 demonstrated that *E. coli* cells had undergone spontaneous mutation and become resistant to T_1.

The technique is as follows. *E. coli* cells are grown on a master plate of nutrient agar until they form a continuous lawn of cells. A sample of cells is transferred by pressing on it a sterile velvet surface supported on a circular block (replicator) to a fresh plate of agar medium already coated with T_1 phage particles. The method produces an identical pattern of colonies on the second plate. Some mutant phage resistant colonies also grew on this plate.

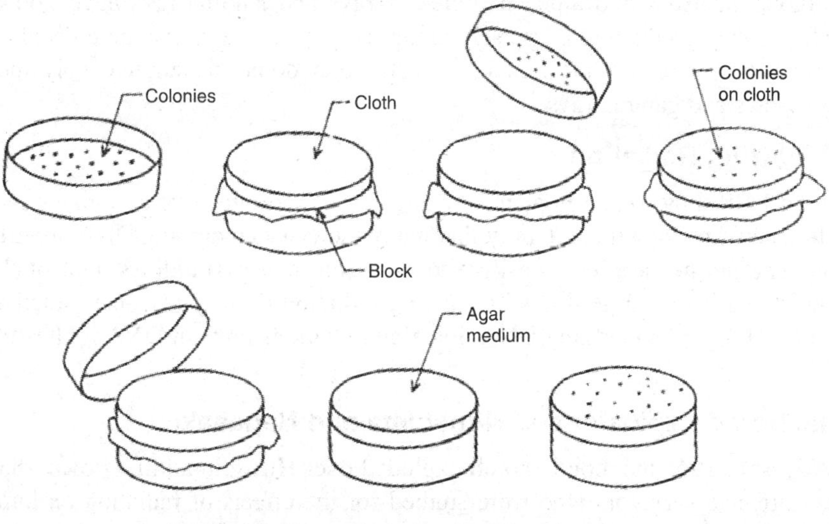

Fig. O The technique of replica plating.

The question arises about the origin of these mutant colonies. If the mutant colonies had originated from the master plate, then it would be possible to obtain a culture of such cells by inoculating from locations on the master plate corresponding to positions of resistant colonies on the second phage—containing plate. Lederberg therefore repeated the procedure by preparing a second master plate and replica plating on to a phage coated plate. This was done 3 to 4 times, each time plating cells thinly. In this way single resistant colonies could be identified (figure O) and a true breeding phage resistant strain would be developed. The experiment proves that the resistant cells arose by mutation on the first master plate and were not induced by the phage to become resistant.

Induced Mutations (Mutagens)

In 1927 H.J. Muller showed for the first time that mutations could be induced in *Drosophila* by use of external agents or mutagens. He was awarded Nobel Prize in 1946. When flies were irradiated with X-rays results in a linear increase in the frequency of mutations. By 1930s, it became established that physical agents such as X-rays, gamma rays, UV radiation, and some chemical agents are all effective as mutagens. At the same time L.J. Stadler demonstrated that X-rays could produce gene mutations in plants of barley. Mutagenesis by radiation and chemicals are discussed separately below.

Ionizing Radiation

Due to their shorter wavelength, X-rays and gamma rays penetrate tissues deeper than visible and UV light. They can impart enough localized energy to absorbing tissue to ionize atoms and molecules. When a highly energetic wave moving at high speed is stopped, it releases energy. This energy makes an atom lose an electron and become a charged particle or ion. The process is called ionization. The free moving electron causes other atoms to lose electrons and become positively charged ions. The two processes generate pairs of positively and negatively charged ions. A number of ions may be clustered together to form an ion track. Ions undergo chemical reactions to neutralize their charge to reach a more stable configuration. While doing so they (ions) produce breaks in chromosomes (DNA) thereby inducing mutations. The free ions moreover, may combine with oxygen and produce highly reactive chemicals which may also react with DNA and cause mutagenesis. Some ionizing radiation is electromagnetic such as X-rays and gamma rays and some consists of subatomic particles such as electrons, protons, neutrons and alpha particles. X-rays and gamma rays have a low rate of linear energy transfer as they produce ions sparsely along the ion track and penetrate deeply into the tissue. Charged particles have a higher linear energy transfer, they do not penetrate deeply and produce more damage than X-rays and gamma rays.

Effects of Ionizing Radiation

Zirkle in 1930 showed that in plants the nucleus is more sensitive to ionizing radiation than the cytoplasm. It is now known with certainty that many molecules including DNA are affected by ionizing radiation. The purines are less sensitive to radiation than pyrimidines. Out of the pyrimidines, thymine is most sensitive. Large doses of ionizing radiation destroy thymine, uracil and cytosine in aqueous solution. By deploymerizing DNA, ionizing radiations prevent DNA replication and stop cell division.

The Atomic Bomb Explosions at Hiroshima and Nagasaki

On August 6, 1945 the atomic bomb exploded over Hiroshima killing over 78,000 people and leaving many affected survivors who were studied for the effects of radiation on human beings. On August 9, 1945 a second atomic bomb was exploded over the city of Nagasaki. The words hibakusha (explosion affected person) and higaisha (victim or injured person) became understood to people in the world. A number of individuals who had received extensive radiation had no children for several years. Many of the survivors have shown visible chromosome abnormalities such as breaks and translocations. Persons over 30 years of age who had received more than 200 rads of radiation were found to be more sensitive to radiation than persons of the younger age group. About two percent of survivors developed leukemia within the following decade. The children of survivors have not shown a detectable increase in genetic abnormalities.

UV Radiation

E. Altenberg was first to show that ultraviolet radiation can induce mutations. UV rays have longer wavelength than X-rays and gamma rays, hence they cannot penetrate tissues as deeply. Their mutagenic action is limited to bacteria, fungal spores or other free cells whose genetic material lies very near the irradiated surface. L.J. Stadler induced mutations in pollen grains of maize plants. He found that UV rays with a wavelength of about 260 nm (nanometer) were more effective in mutagenesis. In fact the wavelength most readily absorbed by DNA also happens to be 260 nm, thus proving a direct correlation between UV induced mutations and DNA. The Sun is a powerful source of UV. As such the Sun's rays would be expected to cause widespread damage through mutations in all living organisms. But fortunately a layer of Ozone in the upper atmosphere absorbs most radiation below 290 nm. The UV rays falling on DNA are absorbed by the pyrimidine bases especially thymine.

Molecular Basis of Point Mutations

Mutations which alter nucleotide sequences within a gene are of two types: base pair substitutions and frame shift mutations. In base pair substitution one base pair, for example AT may be replaced by another such as CG or GC. These are of two further types, namely transitions and transversions. Transitions are base changes in which a purine is substituted by another purine as when A—T pair is replaced by a G—C pair or vice versa; or when a pyrimidine is replaced by a pyrimidine such as when T—A is replaced by C—G, or vice versa. Transversions are alterations in bases in which a purine is substituted by a pyrimidine, that is, when an A—T pair is replaced either by T—A or C—G, and vice versa (Figure P).

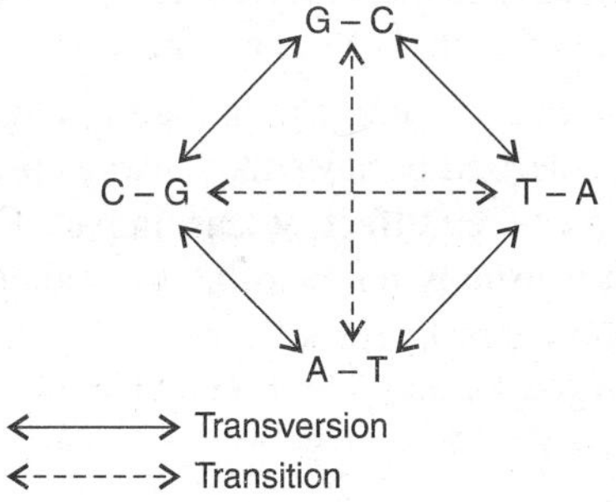

Fig. P

In frame shift mutations, (so called because they shift the normal reading frame of base triplets in mRNA) single base pairs are deleted from or added to DNA in interstitial position. The genetic code requires reading of consecutive base triplets from a fixed starting point. If a single nucleotide is inserted or deleted, it shifts the reading frame, and all the subsequent triplets are read off differently. The entire portion of the polypeptide chain after insertion or deletion is translated wrongly, resulting in a "nonleaky" phenotype. Frameshift mutations therefore differ from base pair substitutions which produce "leaky" phenotypes by altering only one nucleotide in a single triplet, so that only one amino acid is wrongly placed.

Nonsense and Missense Mutations

When a mutation in a triplet changes the codon so that it is recognized by another amino acid, it is called missense mutation. But if the mutation changes the codon for a specific amino acid into one which signals chain termination (non-sense codon), it is called a nonsense mutation. Missense mutations occur more frequently than non-sense mutations, and usually result in single amino acid replacements in the polypeptide chain. Such a chain may still have biological activity. Nonsense mutations result in premature termination of polypeptide chains so that only fragments of chains are formed. The lengths of fragments depend upon the distanace of the nonsense mutation from the starting codon.

Chemical Mutagens

More recently a large number of chemicals have been found to be mutagens (i.e., capable of producing mutations). The most important of these are peroxides. In addition, the list includes formaldehyde, permanganate, urethane (a narcotic), and caffeine, (alcohol has not been found to be mutagenic). Most chemical mutagens other than mustard gas and peroxides produce only a very small percentage of mutations at best (that is, when they are employed in a dose that gives their maximum effect), and many of them are effective in certain organisms but not in others; or they are effective in certain organisms but not in others; or they are effective only at certain stages in mitosis, or at a certain stage in development, or even in one sex but not the other (formaldehyde is effective in the male *Drosophila*, but not in the female).

Some of the peculiar limitations of chemical mutagens are readily understandable. Thus, for example, before a chemical agent can produce a mutation in a resting nucleus, it must penetrate the nuclear wall, and a chemical that cannot penetrate the wall could not, of course, produce mutations during the resting stage, though conceivably it might at stages in mitosis when there is no wall. Moreover, no mutagenic action would be expected of any chemical that is destroyed by the cytoplasm before it can reach the nucleus or the chromosomes, and in some organisms such destruction might occur, but not in others (depending on differences in their metabolism); or, destruction might take place at certain stages in the life cycle of a given organism but not at other stages.

By contrast, X-rays tend to be more uniform in their mutagenic effectiveness, first because they have great penetrating power, and second because they always produce ions. It is this ionizing action, it will be recalled, to which they owe their effect, at least in part. Chemical agents, by contrast, have no such single action in common. It is usually not possible to predict whether a certain substance will be mutagenic on the basis of its chemical properties. In fact, two substances might be chemically similar, such as mustard gas and lewisite, yet the one might be mutagenic and the other not or only slightly so.

Alkylating Agents

This is the most powerful group of mutagens. Chemicals of this group bind invitro to the N-7 position of guanine. In vivo the 0-6 position of guanine and the 0-4 of thymine are preferred. They transfer alkyl groups to the nitrogen atoms of the bases in DNA. Alkylation of guanine causes ionization of the molecule and changes its base pairing specificities. Alkylated guanine pairs with thymine instead of cytosine. Thus on replication there is a transition from a G—C to an A—T base pair. Alkylation of purines leads to hydrolysis of the sugar base linkage so that the purine base is lost from the backbone of the DNA molecule producing apurinic gaps. When DNA replication takes place, almost any base may be inserted in the gap. Insertion of a wrong base in a gap produces transitions as well as transversions. There are some bifunctional alkylating agents which form cross links between guanines on the same or opposite stands of the double helix. This causes more frequent production of apurinic gaps. Examples of alkylating agents are nitrosoguanidine, mustard gas and mustard compounds, ethyl methane sulfonate (EMS) and ethyl ethane sulfonate (EES), alkyl halides, sulfuric and phosphoric esters, ethylene imines and amides. These compounds are also described as radiomimetic because their effects resemble those of ionizing radiation.

EMS is one of the most powerful mutagenic agents known. It can add an ethyl group (-CH2 CH3) to a guanine, and much less frequently to adenine. Due to ethylation guanine pairs with adenine leading to A—T = G—C transitions. EMS also produces apurinic gaps into which any of the four bases may be inserted giving rise also transversions.

Base Substitution by Tautomerism

The purine-pyrimidine base pairs in double helical DNA are determined mainly by the positions of hydrogen atoms which cross link the bases. Normally A pairs with T and G with C. Watson and Crick suggested that if a base was present in its tautomeric form at the time of DNA replication, then a wrong base would be synthesized in the new strand. At the next replication cycle, the tautomer would revert back to its normal form, the two strands would separate, and this time the normal correct base would be synthesized on the new strand. This would result in substitution of one base pair for another, i.e., A—G and T—C transitions would occur.

Base Analogues

These chemicals with structure similar to bases in normal DNA and become substituted for normal bases during DNA replication. The first base analogue studied was 5-bromouracil (5-BU) which has chemical structure similar to that of thymine (5-methyl-uracil) except that the methyl group of thymine is replaced by bromine.

Nitrous acid (HNO_2)

This acts on non-replicating DNA by removing amino groups of nitrogenous bases, converting adenine to hypoxanthine, guanine to xanthine and cytosine to uracil (figure Q). The conversions lead to AT—GC and GC—AT transitions as explained in the figure. In a similar way hydroxylamine (NH2OH) changes hydroxymethylcytosine into uracil leading to AT—GC transition. These mutations are able to reverse their own effects (back mutation) as well as those of other base analogues.

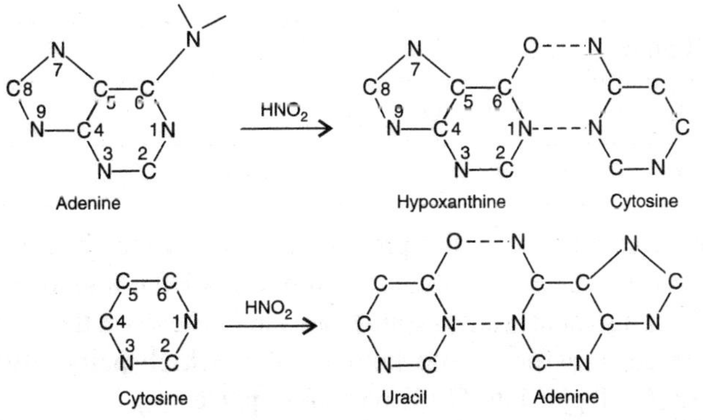

Fig. Q Action of nitrous acid. It converts adenine to hypoxanthine which pairs with cytosine ; cytosine is converted to uracil which pairs with adenine.

Certain chemicals like ethoxycaffeine, urethane and formaldehyde produce organic peroxide and free radicals leading to mutations. They probably cause destructions of nucleicacid bases resulting in breaks in single strands.

Mutagene-type and mode of action (Substitution)

Substitution is a gene mutation where one nucleotide is replaced by another nucleotide. Substitution mutation is of two types, namely,

1. Transversion and
2. Transition

Transversion

It is a substitution mutation where purine is replaced by pyrimidine or vice-versa.

Transition

It is a substitution mutation where a purine is replaced by purine or a pyrimidine is replaced by pyrimidine. Transition is caused by the following factors :

(i) Tautomerization (ii) Base analogs and (iii) Deamination

(i) Tautomerization. Certain chemicals exist in more than one chemical form. These chemicals are called tautomers and this phenomenon is called tautomerization. It is caused by shifts in electrons or protons in the normal nitrogen bases. Owing to tauatomerization the amino (—NH_2) forms of cytosine and adenine may convert into imino (NH) tautomers and likewise keto (C—O) forms of thymine and guanine may convert into enol (COH) tautomer. This leads to unusual base pairs in DNA. Normally, A—T and G—C are the base pairing found in DNA. But owing to tautomerization unusual base pairings such as T—G, C—A and A—C, G—T are noticed.

(ii) Base analogs. Certain chemical substances have similar molecular structure to the DNA bases. They are called base analogs. The chemical structure of 5-bromouracil is very much like that of the natural base thymine. This is the first base analog inserted into DNA molecule. When 5-BU is supplied to the medium of growing bacterial cells it is incorporated into new DNA in place of thymine. The thymine can be replaced by 5-BU. The 5-BU functions within the chromosomes just like thymine. The cells whose chromosomes carry 5-BU shows a higher rate of mutation than DNA with thymine. The 5-BU molecules contain two molecular or tautomeric forms. One form called the keto state is the more common, while the other, the enol state is rare. The presence of 5-BU in the DNA sequence may cause synthesis of unusual mRNA which leads to mutation.

Normal gene				
A	G	G	C	G
T	C	C	G	C

Base analog				
A	G	G	C	G
5 BU	C	C	G	C

(iii) Deamination. Deamination is the process in which some chemical substances such as nitrous acid, cause transitional mutation through the removal of the amino group (deamination) from adenine and cytosine. The removal of amino group from adenine leads to the production of hypoxanthine. The hypoxanthine, later on transforms into ketotautomer which pairs with cytosine. Thus due to deamination transition of A : T pair into G : C pair takes place.

Detection of mutations

In *Drosophila* and most other animals, only those mutations which occur in the germ track can be transmitted to the offspring, and we might, therefore, get the impression that mutations are limited to the germ track. However, in the case of mosaics, it can be seen that mutations occur in the body cells, and it is, therefore, evident that mutations are not limited to the germ track. It has been suggested that the cells of the germ track are more susceptible to mutation than cells in other parts of the body, and that this would apply in particular to germ cells which are undergoing the reduction division. We do not know whether this is true. It is possible that irregularities connected with crossing over might occasionally lead to mutation, but abnormal crossing over would by no means be the usual cause of mutations. For one thing, mutations are just about as frequent in the male of *Drosophila* as in the female, though there is no crossing over in the male. Also when mutation occur outside the germ track as in mosaics, they most likely are in no way connected with a reduction division or crossing over.

The dominant mutations are very rare that can be detected phenotypically. But most mutations are recessive and only in the segregating generations they can be observed.

ClB Technique to detect Mutation

In 1927 H. J. Muller discovered that mutations could be artificially induced by X-rays in all the chromosomes of *Drosophila*. The mutation induced by him caused visible changes and

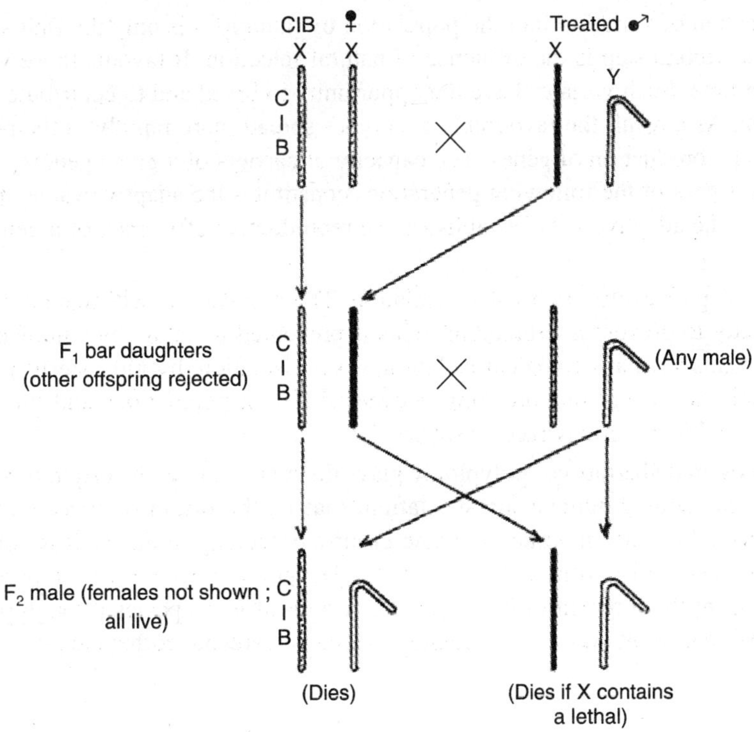

Fig. R The ClB method for detecting lethals.

inherited to the successive generations. He devised a technique to detect mutation in *Drosophila*. The technique is called ClB technique. For this work a special type of female is used. This female carries one normal X chromosome and one X chromosome in which the central portion is inverted. This inversion prevents the crossing over with the normal X chromosome. The X chromosome contains three genes, ClB. The gene C is a dominant cross over suppressor gene. The gene 1 is a recessive X linked lethal gene and B is a dominant gene producing bar eye. Females of this type are called ClB females. A male getting this ClB X chromosome dies, because of the X linked lethal gene 1. Bar eyed ClB females are mated with normal males which have been treated with X-rays. Half of the males died and the sex ratio of female and male is 2:1 among the survivors. The bar eyed males die because of the lethal gene in their single X chromosome. Among the F_1 some females carry ClB chromosomes inherited from the father. This female is selected and mated with a normal male. One half of the male offspring will die because of the presence of the lethal gene on the X chromosome. The other half will live if there is no lethal gene in the X chromosome. If a lethal mutation has appeared in the X chromosome derived from an X-rayed male, the other half of the F2 males will also die, and there will be no males at all among the offspring. Thus it is possible to detect the mutation, (Figure R).

Evolutionary Significance of Mutation

Mutations are the important sources of variations in populations. Variations are the raw materials for evolution. So mutations supply the raw materials for evolution. The mutation process constantly generate new hereditary variants. These variants accumulate in populations. The variation produced by mutation may be useful or useless or harmful or neutral to the population. The population will be benefited by the production of useful mutations.

The formation of variation into the population by mutations is only the first step in the process of evolution. The second step is the influence of natural selection. It favours those variants which are useful at a given time. Such variants have the opportunity to breed and to contribute more young ones to the population. As a result the favourable genotypes spread more rapidly in the population. This is called differential reproduction of genes. The capacity of carriers of a given genotype to transmit their genes to the genes pool of the following generations constitutes the adaptive value or Darwin's fitness of that genotype. The adaptive value establishes the reproductive efficiency of a genotype in a certain environment.

The third step of evolution involves isolation. The populations will remain distinct only when there is some factor to prevent interbreeding. This is prevented in nature by a number of reproductive isolating mechanisms such as ecological isolation, sexual isolation, hybrid - sterility and so on. These isolating mechanisms prevent interbreeding between different populations and thus each population becomes distinct and forms a new race or subspecies or species.

Of chromosomal aberrations, polyploidy plays the major role in the origin of species, especially in plants. It brings about instantaneous speciation, that is, the origin of species at a single stroke. Although this method is very important in some groups, especially in plants, it is not a general occurrence. Polyploid species arise from diploid ancestors. The polyploidy species is reproductively isolated from its parents from the moment of its origin. The origin of new species through polyploidy is very common in wheat, oats, cotton, tobacco, potato, banana, sugarcane, coffee etc.,

CHAPTER 8

Alterations in Genetic Make Up-Changes in Chromosome Structure

Origin

When two segments of a population become isolated from each other and remain isolated for a long time, each tends to accumulate its own combination of mutant genes. Each combination by itself incompatible with fertility, but when the two combinations come together in the hybrid they often form a third combination that is not compatible with fertility. In other words, the hybrid is sterile. Once two populations are no longer able to produce fertile offspring when crossed, they no longer are able to pool their mutant genes through sexual reproduction and genetic recombination, and they then tend to become more and more distinct from each other through their independent evolution. The initial steps in the evolutionary separation of two segments of a population from each other are sometimes referred to as speciation ; and two populations which are not fertile when crossed but still not different enough to be regarded as separate species, are referred to as subspecies.

Why did varieties become different? Three reasons : (1) for one thing, geographical isolation breaks up a species into smaller groups and so it assists in the spreading of mutant genes by accident (random drift). As a matter of chance, the same mutant genes would not necessarily arise and spread in any two groups. Thus, each isolated group would tend to become different from every other - the species would become diversified. (2) Isolation would then maintain diversification by preventing one group from crossing with another. (3) Geographical isolation might also lead to diversity by reducing competition and allowing inferior types to survive. This is well seen in Australia where the lower types of mammals—the marsupials—survive because they were protected from competition with the higher mammals by Australia's isolation. Geographical barriers are not the only means by which animals or plants might be isolated from one another. One race might be able to tolerate a higher temperature than another, and so the two might occupy different localities. They might also be separated by differences in the moisture they can tolerate, by differences in the kind of food they eat, and by other differences of an ecological nature.

Types

Structural changes within chromosomes do not change the amount of genetic material nor total number of chromosomes in a cell. Basically they are visible alterations in the shape of chromosomes and sequence of their segments which change the linear order of genes on a chromosome. Also known as chromosomal aberrations, they are transmitted through mitosis and meiosis and lead to changes in phenotypes and in the expected genetic ratios. Their study is useful for gaining knowledge about the nature of chromosomes and gene mapping. Chromosomal arrangements occur spontaneously due to natural causes from time to time. In grasshoppers about one in a thousand individuals appears to carry a newly arisen cytologically visible chromosomal rearranagement in the germ line. In human beings also their incidence is high and would be deleterious, but luckily most of them are eliminated by natural selection. Structural variations chromosomes are of four distinct types: deletions, duplications, translocations and inversions. There are special types of B chromosomes also detected.

Duplications or Repeats

A small segment of a chromosome might sometimes be duplicated as a result of unequal crossing over, as shown in (figure S). The duplication in this case is known as a repeat (bb in chromosome abbc). Further unequal crossing over might then take place within the repeat stock. This might result in two chromosomes : one having a second repeat and the other a reversion to normal.

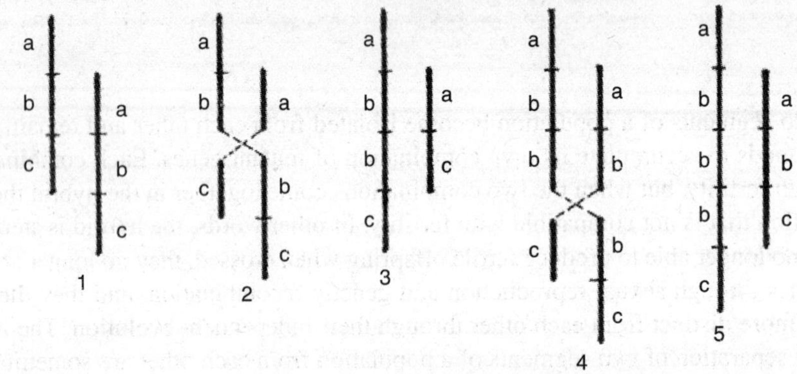

Fig. S Unequal crossing over.

In *Drosophila,* the mutant Bar (narrow eyes) is due to a small repeat. This cannot be seen in the chromosomes as they are found in most parts of the fly, but in the salivary glands it happens that the chromosomes are unusually large; in them, it is possible to see that a certain segment of the normal chromosome is repeated in Bar flies (figure T) left and middle. Further unequal crossing over within the Bar stock might cause the repeat to be deleted, giving a reversion to the normal condition (called Bar reverted); or still another repeat might be added to the chromosome, giving two repeats and resulting in the mutant known as "Bar double". Other mutants besides Bar are known to be due to small repeats and perhaps some mutants are due to repeats that are too small to be visible.

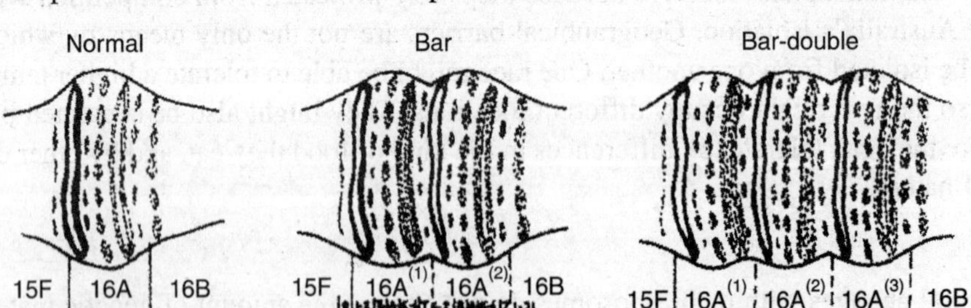

Fig. T A repeat in Drosophila (bar eyes). (From Eileen Sutton in genetics.)

Position Effect

Translocations and inversions cause no loss or gain in amount of chromosome material but merely a segmental rearrangement of genes. Very often the rearrangement has no visible effect on development. But sometimes it causes a visible change, or position effect, discovered by Sturtevent, who found that two repeats together caused narrower Bar eyes than two in separate chromosomes. We do not know how rearrangements cause a position effect. Perhaps the chemical products of adjacent genes sometimes react immediately upon being formed but fail to interact when they are separated, or they interact with the chemical products of other genes, with a resultant position effect. However, in some cases, a rearrangement, such as an inversion, transfers heterochromatin from one part of a chromosome to some other part, as shown in (figure U). Here, the heterochromatin is brought next to some gene which normally is not next to the heterochromatin and there is evidence which indicates that heterochromatin, when abnormally located, can suppress the expression of genes adjacent to it. The position effect might very well represent this sort of heterochromatic suppression.

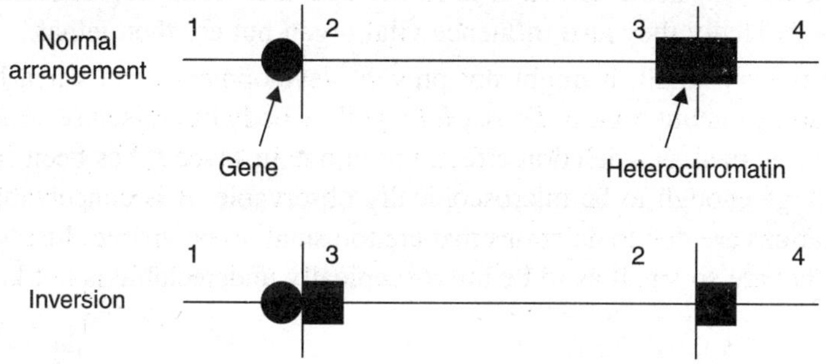

Fig. U A segmental rearrangement involving heterochromatin.

Deletions

When a chromosome is broken into three segments, it is a mere matter of chance whether or not the internal segment will again join up with the two external segments. Sometimes the two end segments become directly connected, and the middle segment is omitted or deleted, as shown in (figure V). A chromosome from which an internal segment has been removed is said to have a deletion. X-rays readily break up chromosomes into segments; hence, they readily produce deletions (along with other chromosomal rearrangements). But deletions also turn up occasionally in the absence of any treatment in both plants and animals. Deletion heterozygous are genically unbalanced because they have only one dose of every gene in the region of the deletion as compared to two elsewhere. In case the deletion is large, the heterozygote cannot develop because of genic unbalance. Offspring that are homozygous for a large deletion cannot develop because they completely lack the genes in the region of the deletion, and any large deletion as a rule contains one or more genes necessary for life.

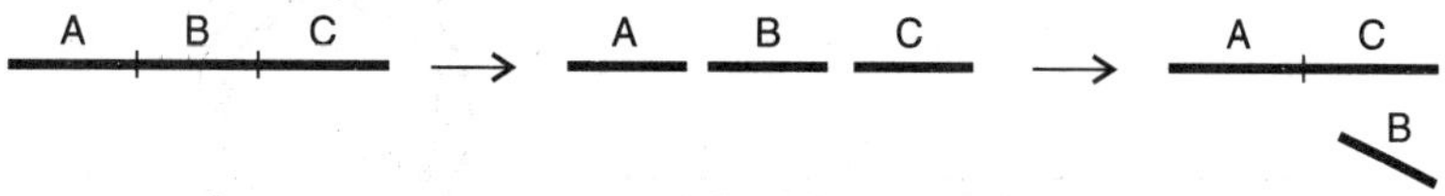

Fig. V A deletion.

Small deletions might not prevent development but often they cause the adult to have some mutant trait. Thus, in *Drosophila*, a small deletion in a certain part of the X chromosome causes the tip of the wings (in the heterozygote) to be notched and is referred to as "Notch". This deletion has been observed to arise rather often (over 50 times) and varies somewhat in length in deleted X's of independent origin, but usually it is about 1.5 map units in length; it always includes the locus of "facet" (rough eyes, located at point 3 on the chromosome map) and sometimes also includes the locus of white (located at point 1.5). Notch flies are never males because the males carry only one X and, if this should contain the Notch deletion, the fly would completely lack the deleted segment and would fail to develop.

In the female, however, there are two X's and if one should have the deletion, the other might be normal and the fly would develop. Females heterozygous for the deletion are Notch and the deletion, therefore, is dominant as regards its influence on wing development.

In *Drosophila* again, small deletions often cause a reduction in bristle size (minute bristles) and are known as "Minutes". Over 75 different Minutes have been found, distributed in all four chromosomes of *Drosophila*. Homozygous minutes cannot live. The heterozygous appear Minute, so that the deletion is dominant for its visible effect. It is rather odd that small deletions should influence the bristles so often. Undoubtedly they also influence vital organ but are then lethal.

If a deletion is very small, it might not prevent development even when homozygous and it might then produce some mutant trait. In *Drosophila,* yellow body has arisen several times by mutation and ordinarily is not regarded as a deletion effect, but in one instance it has been found to be due to a small deletion, yet large enough to be microscopically observable. It is conceivable that, in some instances, yellow mutations are due to deletions that are too small to be visible. Just how often mutations represent deletions that are so small as to be microscopically undetectable is not known.

Inversion

Suppose that we picked out four genes along the length of a chromosome and lettered them ABCD. Then, by means of X-rays, it would be possible to break the chromosome into three segments A, BC, and D. The middle segment might then become reversed and attached to the end segments, giving rise to the rearrangement ACBD. This is known as an inversion.

When a chromosome is broken into just two segments, it never happens that one of the segments then becomes inverted. Inversions always involve two breaks, and it is always the internal segment which becomes inverted, never one of the external segments. Only the broken and sticky ends of chromosome segments can become attached after breakage as previously mentioned. The unbroken ends contain a covering or cap (the telomere) which prevents them from sticking to one another of to a broken end. Inversions vary greatly in size. Some involve just a small segment of a chromosome, others a very large segment, and there are all intermediate sizes. The very large ones might stretch almost from one end of a chromosome to the other end, leaving only a small external segment at each end (figure W).

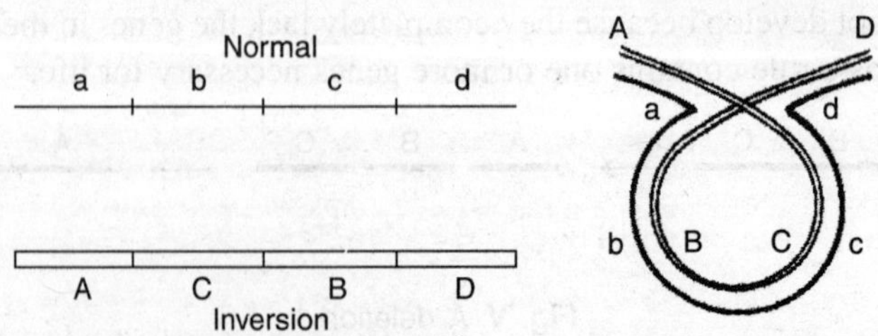

Fig. W An inversion heterozygote. Left, the chromosomes before synapsis ; right, synapsis.

Inversion in Nature

Inversions occur in a state of nature and lead to racial differences. This has been clearly shown by studies made by Dobzhansky and Sturtevant on *Drosophila pseudoobscura*, a species of *Drosophila* which inhabits western North America. Specimens of *D. pseudoobscura* collected from different regions in western U.S. and Mexico often differ in the arrangement of genes in a given chromosome in such a way that, if one arrangement of genes is regarded as normal or standard, it is possible to derive the other arrangements from this by one or more inversions. These inversion races are limited to definite localities, and they have been named, for the most part, after the locality which they inhabit, as Pikes Peak, Estes Park, Cuernavaca. It is possible to arrive at the evolutionary relationship of the various inversion races by means of overlapping inversions, as explained above. For example, if we proceed upwards from standard in the figure X we come first to "hypothetical A", which has not yet been found (perhaps it died out); then we come to "Santa Cruz", which when crossed to standard gives a compound loop. Therefore, differs from standard by the inversions, the first of which according to Dobzhansky and Sturtevant was the one that gave rise to "hypothetical A". Santa Crus in turn gave rise by single inversion, to four separate races. One of these four (Tree Line) next gave rise to four other races (Olympic, Estes Park, Oaxaca, Hidalgo) by means of a second inversion (after the Santa Cruz stage), the second in each case overlapping the first inversion. This is shown by the fact that each of the

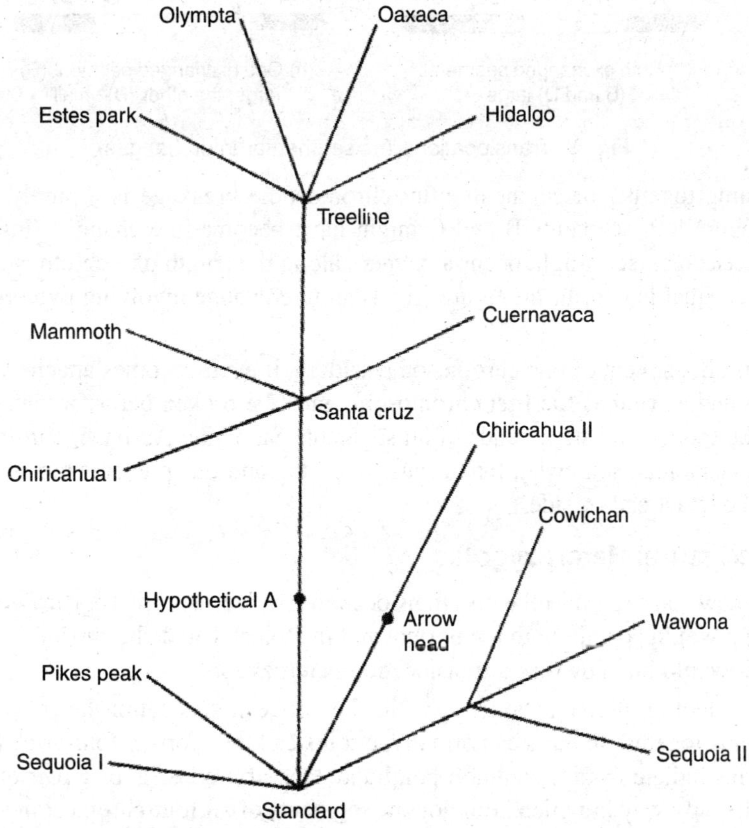

Fig. X The evolutionary relationship of races produced by inversions (in the third chromosome) of Drosophila pseudo-obscura. (After Dobzhansky and Sturtevant in Genetics.)

four last races, when crossed to Santa Cruz, gives a compound loop, one of the components of which is the inversion that produced Tree Line (the race from which the other four were presumably derived). In like manner, the rest of figure is derivable from the standard race.

Translocations

In the figure Y (left), two non-homologous chromosomes are shown above and each is arbitrarily considered as made up of two segments, labeled AB and CD, respectively. By means of X-rays it would be possible to break up each chromosome accidentally into the two segments in question. Next A and D might become attached and, also, C and B. This would give rise to AD and CB, shown in figure left. We might think of AB and CD as giving rise to AD and CB by the interchange of segments B and D (or A and C). These segments are not homologous. An interchange of nonhomologous segments of two chromosomes is known as a segmental interchange or a mutual translocations. The first known translocation was discovered by Bridges in *Drosophila*. It arose spontaneously (apart from artificial treatment), but spontaneous translocations are comparatively rare.

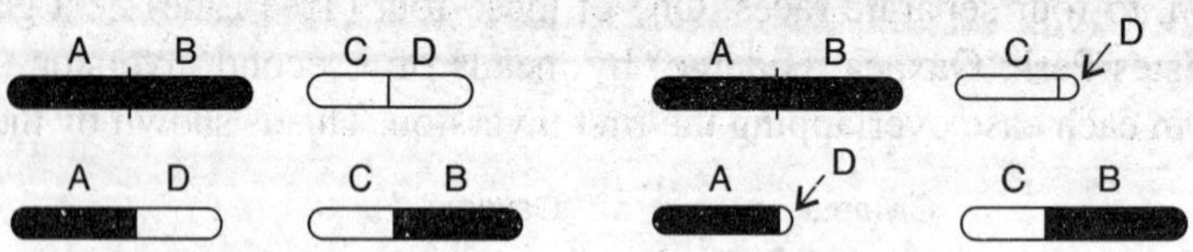

(a) Both exchanged segments (B and D) large

(b) One exchanged segment (B) large, the other (D) small

Fig. Y Translocations (or segmental interchanges).

The coming together of segments after chromosome breakage is a purely accidental matter. Hence, in the figure left, segments B and C might have become interchanged (instead of B and D). Moreover, the breakage itself might occur anywhere along the length of a chromosome and thus result in segments of unequal length. In the figure (right) an interchange involving two segments of unequal length is shown.

A broken-off segment of one chromosome seldom, if ever, becomes attached to an intact end of another. The second as well as the first chromosome must be broken before a translocation can occur and then only the broken and sticky ends of all segments can unite. An intact chromosome apparently is enclosed in an envelope somewhat like a sausage casing and this prevents the attachment of a fragment either to the intact end or side.

Meiosis in Structural Heterozygotes

Suppose now that a sperm cell with a translocation fertilized a normal egg in *Drosophila* (figure Z). The fertilized egg would contain both the normal and the translocated chromosomes, and the fly which develops from it would be known as a translocation heterozygots.

When a reduction division takes place in the above translocation heterozygots, the chromosomes do not come together in pairs as normally, but instead they form a four-armed figure (figure left middle). The arms radiate from a common point and each arm consists of a pair of homologous segments. This is the only way in which homologous segments of all four chromosomes can pair throughout their length. The chromosomes next separate at their mid regions (because of termianalization of chiasmata), but they remain attached for a time at their ends. As a result they form a circle. Within the circle the normal chromosomes (AB and CD) alternate with the translocated (AD and CB). Next, the circle might get twisted in the form of an 8, and when this happens alternate chromosomes of the former ring face the same pole at metaphase. As a result, the two normal chromosomes (AB and CD) go to one gamete and the two translocated (AD and CB) to the other (figure bottom left). However, the circle might not twist into an 8. In this event, adjacent chromosomes face the same pole at metaphase. It might happen that AB and CB face one pole and AD and CD the other (figure middle bottom). Accordingly, AB and CB would go to one gamete; AD and CD, to the other. On the other hand, AD and AB might face one pole; CD and CB, the others; and the two corresponding classes of gametes would be formed (figure Z bottom right).

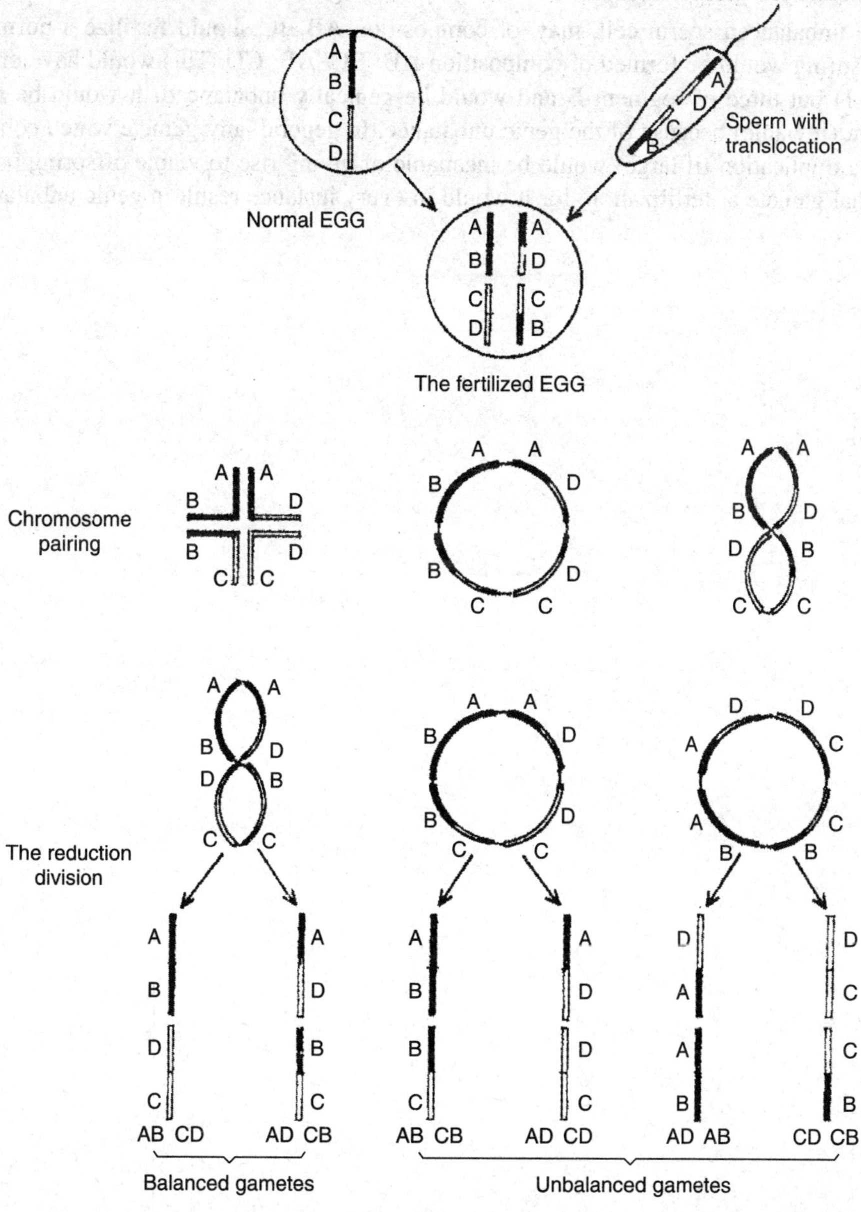

Fig. Z A translocation heterozygote.

Whenever two adjacent chromosomes in a ring to the same pole, the resulting gametes contain one normal and one translocated chromosome, as AB (normal) and CB (translocated). A gamete with AB and CB lacks segment D but has a double dose of segment B. Such a gamete is said to be a deficiency for D and a duplication for B. The translocation heterozygote forms six classes of gametes, but only two of these contain neither a deficiency nor duplication: namely, the class with the two normal chromosomes (ABCD) and the class with the two translocated (AD CB). We shall refer to these two classes a balanced gametes, and to the remaining four (the deficiency duplications) as unbalanced gametes, and to the remaining four (the deficiency duplications) as unbalanced. It should be noted that a balanced gamete contains one and only one segment of each kind.

If an unbalanced sperm cell, may, of composition AB BC should fertilize a normal egg (AB CD), an offspring would be formed of composition AB BC/ AB CD. This would have only one dose of segment D but three of segment B and would be genically unbalanced. It would be incapable of development (inviable) because of the genic unbalance. In general, any gamete which contains a deficiency and a duplication (if large) would be incapable of giving rise to viable offspring on combining with a normal gamete at fertilization, for it would in every instance result in genic unbalance.

CHAPTER 9

Alterations in Genetic Make Up-Changes in Chromosome Number

Changes in chromosomes that are easiest to identify cytologically are those where an increase or decrease in chromosome number has occurred. Such numerical variations are known as polyploidy. When there is addition of one or more entire sets of chromosomes (a set means all the chromosomes which constitute a haploid complement), the condition is known as euploidy. When one or more single chromosomes are added or deleted we call it aneuploidy.

Euploidy

Most diploid sexually reproducing organisms have an alternation between a haploid and a diploid state in the life cycle. The haploid state, mostly confined to germ cells, is characterized by the presence of a single set of chromosomes (n). When two haploid germ cells produced by a male and female parent unite during fertilization, the zygote formed contains two haploid sets of chromosomes and it becomes diploid (2n). Mitotic divisions of the zygote produce a diploid adult organism of which each cell contains one haploid set of chromosomes from the male parent, the other from the female parent. Since both sets of chromosomes are morphologically identical, pairs of identical chromosomes exist in a cell nucleus. The two members of a pair are said to be homologous to each other. When meiosis starts in the germ mother cells, homologous chromosomes attract each other and start pairing. The exactness of pairing depends upon the identical nature of the homologues.

Aneuploidy

Aneuploids are individuals with an uneven number of individual chromosomes (Greek aneu meaning uneven). The cell has one or a few chromosomes more or less than the normal diploid number. In a diploid cell one complete set of chromosomes is present twice so that two homologues are present for each chromosome. If there is one additional chromosome it leads to a particular chromosome being represented by three homologues instead of usual two, and the condition is known as trisomy (symbolized as 2n + 1). If there are two additional chromosomes in a cell so that one chromosome comes to be represented by four homologues, it is called tetrasomy (2n + 2). It can also happen that the two additional chromosomes are homologues belonging to two different chromosomes of the complement so that two chromosomes are represented each by three homologues. Such an individual is said to be trisomic for two different chromosomes (2n + 1 + 1).

The alternative situation also exists involving loss of a chromosome from a diploid complement. This results in the presence of one chromosome without a homologous partner called monosomy (2n – 1). If both homologues of a chromosome are missing, it is called nullisomy (2n – 2).

Origin

All types of ploids arise due to cytological accidents. Monosomics and trisomics arise sporadically from gametes with n-1 or n+1 chromosomes. These gametes occur as a result of non-disjunction in one or another pair of chromosomes. Aneuploids appear to arise most commonly, as a result of the troubles at meiosis that are caused by asynaptic genes. Euploidy seems to arise most commonly through misadventures in meiosis that caused 2n gametes to form. In mitosis also failure of spindle fibre mechanism may result in doubling of chromosome complement. If the doubling occurs early in development of the plant, it may produce sectoral chimera.

Grant (1956) had presented some interesting observations on the relation between the life cycle and breeding behaviour of the parental species. The author presents the hypothesis that the result of natural hybridization in annuals will be a tendency towards polyploidy when self-fertilizing species are involved and introgression when cross fertilized species are involved. In biennials and short lived perennials without asexual propagative feature, where the time available for reproduction by a natural hybrid is also crucial, there is a tendency towards polyploidy, although they frequently exhibit less polyploidy than the most closely related groups of annuals. Many biennials are short lived perennials are predominantly out crossing species which may account for the reduced frequency of polyploidy. Some authorities estimate that atleast 1/3 of the angiosperms are polyploids. Among cultivated species the proportion is equally great if not even greater. This supports strongly that the cytological accidents leading to polyploidy are common.

Types

Euploidy. This is general term covering situations in which the result is the number of repetitions of entire chromosomal sets.

x — Monoploid
3x — Triploid
4x — Autotetraploid
2x + 2x — Allo tetraploid

Aneuploids

The term aneuploidy is a general term to describe organisms whose chromosome number is not a whole number multiple of the basic number of the group.

2x – 2 (one pair) — nullisomic
2x – 1 — monosomic
2x – 1 – 1 — double monosomic
2x + 1 — trisomic
2x + 1 + 1 — double trisomic
2x + 2 (one pair) — tetrasomic
2x – 1 + 1 — monosomic and trisomic

In general, deficiency or repetition of particular chromosomes results in unbalances in the genotype. Aneuploids are usually less vigorous than their diploid progenitors. Nullisomics and monosomics are ordinarily viable only in species with polyploidy ancestry where previous duplication of chromosomes appears to cover for missing chromosomal materials. In some polyploidy species—wheat for example, the degree of tolerance to repetitions of particular chromosomes is great and even tetrasomics may be nearly indistinguishable from normal diploid plants. A common effect of autopolyploidy is to increase the size of the vegetative portions of the plant, making autopolyploids more vigorous than diploids.

Effects of Auto and Allopolyploidy

Autopolyploids arise when the additional sets originate from the same species. For example, if the haploid set of a species is designated A, the diploid is AA, triploid AAA, tetraploid AAAA, and so on. Autotetraploid can arise through one of the following ways (1) fertilization of an egg by two or more sperms giving rise to a zygote with three or more sets of chromosomes; (2) normal mitotic division in the diploid zygote in which chromosomes duplicate but cell division fails to occur so that four haploid sets of chromosomes produce a tetraploid nucleus; if the same mitotic error occurs during embryo development it results in some tetraploid tissues in an adult diploid individuals ; (3) failure of meiotic division in germ mother cells so that unreduced diploid gametes are formed instead of haploids. Although autopolyploids have homologous genomes, yet those having odd numbered sets of chromosomes such as 3n, 5n, 7n, 9n and so on, show a high degree of sterility. This is because during meiosis, pairing between two homologous chromosomes only results in normal segregation of the haploid set into gametes. If three homologues are present they may or may not become paired to form a trivalent. Since pairing in any region is restricted to only two homologues at a time, third homologue may fail to pair and remain as univalent, or may pair at some places to form a trivalent. During anaphasic segregation at meiosis I, two of these homologues may move into one daughter cell, and the third into the other daughter cell. Since all chromosomes of the haploid set have three homologues each, their random distribution or independent assortment will cause the resulting gametes to have varying numbers of different homologues. In this way a true haploid or a true diploid gamete would be rarely formed. Instead, unbalanced gametes with chromosome numbers ranging between n and 2n would be formed. Such gametes are not viable and triploids consequently are sterile.

Autotetraploids are either fertile or only partially sterile. As there are no univalents and trivalents formed, the four homologues can segregate to form viable diploid gametes, so that tetraploids are often fertile. Groundnut, potatoes and coffee are well known examples of autotetraploid species. Among cereals, autotetraploid rye is grown in Sweden and Germany. Some of the giant sized plants of *Oenothera lamarckiana* in which De Vries first noticed in Holland and attributed to a mutation, had later turned out to be autotetraploids.

Allopolyploidy

This is the second type of euploidy where the additional set of chromosomes comes from a different species. For example, suppose a diploid species with two chromosome sets AA crosses naturally or artificially with another species BB. The off-spring produced would be AB which is viable but sterile. This is because during meiosis the chromosomes belonging to the set A do not find homologous partners in chromosomes of B. Due to failure of pairing at anaphase I, the chromosomes move at random towards the two poles. Thus each gamete gets an unabalanced mixture of A and B chromosomes and sterility results. There is one way of restoring fertility to a sterile hybrid (AB). If during mitotic division in the AB hybrid all the chromosomes are allowed to divide but cell division is inhibited, the result would be tetraploid nucleus with two sets of A and two sets of B chromosomes (AABB). Therefore, when meiosis starts, all chromosomes belonging to one set of A will find homologous partners with the remaining A chromosomes and perfect pairing will result. Similarly, the two sets of B chromosomes will pair with each other and viable fertile gametes would be formed. Such an allopolyploid individual is called amphidiploid. If possible to induce amphidiploidy artificially by treating young buds or seeds with the alkaloid colchicines, a mitotic poison which inhibits, spindle formation, consequently cell division. This leads to all the duplicated chromosomes becoming included in a single tetraploid nucleus.

Raphanobrassica is an interesting example of a newly synthesized genus for illustrating allopolyploidy. In 1927 a Russian Geneticist Karpechenko 1935 made a cross between *Raphanus sativus* (Radish) and *Brassica oleracea* (cabbage) with the aim of producing a new plant that would have the roots of radish, and in the aerial portions would bear cabbage. This hybrid that was actually formed had the roots of cabbage and tops of radish plant. The hybrid produced between radish and cabbage proved useless economically. But it proved very important genetically. Both radish and cabbage plants are diploid with 18 chromosomes. Thus gametes from each parent had 9 chromosomes and their union produced the F_1 hybrid with 18 chromosomes. This hybrid was sterile because the 9 chromosomes of radish did not pair with the 9 chromosomes of cabbage. Sometimes however, viable pollen and ovules were produced having all 18 chromosomes. Fusion of such unreduced gametes produced tetraploid (4n=36) plants with 18 chromosomes of radish and 18 of cabbage. Pairing took place amongst the radish chromosomes to form 9 pairs; similarly the cabbage chromosomes also formed 9 pairs. Normal segregation gave rise to viable gametes. The hybrid therefore became fertile and was given the name of a new genus *Raphano brassica*. This is a beautiful demonstration of how a new genus can be artificially synthesized through allopolyploidy.

.Segmental Allopolyploids

Sometimes not all hybrids between diploid species however show a complete absence of meiotic pairing and some indeed may have a reasonable degree of fertility. When doubled these form mostly bivalents but the formation of some quadrivalents indicates that some chromosomal segments are homologous in the two sets. Stebbins referred to these as segmental polyploids. For example segmental tetraploids had two genomes. Pairing is largely in segments between the genomes. This means the chromosomes are not 100 per cent homologous. A portion of one genome chromosome attaches with a portion of another genome chromosome and they pair.

Allosyndesis

This means pairing between non-homologous chromosomes of two species in a alloploids (example pairing between a few A and B genome chromosomes). This shows both the genomes A and P had common ancestor during evolution.

Autosyndesis

This means a few chromosomes of A genome pair within themselves (in autoploids) and a few chromosomes of B genome pair within themselves.

Origin and meiosis in trisomics (2n + 1) ; monosomics (2n-1) and nullisomics (2n – one pair)

The union of an n+ 1 gamete and one n gamete, the offspring is 2n + 1. It contains three chromosomes of one kind in the diploid (figure I).

Triploids will produce some n + 1 gametes and, if the triploid were crossed to a normal diploid, the n + 1 gametes of the triploid would combine with the n gametes of the diploid and produce 2n + 1 offspring. Thus, it is readily possible to get trisomics by crossing a triploid and a diploid. But trisomics might also be produced accidentally by normal diploids. When reduction division takes place in a diploid, it sometimes happens that both members of a pair of chromosomes pass to one pole and neither to the other. This would be non-disjunction. If only one chromosome pair undergoes non-disjunction, then the two resultant gametes are n + 1 and n – 1 respectively. The n + 1 gamete and a normal (n) gamete would, by combining, give rise to a 2n + 1 offspring (a trisomic) and the n – 1 gamete would give rise to monosomic (2n – 1). Sometimes alternative situation also exists involving loss of a pair of chromosomes due to nondisjunction, and it is called nullisomy (2n – 2). On selfing a monosome, nullisomics, monosomics and normal disomes (diploid) are obtained.

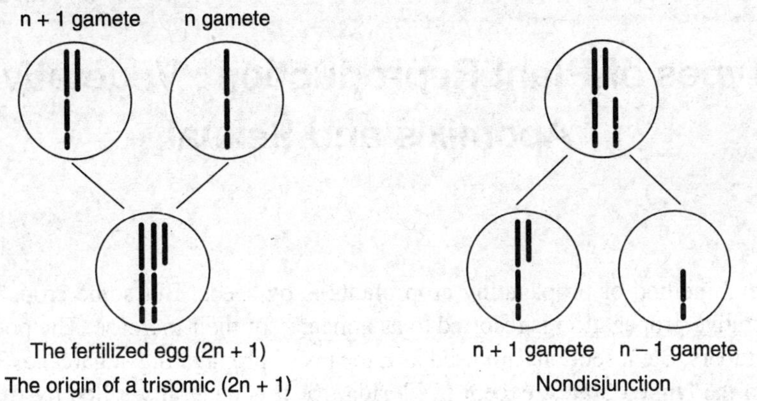

Fig. I

CHAPTER 10

Types of Plant Reproduction : Vegetative, Apomixis and Sexual

Vegetative

The normal method of propagating crop plants is by seeds. But some crops produce seeds so poorly that vegetative propagation is resorted to as a means for their increase. The potato is propagated vegetatively by tubers. Stem sections are used as a means of propagating sugarcane. Sugarcane seldom produces seed in the United States, except in Florida, for it is generally killed by frost in other areas. Coastal and certain other varieties of Bermuda grass set seed so poorly that they are propagated entirely by vegetative sprigs. Zoysia, a grass widely promoted for lawns in the southern states of U.S.A., is propagated by vegetative sprigs also. Many species with succulent stems, such as sugar beets, alfalfa, red clover, birds foot trefoil or lespedeza may be vegetatively propagated from stem cuttings. A group of plants which is propagated vegetatively from a single plant is called a clone. All the plants within a clone are identical in heredity and bear the characteristics of the original parent plant. Plant breeders may use clonal propagation to establish lines when the parent stocks will not breed true from seeds, or cannot be established easily from seeds.

Apomixis

Apomixis is a type of reproduction in which the sexual organs or related structures take part, but in which seeds are formed without union of the gametes. Seeds formed in this manner are vegetative in origin. Various forms of apomixes may occur. In one form, termed parthenogenesis, the embryo develops directly from an unfertilized egg. If the chromosome number of the gamete has been reduced in the normal manner at meiosis and chromosome doubling of the unfertilized gamete does not occur, the apomictically produced embryos and the plant developing from it will be haploid. If the chromosome number of the gamete is unreduced as a result of some abnormal occurrence during meiosis, the apomiotically produced embryo and plant will be diploid. Apogamy is another form of apomixis in which the embryo develops from haploid nuclei other than the eggs. Commonly, apogamy results from the fusion of two cells of the embryo sac. In apospory, the embryo sac is formed directly from a somatic cell, without reduction and formation of spores. The embryo then develops directly from the diploid egg without fertilization. Although sexual union does not occur in the development of apomictically produced seeds, pollination may be necessary to initiate the development of the endosperm. In some common species of grasses, particularly Kentucky blue grass and dallisgrass, seed is produced largely by apomiotic means. It is important that a plant breeder be informed of a tendency of a species to produce seed by apomixis to avoid confusion and error in breeding experiments. Crosses attempted in apomictic species would generally produce progenies like the mother plant. This phenomenon may also have advantages for the breeder. A superior plant type which produces seed by apomictic means will usually breed true for the characteristics of the mother plant.

Darlington defined apomixis as "a system of reproduction having the external character of sexual reproduction but omitting one or both of its essential cellular processes i.e., meiosis and fertilization". Seedlings produced by apomixis are known as apomicts. Darlington classified apomixis in angiosperms into

1. Parthenogenesis. Development of sporophyte from a female gamete or sometimes a male gamete without fertilization. There are two kinds depending on

(a) Failure of fertilization—haploid parthenogeny, e.g. Bees

(b) Failure of meiosis—diploid parthenogeny. e.g. Aphids.

2. Apogamy or Apogamety. Development of cells other than egg cell into the embryo sporophyte. Synergids and antipodals may form supernumerary embryos.

3. Apospory. Development of cells other than embryo sac cell. Nucellus or integument develops into diploid embryo sac. These compete with the normal embryo sacs. This means, here the normal embryo sacs also may function. Apospory may occur along with apogamy. e.g. Citrus, Mango etc.

According to Stebbins "Apomixis is the substitutive for sexual reproduction and it is asexual reproductive process that does not involve nuclear fusion i.e., fertilization". Stebbins classified apomoxis into:

(i) Agamospermy. This includes all types of apomictic reproduction in which embryos and seeds are formed by asexual methods.

(ii) Recurrent apomixis. The seedling may originate from the egg mother cell or from another cell of the nucleus.

(iii) Non-recurrent apomixis. In this case, an embryo arises directly from the egg nucleus without fertilization. Since the egg is haploid the resulting embryo will also be haploid. It does not consistently occur in any particular kind of plant as do recurrent apomixis and adventitious embryony. e.g. Corn.

(iv) Adventitious embryony (also known as nucellar embryony or nucellar budding). The apomictic embryo arises from nucellus or the integuments. A number of apomictic embryos may develop within the same ovule. A sexual embryo may also develop at the same time by the normal reproductive process of reduction division and fertilization. Consequently at maturity the seed may contain one, two or more embryos enclosed within a single seed coat, only one of which would be true sexual embryo and the others nucellar embryos. e.g. Citrus, Mango etc.

(v) Vegetative apomixis. In some cases vegetative buds or bulbils are produced in the inflorescence in the place of flowers. e.g., Some *Allium* and grass species. Vegetative apomixis is the reproduction by means of bulbs, bulbils, runners, layers, etc., e.g., agave, Lilium, Roses, ferns, *Bryophyllum* etc.

(vi) Diplospory. Primary egg mother cell gives embryo sac without meiosis.

(vii) Pseudogamy. Production of a diploid embryo like the mother. This is induced by pollination but without complete gametic fusion. e.g., Potato, Strawberries, Datura etc.

(viii) Apomeiosis. Produciton of diploid gametophyte without meiosis.

(ix) Generative apospory. Production of diploid gametophyte from egg mother cell without meiosis.

(x) Somatic apospory. Production of diploid gametophyte from the non-sporogeneous tissue.

(xi) Agamogony. This refers to types of apomixis in which the daughter sporophyte develops from a diploid gametophyte in a manner suggesting the presence of an alternation of generation.

(xii) Polyembryony. This is an occurrence in the seed more than one embryo. Twin seedlings may occur which may be genetically identical or different depending upon the manner of origin.

Types of apomixis

The apomicts are said to be Pseudogamous. Apomixis is said to be obligatory when no other method of reproduction is available to the plant. e.g. Agave, mangosteen. In facultative it can reproduce both by sexual and apomictic methods. e.g. Sugarcane, Mango, citrus etc.

It has been observed that in all these cases the tendency to parthenocarpy is gene controlled. In the bananas, seedless grapes, oranges etc., diploids without seeds are found. In these plants the meiosis is normal but the break down of the seed formation occurs in embryo sac formation or post—fertilization stages. Here also parthenocarpy is Mendelian. Parthenocarpy has its greatest application in horticulture since seedless fruits are always the ideal of the consumer. Further it may also increase the proportion of edible parts. Apomixis has both advantages and disadvantages from the evolutionary point of view. The advantages are that the good phenotypes remains stable by apomiotic type of reproduction. The disadvantages are apomixis puts a limitation to genetic recombination and segregation and become a dead end in evolutionary progress.

Sexual Reproduction

A wide degree of genetic variability cannot be created or maintained in plants, except through their sexual reproduction. The flower contains the sexual reproductive structure of the plant. It commonly consists of four floral organs, *viz.,* sepals, petals, stamens, and pistil. Typically, the petals are large and brightly coloured; the sepals are small and inconspicuous. Petals and sepals are not necessary for reproduction. Only the stamens and the pistil function in the production of seeds. The stamens usually consists of a slender stalk or filament which supports the anther. Within the anther the pollen grains develop. The pistil commonly consists of an enlarged base or ovary in which the seeds are formed, an elongated stalk or style, and a stigma on which pollen may be deposited. Within the ovary are found the ovules or immature seeds which after fertilization develop into the mature seeds. The number of ovules within an ovary may vary from one, as in wheat or barley to several hundred as in tobacco.

Complete flowers contain all four floral organs (sepals, petals, stamens, pistil). Incomplete flowers lack one or more of these floral organs. Complete flowers are borne on cotton, flax, tobacco, rape, potatos, soybeans, red clover, white clover, alfalfa, birdsfoot trefoil, vetch and many other crop plants. Crops belonging to the grass family, including corn, sorghum, wheat, barley, oats, rice and the common hay and pasture grasses, have incomplete flowers in which the petals and the sepals are lacking.

Seeds are formed within a plant by a succession of processes in which the stamens and the pistil play important roles. Within the immature anther are four cavities containing many microspore or pollen mother cells. Each mother cell undergoes two successive nuclear division and forms a tetrad of four microspores. Each microspore may develop into a pollen grain. The microspore is transformed into a pollen grain by a thickening of the spore wall and a division of the microspore nucleus to form a tube cell nucleus and a generative nucleus. As the anther matures the pollen sacs open and the pollen grains are dispersed. The pollen grains are produced in great numbers from 20 to 50 million may be produced on the tassel of a single corn plant. Pollination is the transfer of the pollen grains from the anther to the stigma. The female germ cell, or gamete, is produced within the ovule by a succession of

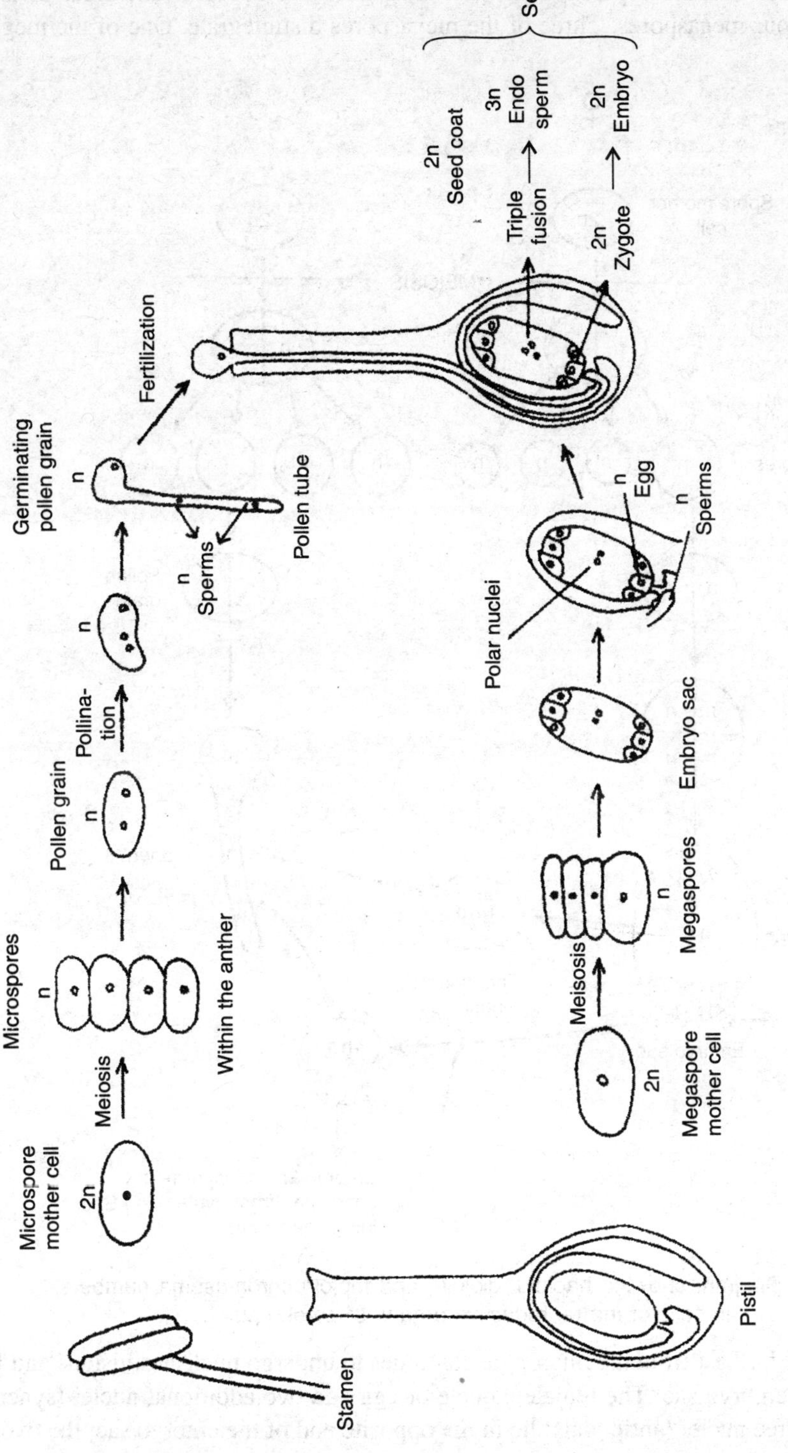

Fig. II Steps in the reproduction of a seed plant. Starting with the spore mother cells in the anthers in the ovules, a succession of events takes place which leads to fertilization and eventual formation of a seed.

steps similar to those which led to the production of the sperm. Within each ovule is a single megaspore mother cell, like each of the microspore mother cell undergoes two successive nuclear divisions and produces a tetrad of four megaspores. Three of the megaspores distintegrate. One of the megaspores,

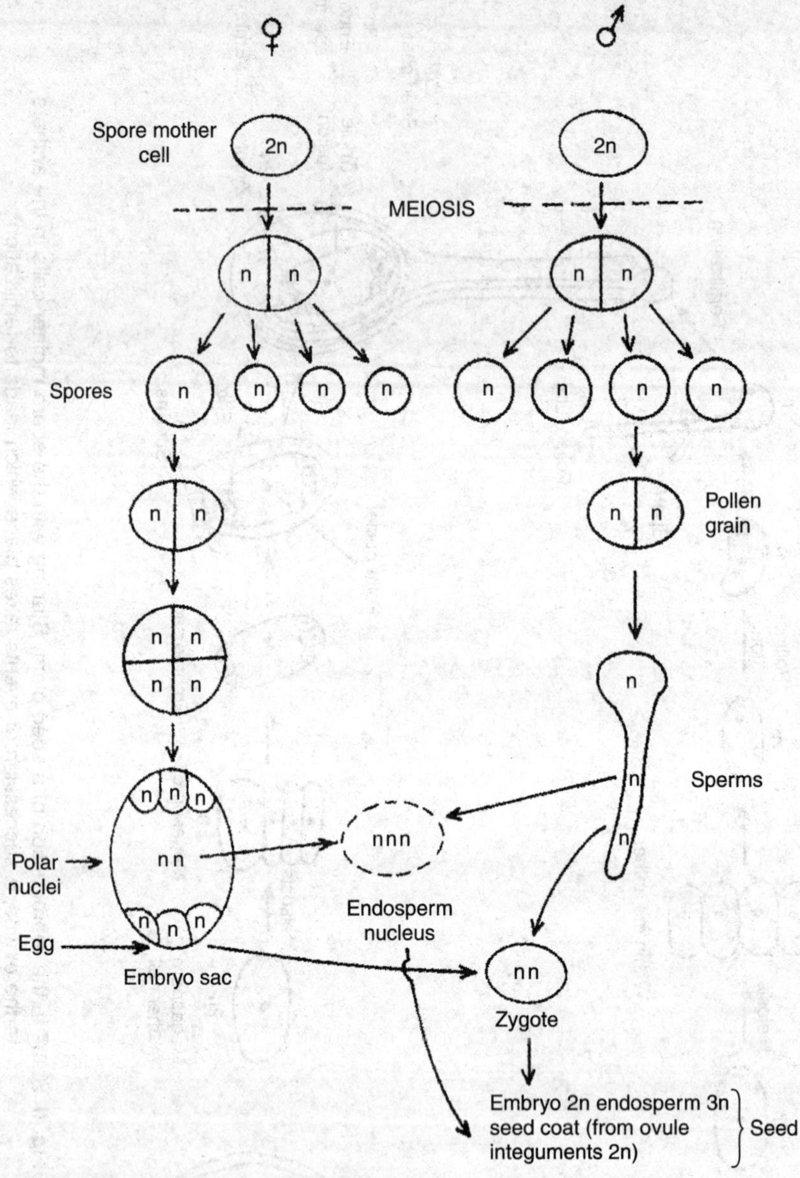

Fig. IIA Diagram showing haploid, diploid, and triploid chromosome numbers in cells of the reproductive organs of a seed plant.

usually the megaspore farthest from the micropyle continues to undergo nuclear divisions and forms an ovoid, eight-nucleate embryo sac. The female gamete or egg and two additional nuclei (synergids) lie near the micropyle; three nuclei (antipodals) lie in the opposite end of the embryo sac; the two remaining nuclei, termed polar nuclei, lie in the central area. After two sperms are emptied from the pollen tube into the embryo sac, one sperm fuses with the egg to form a zygote, a process known as fertilization. The second sperm unites with a nucleus which was formed by the earlier fusion of the two polar nuclei, or all three of these nuclei may fuse simultaneously. The nucleus resulting from this triple fusion is called the primary endosperm nucleus. These processes, in which both sperm nuclei function, is referred to as double fertilization (figure II and IIA).

The seed has its beginning with the fertilized egg (zygote) and the endosperm nucleus. The fertilized egg develops into the embryo which, on germination of the seed, grows into the new plant. The primary endosperm nucleus divides many times to form numerous nuclei. These nuclei become enclosed by cell walls to form the endosperm, a tissue in which starch, oil, or protein is stored. This stored food supplies to the germinating embryo and the early stages of seedling growth. In the cereals the larger part of the seed is endosperm. In seeds of the soybean, peanut, and other legumes the endosperm is absorbed by the developing embryo and the food materials are stored in cotyledons. The seed coat develops from integuments surrounding the ovule.

Effect on Generating and Fixing Genotypic Variation

Continuity is of the essence of life. The continuity must be maintained by the transmission of life from one individual to a succession of new ones, its offspring. Heredity is the transmission of characters from one generation to the next, i.e., from parent to the offspring. Muller had estimated that all the spermatozoa from which the present population of the world arose (more than two billion individuals) would make no larger bulk than half of an ordinary aspirin tablet. The essential part that the chromosomes of two billion eggs would occupy about the same space. Heredity does not necessarily mean that parents and offspring are completely identical. Probably no two human beings or two plants of any species are ever exactly alike. Hereditary substance is composed of discrete parts, called genes, and that different genes may undergo mutation independently. Moreover, recombination is one of the commonest sources of hereditary variability. A fertilized egg contains all the hereditary factors which will interact with the environment to produce a body of its own species and variety. Bateson in 1906, proposed the name "Genetics" for that branch of biology concerned with "the elucidation of the phenomena of heredity and variation". In the plant kingdom the fact of sexuality was confirmed in 1694 by Camerarius. Plants that are normally self-pollinated differ in genetic make-up from plants that are normally cross pollinated. In a crop that is self pollinated, it is the rule that plants will be homozygous. This assumption may be made since 1) gene pairs homozygous AA or aa will remain homozygous with self pollination. 2) gene pairs heterozygous Aa will segregate producing homozygous and heterozygous genotypes in equal proportions. With self pollinations, heterozygosity is reduced by one-half with each successive self-fertilization.

Heterozygous plants may arise in a homozygous population of a self pollinated crop by natural cross—pollination or by mutation. Continuous self pollination or inbreeding for several generations in normally cross pollinated species is generally accompanied by a loss of vigor and productiveness.

Environmental variations may be discerned by growing plants with similar heredity in different environments. Heredity variations are the result of plants possessing different genetic characters. Generally they may be observed if different varieties or species are grown under similar conditions. Hereditary variations may be simple and easily observed on plant characters such as differences in colour of the plant or seed etc., The variations may be more complex such as vigour of growth, tillering capacity, disease resistance, height of plant, date of maturity etc., If a mixed lot of wheat is grown in an environment suited to the development of stem rust and all the plants become infected with the disease, except one, we may assume that the healthy plant differs from the diseased one by being inherently resistant or immune and that the diseased plants are susceptible. The assumption may be verified by growing the progeny of the healthy plant in an environment favourable for the development of rust and observing whether or not the progeny becomes diseased. The breeders particular task is to sort out those heritable variations which will be useful for the improvement of crop plants and concentrate them in a variety.

Heritable variations in plants originate through gene recombinations, mutations and polyploidy. The mechanism of heredity is dependent upon the behaviour of chromosomes and the genes, they carry. Some facts regarding characters, genes, and chromosomes are as follows :

1. In a mixed population of a plant species, the breeder selects plants with traits important for the development of improved varieties. Example of such traits are seed colour, plant height, earliness of flowering, maturity etc., The contrasting traits are determined by alternative contrasting genes.
2. The genes are located on the chromosomes. They are the determiners of the characters of a plant. The influence of each gene may be exerted individually or in combination with other genes and in conjunction with the environment. The genes occur in alternative kinds called alleles, which result in the development of contrasting forms of the characters they determine. Genes express themselves to the exclusion of their alleles are referred to as dominant. The contrasting form of gene which is not expressed in the presence of dominant is referred to as recessive. Genes sometimes change in nature so that they produce different forms of the character, the new form being reproduced in succeeding generations. Such changes in a gene is known as mutation.
3. Chromosomes are important in heredity because they carry the genes. The distribution of the chromosomes and of the genes they carry to the germ cells determines the specific distribution of the genes to the progeny. The simplest hereditary character is one that develops under the control of a single gene. However, the plant breeder works such as size, yielding ability, winter hardiness, lodging resistance are each influenced by numerous gene which may be scattered about on several chromosomes. The breeding behaviour of an individual plant is studied by growing its progeny. Only by this procedure we can know whether a plant is homozygous or heterozygous for a particular dominant character. Progeny testing is a basic procedure in plant breeding. Selection of superior plants from a mixed population is usually made on the basis of appearance of phenotype. The plant breeder may select a shorter plant, a plant with more vigour etc., The progeny test provides an opportunity to evaluate the genotype of the selected plant.

The gene recombination may be studied in a dihybrid cross, the distribution of the genes concerned to the progeny will be determined by the distribution of the particular chromosomes that carry the genes. In barley, suppose if the breeder describes white and smooth awn and if both are recessive and found in different plants and by hybridization, he can combine both and in the F_2, it may be expected that one out of each 16 plants will be "white, smooth awn", the combination of the traits desired by the breeder in this particular cross. The genotype of the white, smooth awn, plant differs from that of either parent, hence it represents a recombination of genes.

Back crossing also may be used for other breeding problems. For example, in improving inbred lines of corn, under conditions of controlled pollination, by a cross between two inbred lines that contain complementary characters, often one parent is more desirable than the other in a majority of its character. One to three back crosses may prove advantages using the more desirable inbred as the recurrent parent. Richey (1927) gave the percentage of plant homozygous for the 'n' factors, in each 'r' successive generations, calculated from the formula,

$$\left(\frac{2r-1}{2r}\right)^n$$

r—number of successive back crossing

n—number of genes homozygous

For example, percentage of homozygous of five genes entering homozygous condition, after 4 back crossing.

$$\left(\frac{(2\times 4)-1}{2\times 4}\right)^5 = \frac{7}{8} = (0.88)^5 = 53\%$$

If a character in the non-recurrent parent is controlled by two genes, then the breeder will get one homozygous plant for every 16 plants in the F_2 generation. So he has to raise large number of F_2 population. But in back cross, out of 4 plants 1 will be homozygous, so the breeder need not raise large population since the chances are more, as the genotypic and phenotypic ratio in back cross is 1:1:1:1 in first back cross population.

CHAPTER 11

Methods of Plant Improvement : Pure Line and Mass Selection

The Pure Line Theory

The theory of pure line was established by a Danish botanist, Johannsen, in 1903, Johannsen conducted selection experiments with a mixed seed lot of the Princess bean. He selected from this random lot large and small seeds. These were planted and the seeds were harvested from each of the growing plants. The seeds harvested from each plant varied in size, but the average weight of the progeny from the large seed was larger than the average weight of the progeny from the small seed. This indicated that the selection had been effective in separating bean seeds that possessed different genes for seed size. Since beans are self-fertilized, the seeds were pure from the start. The original selection was from a mixture of pure lines and was therefore successful in separating beans with different genotypes for size.

To test the efficacy of additional selection, Johannsen established 19 purelines by selection of individual beans from the mixed lot. Within each of these pure lines he again selected a large and small seed. The progenies of the large and the small seeds from a single pure line varied in the weight of the individual seed again, but the average weight of the progeny from the large seed was quite similar to the average weight of the progeny from the small seed within the same pure line. These results indicated that selection with in a population of mixed genetic types, such as the original mixed lot of beans, may be effective in isolating lines that are inherently different. But once the pure line has been isolated, further selection within the line is ineffective. In the original mixed lot of beans, variation in seed size were both hereditary and environmental. Within the pure lines, variations in seed size were due to the environment only.

Pure Line Selection

A progeny descendent solely by self-pollination from a single homozygous plant is known as a pure line. A pureline variety is developed by increasing the self-fertilized progeny from a single, true breeding plant. A variety developed by pure line selection is more uniform than a variety developed by mass selection, since all the plants in the pure line variety will be exactly alike. This is assuming, of course, that the plant originally selected is homozygous for all gene pairs, an assumption which plant breeder often make, but condition which is seldom, if ever, completely realized. Two examples illustrate how varieties may be developed by pure line selection from older varieties which are genetically mixed populations.

Example 1. Kanred is the product of a single head of wheat selected in Kansas in 1906, from Crimean, a variety introduced from Russia by the United States Department of Agriculture. One of 554 different plant selections, it was found to be resistant to rust and was distributed to Kansas farmers in 1917.

Example 2. The Columbia variety of oats was developed from a single plant selected form the Fulghum variety at the Missouri Agricultural Experiment Station in 1920. The strain increased from this plant was earlier, taller, more uniform and higher in yield than the Fulghum stock from which it was selected.

Pure line selection may be practiced by farmers who may observe off-type plants in their fields. Many useful varieties have been produced in this way. Pure line selection is practiced within segregating populations after artificial hybridization of two varieties. New genotypes are not created by pure line selection. Improvement by this method of selection is limited to the isolation of the best genotype already present in the mixed population. Once the superiority of a selected strain has been proved by thorough testing procedures, it may be increased, named and distributed as a new agricultural variety.

Mass Selection

If a group of similarly appearing plants is selected and harvested, and the seed is composited, the resulting mixture is known as a mass selection. A mass selection of a self-pollinated crop will be a composite of more or less similar and supposedly true breeding genotypes. A variety developed by mass selection will generally be more or less pure for those physical features which may be easily seen and used as the basis for purification, such as presence or absence of awns, colour markings, or maturity. But its component lines may differ in quantitative characters, such as yield, size or quality, since small differences in the quantitative characters cannot be visibly distinguished.

Two examples of varieties of winter barley developed by mass selection are Purdue 21 and Missouri Early Beardless. Their development illustrates procedures by which varieties of self pollinated crops originate as mass selections.

Example 1. Purdue 21 originated in Indiana as a composite of seed from six plants that survived the winter in a twentieth acre plot seeded to Tennessee winter barley at Lafayett, Indiana. Seed was first released to farmers in Indiana in 1921.

Example 2. Missouri Early Beardless originated from a lot of seed purchased in the fall of 1931 by a Missouri farmer from a St. Louis seedsman. It was observed that about 25 per cent of the plants in the field were shorter and earlier than the remainder of the field. Many of these early plants were harvested by hand, and the seed from them was bulked. An increase of this bulked seed was distributed to Missouri farmers in 1934.

In mass selection, plants are chosen on the basis of the phenotypes and the harvested seed is composited without progeny testing. The object is to improve the general level of the population by selecting and bulking the superior genotypes already present. It is not possible to know whether the plants being grouped are homozygous or heterozygous. Since the heterozygous plants will segregate in the following generation, phenotypic selection may need to be repeated. The environments in which a plant grows affects its development and purity. With mass selection it is not possible to know whether the selected phenotype is superior in appearance owing to hereditary characters or environment. Mass selection is often used to purify mixed variety, testing may be terminated and seed increase started any time after it has been verified that the new strain does not differ in adaptation and performance from the mixed variety, and that it is superior to the mixed variety in uniformity.

Hybridization in Self and Cross Pollinated Crops

The aim of hybridization is combining the properties or favourable genes found in different individuals. Parents for hybridization should be chosen from varieties grown in tracts with similar climate or from varieties cultivated in the neighbouring tracts. The parents should differ only in a few characters. Crossing between parents that are morphologically similar would offer great possibilities for selection as most of the genes would be of dominant type.

Selection Procedure with Hybridization (Self-Pollinated Crop)

Two selection procedures are commonly used after hybridization to sort the desirable genotypes from the segregating progeny.

(a) Pedigree selection in which plants with the desired combination of characters are selected in the F_2 generation and the progenies of each selected plant reselected in succeeding generations until genetic purity is reached.

(b) bulk population method, in which selection is delayed until a later generation usually to the F_5 and F_6 after hybridization at which time segregation will virtually have ceased.

Example : Assumed Cross

A variety of wheat Pawnee which is short, early plant type, smut resistance and high yielding. Another variety stem rust resistance.

Example of Pedigree Method of Selection

First year—Cross pawnee x stem rust resistant variety

Second year—Grow 10 to 25 F_1 plants

Third year—Grow 2000 to 6000 F_2 plants

Spraying with inoculam stem rust and smut diseases. Select several hundred short, early, vigourous, Pawnee like plants with stem rust resistance.

4th Year

Grow F_3 progeny rows from 300 to 500 selected F_2 plants. Space seeds in row so that individual plants may be studied. Resistant plants may be selected from short, early maturing families or families pure for rust resistance may be reselected. Normally 50 to 100 families may be retained at the end of the 4th year.

5th to 8th Year

Reselect superior families in F_4 to F_7 until each is uniform. Only the best appearing and most uniform rust-resistant with short, early maturing lines should be reselected and carried forward to the next generation. The total number of lines carried at the end of this period may be reduced to not more than 25 to 50.

9th Year

Grow preliminary yield test.

10th to 13 Year

Lines remaining are tested for yield in comparison with standard commercial varieties. Only the highest yielding lines are retained in the yield tests each year. During the testing period, observations are made on height, straw stiffness, winter hardiness, maturity, disease resistance and quality. Superior lines may be grown in tests over the state or entered in regional tests to learn the range of adaptation. By the end of 13th year, only 2-5 superior lines will generally remain. If superior to the commercial check varieties, one line may be chosen for increase and distribution.

14th and 15th Year

Increase seed and distribute as new variety.

(c) Back cross. The back cross is a form of recurrent hybridization. The plan of back cross is relatively simple. The parent varieties are selected and crossed. The F_1 hybrid is successively back crossed several times to one parent variety which possesses maximum number of favourable characters. After each back cross, selection is made for the superior character and the hybrid plants possessing the superior characters are used in the back cross. The parent which is repeatedly used for the back cross is called "recurrent parent". The other parent is called "donor parent" for the superior character it possessed. Since the donor parent is not repeatedly used in the back cross, it is also called "non-recurrent parent". The purpose of back cross is to recover the genotype of the recurrent parent. The back cross is a form of inbreeding and the features of the recurrent parent are automatically recovered after successive backcrosses. The only selection practiced is for the one superior trait contributed by the non-recurrent parent. The number of back crosses may vary from 1 to 8, depending upon how completely the breeder wishes to recover the genes from the recurrent parent. The back cross procedure is most easily carried out if the superior character of non-recurrent parent is simply inherited, dominant and easily recognized in the hybrid plants.

Example : Assumed cross in wheat

	Rust resistant variety (dominant genes)	×	Pawnee (cultivated) (recessive genes)
	F_1		
1st Back cross	50% Pawnee genes	×	Pawnee
2nd back cross	75% Pawnee genes	×	Pawnee
3rd back cross	87.5% Pawnee genes	×	Pawnee
4th back cross	93.75% Pawnee genes	×	Pawnee
	96.875% Pawnee genes		

Self the plants from B4 to obtain plants homozygous for rust resistant.

(d) Multiple crossing (Multiline). A complex system in which 8 to 16 varieties are systematically crossed has been used in the production of new varieties of some self-pollinated crops, particularly barley. These multiple crosses are produced by crossing pairs of parents and then crossing pairs of F_1's until all the parents enter into a common progeny according to the scheme outlined below.

$$A \times B \quad C \times D \quad E \times F \quad G \times H$$
$$AB \times CD \quad EF \times GH$$
$$ABCD \times EFGH$$
$$ABCDEFGH$$

This system of crossing has the advantage of bringing together quickly combinations of genes from several parents. Many possibilities of combinations exist since every seed produced after the initial cross is essentially a new hybrid. A disadvantage of this system is that many undesirable combinations may be brought together since such a large number of parent varieties is involved. The possibility of obtaining desirable combinations would be enhanced by selection within each progeny before the next cross is made, but this procedure will require a longer time to reach the final cross.

Cross Pollinated Crops

Selection procedures used in breeding cross-pollinated crops differ from those used in self-pollinated crops. In the self pollinated crops individual plant selections are used to establish uniform pureline varieties. But in cross-pollinated crops which are highly heterozygous, individual plants are seldom used to establish a variety for the simple reason that segregation and cross pollination make it difficult to maintain the parent type within the progenies. In cross pollinated crops, mass selection is a more common type of breeding. Selection procedures more commonly used with cross pollinated crops, in addition to mass selection include progeny selection, line breeding and recurrent selection. Mass selection was practiced in sugar beet, cotton etc., for improvement.

(a) Progeny selection : or Line breeding (plant-to-row). Progeny selection is a procedure in which progenies are grown in individual plots in order to determine the breeding behaviour of selected plants. By the progeny test plants whose superiority is due to genetic variation may be distinguished from plants whose superiority is due to the environment.

In cross pollinated crops, individual plants are more or less heterozygous and the progeny will segregate for the heterozygous characteristics. By growing a progeny of 25 to 50 plants, the range of variability of any particular line may be established. Progeny selection is most easily carried out with crops that may be evaluated and harvested as individual plants like sugar beets, cotton etc., Progeny selection is more difficult with crops such as grasses and legumes which grow in dense stands so that individual plants are hard to separate.

Varieties of cross pollinated crops are seldom developed from the progeny of a single plant because inbreeding reduces vigor. More commonly, a group of progeny lines which are similar in phenotype is composited. This procedure, as used with cotton or sugar beets, is sometimes referred as to "line breeding". Various modifications of the procedure are practiced by different breeders. For example, each group of phenotypically similar lines may be grown in isolation and open—pollination permitted within the group. This procedure helps to maintain the vigour that might be lost by more rigid control of pollination.

(b) Recurrent selection. Recurrent selection is used with cross pollinated crops to concentrate genes for a particular quantitative characteristic in a population. The characteristic under consideration should be one that can easily be recognized by the phenotype. Recurrent selection has been used to improve oil content in corn, fiber strength in cotton, sugar content in sugar beets etc., The procedure is to select from a mixed population plants that are superior for the character under consideration. The plants are selfed and the selfed seed is used to grow plant-to-row progenies. The plant-to-row progenies are then crossed in all possible combinations. Hybrid seeds from these crosses are composited and a bulk population is established to start the first recurrent selection cycle. From the bulk population, plants superior for the character under consideration are again selected and used to establish new plant-to-row progenies. The plant-to-row progenies are crossed in all possible combinations as before. The hybrid seed is composed to grow a bulk, which is used to start the second recurrent selection cycle. The process may be repeated as long as improvement is shown in the character being selected.

(c) Compounding Synthetic Varieties. Synthetic varieties are used, or their use has been suggested, for the improvement of forage crops, sugar beets, corn and other cross - pollinated crops. Many new varieties of forages are being developed by compounding seed of individual plants, or strains, into a synthetic variety. The synthetic varieties may range from mixtures of seed harvested from a few carefully selected plants to a uniform blend of seed from several distinctly different strains, inbred lines, or clones.

(d) Hybridization. Two basic hybridization procedures are used in the breeding of cross pollinated crops. These involve intervarietal or interspecific crossing.

(i) Intervarietal and Interspecific crossing. Crosses between varieties, or between species, may be used to combine genes for desirable characteristics from different parents, as with self-pollinated crops. In cross-pollinated crops, each plant may itself be an individual hybrid, in which case segregation will occur within the F_1 generation. Hybrid plants in the progeny of the cross, if pollination is uncontrolled, will in turn cross freely with other hybrid plants within the population so that the progeny from the cross is not resolved into homozygosity as with self-pollinated crops. For this reason, after hybridization, selection procedures will differ from those used with self-pollinated crops. Phenotypically desirable hybrid plants will usually need to be selfed for one or more generations to fix the desirable characters in a homozygous condition. From the hybrid population, by progeny selection lines are then established which combine the desirable characteristics of the parent varieties. Some form of out-crossing the selected lines eventually may be necessary to restore the vigour lost during inbreeding.

Introduction and Acclimatization

Introduction may be used as sources of new varieties in self-pollinated crops. Some varieties are grown as originally introduced. The Balbo variety of rye originated from a small seed lot received by the Tennessee Agricultural Experiment Station from Italy. The Ladino variety of white clover and Ladak variety of alfalfa also originated as introductions. In cotton, the Acala variety stems from seed stock introduced from Mexico in 1906, although the present Acala varieties have been changed from the original introduction by selection and acclimatization. Introductions may also be used as sources of desirable genes for disease and drought resistance, winter hardiness, and other valuable characteristics which may then be incorporated into synthetic varieties. The early immigrants to America brought with them seed of the crops grown in their native lands. Most of our important field crops including wheat, oats, barley, rice, sorghum, flax, soybeans, alfalfa, clovers, blue grass, timothy and sugarcane were first introduced in this way. In India, *Gossypium hirsutum* (Cambodia cotton) was introduced first into Pondicherry state from Cambodia (America) and from Pondicherry into Madras state. Similarly P.216 F cotton is an introduction from Punjab state into Madras state.

Acclimatization

When a crop plant is introduced into a totally new production area, it may be less adapted than in the climatic area where it was accustomed to being grown. In certain case newly introduced species, seemingly not well adapted, have after a few seasons established themselves and have become more productive. The ability to become adapted to a new climate is referred to as acclimatization. To what extent can the acclimatization process change a newly introduced crop, or variety, so that it becomes habituated to its new environment? This will be influenced by (a) the mode of pollination (b) the range of genetic variability within the crop (c) the longevity of the crop. A crop or a variety of a crop becomes acclimatized only by an increase of the genotypes within the population that are better suited to the new environment than are the average of the genotypes originally present. Acclimatization is natural selection operating in a heterogeneous population of plants. It proceeds more rapidly in a cross-pollinated crop than in a self-pollinated crop, since gene recombinations will occur with greater frequency owing to the frequent cross-pollination, and some of the recombinations may be more favourably adapted in the new environment. In annual crops, gene recombinations occur more frequently than in perennial crops, and thereby increase the possibility that favourable combinations will arise. On the other hand, a pure line would change very slightly if at all, and hence would not generally be subject to acclimatization. The rate of mutation within the crop is another genetic force which may influence acclimatization. Also, gene combinations may react differently in different environments. Example: Egyptian cotton was tried most successfully in the Soviet Union, particularly in Azerbaijan where it is now stated to grow in a large area.

Introgressive Hybridization

Anderson's (1949) observations on naturally occurring populations of Iris in two different habitats, for example species A at the top of the hill and species B at the bottom of the hill, forming hybrid formations of these two species at the place where the two habitats meet, adapted to a new intermediate habitat led him and Hubricht (1938) to the theory of introgression on hybridization.

Hybrid Vigour

It is common observation that the F_1 generation in many crosses is more vigourous than the parent stocks. The increase in vigour, growth, size, yield or function of a hybrid progeny over the parents is known as hybrid vigour or heterosis. A breeding system based on the utilization of hybrid vigour was first applied successfully to the development of hybrid corn. Hybrid vigour is now utilized in the breeding of sorghum, sugar beets, onions, tomatoes, cucumbers, and many other field and vegetable crops. To utilize hybrid vigour, uniform F_1 populations are produced in such quantities that the F_1 seed can be grown directly. As applied to breeding hybrid corn, the utilization of hybrid vigour involves three steps. (a) production of uniform homozygous inbred lines, (b) crossing inbreds in combinations that gives uniform and productive single cross (F_1) hybrids (c) crossing single crosses in combination to give productive double cross hybrids.

The procedure of producing hybrid corn, as originally devised involves detasseling the female parent inbred lines and permitting open cross pollination in isolation. In commercial hybrid seed production, utilization of cytoplasmic male sterility, has eliminated much of the work of detasseling the female parent lines. By utilization of cytoplasmic male sterility, the scheme of hybrid seed production used with corn has been extended to sorghum, sugarbeets, onions, and other crops. The utilization of hybrid vigour is not restricted to the F_1 of crosses between homozygous plants, but may be exhibited also by the F_1 of crosses between heterozygous plants. In asexually propagated crops such as sugar cane, the F_1 hybrid plant may be propagated by vegetative means without the necessity of producing hybrid seed on a commercial scale. This procedure was utilized in the propagation of the Coastal variety of Bermudagrass. The Coastal variety originated as a vigourous hybrid plant in a cross between two strains of Bermudagrass.

Theories on Causes of Heterosis

Definition. "Hybrid vigour is the increase in size, yield, general vegetative growth etc., observed in the F_1 of certain crosses as compared to the parents".

Heterosis is more marked in naturally cross-fertilized plants for e.g., *Zea mays*. On the other hand inbreeding brings out a number of abnormalities and lethal factors. Heterosis is observable in many characters such as height, size, yield, earliness, viability etc., The superiority of the hybrid is soon lost if the hybrid and the progeny are self-fertilized. In vegetatively propagated plants like sugarcane, potato, fruit plants etc., the problem of loss of hybrid vigour does not arise. Mules show hybrid vigour. They are obtained by cross of horse x donkey (*Equus equus* X *Equus hemionus*). The mules are sturdier than the parents and they are largely used in military transport.

1. Physiologic stimulus or heterozygosity hypothesis. Shull (1908) proposed this first theory according to which heterosis is due to some kind of physiologic stimulus resulting from the union of unlike gametes. The divergent alleles when present together produce an effect which exceed either of the alleles when present in homozygous condition so that A1A2 is more efficient than A1A1 or A2A2. When like genes are present in the zygote the stimulus is less than when unlike genes are present.

2. Dominant factor hypothesis. The dominant factor hypothesis is the most accepted one. This was first proposed by Bruce (1910). According to him one favourable line has a certain set of favourable genes. Another line contains another set of favourable genes which complement those in the first line. The hybrid vigour results from bringing together these dominant genes from the two parents. The deleterious effects of the recessives are covered up by the dominant favourable genes. This theory has not met with general approval since vigorous tall hybrids have been obtained by intercrossing dwarfs from two inbred lines in maize. The chief objections are 1. If only pairs of factors are concerned, a line could be obtained in the F_2 generation which would be homozygous for the two dominant factor pairs and approach the F_1 vigour. Such cases are not met with.

ABCdef × abcDEF

F_1 Aa Bb Cc Dd Ee Ff

Hence, the objections are (a) Why not true breeding heterotic lines? If above hypothesis is correct, it should be possible to recover a plant with genotype AA BB CC DD EE FF which is as highly vigorous as F_1 and true breeding. The objection to Bruce (1910) was removed by Jones (1917) in his modified theory entitled "Dominance of linked genes hypothesis" (b) Why symmetrical distribution? If heterosis is due to dominance of independent factors, the F_2 distribution curve for heterotic character should be skewed rather than symmetrical. Because in such a trait the dominant and recessive loci would be distributed according to the expansion of binomial (3/4 + 1/4)n, where 'n' is the number of genes involved in the expression of the heterotic character. (Figure III).

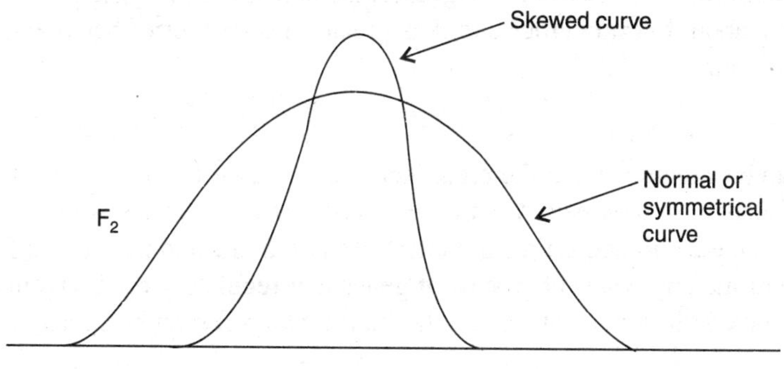

Fig. III

3. Linked favourable dominant genes hypothesis. Jones (1917) who advanced this hypothesis stated that heterosis was due to the complementary action of linked dominant growth favourable genes brought from both the parents. Genes affecting growth are found at numerous loci and have cumulative action. Since any given chromosome may contain X dominant genes and Y recessive genes distributed at random. Heterosis will be manifested in F_1 hybrids owing to the opposition of dominant to recessive at various loci. The possibility of recovery homozygotes having the characteristics of the F_1 is impossible as certain combinations of genes would pass from one generation to the other without recombination and linkage. Further when a large number of factor pairs are involved the chances of obtaining a homozygous for all factors are very remote.

4. Cumulative genes hypothesis. The objection was removed by Collins (1921) who pointed out that heterosis is governed by a large number of genes, so that even in the absence of linkage, skewness would not appear.

5. Greater initial capital hypothesis. Ashby (1930) concluded that :

(a) the hybrid does not differ from the more vigorous parent as regards growth rate, cell size, photosynthetic efficiency of leaves.

(b) The only physiological differences observed are an increased percentage of germination of the hybrid seeds and a greater initial weight of embryo which gives an advantage that is maintained throughout the period of growth.

The phenomenon of hybrid vigour, he added, is due to some process between fertilization and the setting of seed. According to him, hybrid vigour, is nothing more than the maintenance of an initial advantage in embryo size. This hypothesis was disputed by other investigators that in tomato, peas, tobacco, maize that embryo size had little or no permanent significance in causing heterosis. East (1936) and Wang (1947) also supported this hypothesis and concluded that increase in endosperm and embryo size may itself be one of the cause of heterosis, but seed size or the size of any part of the seed cannot be the true cause of heterosis. The size of the seed depends on many factors such as size of the fruit, number of seeds matured etc.

6. Super dominance or Over dominance hypothesis. According to this theory, heterozygous genotype is superior over any homozygote genotype at the locus level, has been put forward by Fisher (1930), East (1936) and Hull (1945), i.e., interaction of alleles at a single locus if A1 A2 A3 are three alleles, A1 A2 or any such combinations of the two factors would be more effective than A1A1, A2A2, A3A3 the homozygous condition i.e., heterozygous exceed in effect than homozygous. Various names had been given from time to time as superdominance by Fisher (1930), interaction of alleles at a single locus by East (1936), overdominance by Hull (1945) etc., However, the term "over dominance" is most commonly used among all. This theory suggests that there is something inherent in the heterozygous condition that brings about hybrid vigour and thus greater the number of heterozygous alleles, greater will be the hybrid vigour.

Conclusion

Finally it can be concluded that the great increase in vigour from crossing inbred lines of maize is of common knowledge. It is evident that the interaction of many genes is involved in such crosses. Heterosis result from the combined action and interaction of allelic and non-allelic factors. So it seems, that heterosis is controlled by genes or groups of genes that regulate growth vigour. The results of the heterosis lead to conclude it is a quantitative inheritance which furnishes a logical means of breeding for heterosis.

Heterosis is observable in many characters such as height, size, productiveness, earliness, viability, resistance to diseases, pests and drought etc., Hybrid vigour has been used for practical plant and animal improvement. Increased vigour of the hybrids have been demonstrated in cereals like maize, pearl millet, sorghum, wheat, barley and vegetables like onion, tomato, brinjal, squashes and grasses, cotton etc., In animals also poultry, mule, dairy cattle and silk worms. In vegetatively propagated plants, hybrid vigour is maintained unless it changes by somatic mutations. Many of the fruit plants like apples, pears, potatoes, strawberries have been selected originally for their hybrid vigour, and multiplied vegetatively.

Irradiation breeding

It has been known since 1928 that mutations may be induced in plants by various forms of radiations. This knowledge led to the extensive use of radiation—induced mutations in genetic studies with plants. Reports of experiments in Sweden in which mutations for straw stiffness, earlyness, quality and other agronomically useful characters in barley had been induced by X-rays stimulated interest in the possible use of radiations as a tool of the practical plant breeder as well as the theoretical geneticist. Atomic investigations have made available new and powerful sources of radiations which are also being investigated in attempts to find mutations useful in crop improvement. The method of breeding is based on the principle that the rate of mutation can be increased exposure of the plant and seeds to radiations. Since useful mutations are produced in nature, it is assumed that valuable mutations also may be produced experimentally.

Vijendra Das and Sree Rangasamy (1973) studied the effect of irradiation with Cobo gamma rays on embryo sac in M_1 generation of rice and found aberration in embryo sac structures, abortion of embryos and instability of endosperm divisions. In a study on effects of radiations on the R_1 and R_2 progenies of *Pennisetum typhoides*, Vijendra Das (1978) observed gamma rays were more effective than X-rays with respect to R_1 lethality, growth reduction, pollen sterility and R_2 mutant frequency.

CHAPTER 12

Mutations as Method of Plant Improvement

A mutation is a sudden change in the hereditary material of a cell. Mutations may result from (a) a change in the gene from one allele to the other (b) rearrangement of chromosome materials, or (c) loss of duplication of chromosome segments. An overwhelming majority of the mutations are harmful or deleterious to the individual. Gene mutations may be either dominant or recessive, but the recessive mutation is by far the more common. Dominant gene mutations generally produce an immediate effect in an individual. The effect of a recessive gene mutation is not usually expressed until two recessive genes come together in an individual as a result of segregation. The mutation rates of individual genes vary, and a specific gene may tend to mutate in one direction more frequently than in another. As a result certain mutant types may be found in a population of plants with greater frequency than others. Of some 300 induced mutant seedlings observed in barley about 60 per cent were white seedlings, 5 per cent were virescent seedlings, 15 per cent were yellow seedlings of various shades, 10 per cent were greenish yellow seedlings, and 10 per cent were miscellaneous; striped, banded, spotted etc., Most of these observed mutant seedling characters were lethal and almost all were unfavourable to growth. The cumulative effect of the small variations have made mutation an important force in the evolutionary changes that plants have undergone. How much mutation may change the performance of a crop variety grown continuously for a number of years has been the subject of much speculation. Many important varieties of crop plants have originated from spontaneous mutations which appeared as striking variants in older and widely grown varieties. One of the more spectacular examples is the mutant dwarf forms of grain sorghum. A dwarf mutant plant of standard yellow milo gave rise to the dwarf yellow milo variety. From a still further dwarfed mutant plant of dwarf yellow milo, a double dwarf yellow milo variety originated. These two dwarf mutants and dwarf mutant plants later found in White milo and other varieties formed basic breeding stocks from which the vast combine grain sorghum industry of the southwest has been developed.

In 1928, Dr. L.J. Stadler reported that mutations are increased in plants after exposure of the seeds to X-rays and radiations from radium. In the year 1927 Dr. H.J. Muller had reported that X-rays have the power to induce mutations in Drosophila. Stadler also reported radiation induced mutations in wheat, oats and maize. In recent years many types of ionizing radiations, including thermal neutrons (atomic radiations), have been used to induce mutations in many kinds of crop plants. In 1934 and 1935, Nilsson-Ehle, at the Svalof (Sweden) Plant Breeding Station, observed some desirable X-ray induced mutations in barley. One mutation, which was observed several times, became known as an "erectoides" mutant. The erectoides mutant is characterized by a compact head type and stiffer straw. Mutations for earliness, kernel weight and tillering capacity were also observed in barley. Other Swedish workers later isolated useful mutants from other agricultural crops. Their findings stimulated much interest among breeders in the possibilities of inducing useful mutations in crop plants by subjecting plants or seeds to ionizing radiations. This system of breeding has since been designated "irradiation breeding". The mutation rate is at best very low, and very large numbers of plants must be examined to find desirable mutants.

For the efficient mutation breeding programme, first one must consider the kinds of parental material, level of ploidy, structural hybridity and reproductive habits etc., The idea is that increasing degrees of polyploidy and consequent chromosomal and genic duplication lead to a sharp decrease in the frequency of mutations. Among vegetatively propagated material the opportunities for induction of mutations are promising. In mutation breeding it should be noted that most mutations are recessive. Hence a very large F_2 population is necessary, if the desired improvements are to be obtained and detected. At Savlof (Sweden) about 2000 X_2 rows are grown each year in the barley improvement programme. During growing period in the X_2 generation, careful inspection is made and any abnormal plants are recorded. If an X_2 row is suspected of containing a new and valuable mutation, then the X_2 plants are harvested separately and X_3 rows are grown. Here the mutation can be verified and segregation ratios can be determined. At Savlof another technique was used to induce and detect beneficial mutations, involving repeated irradiation of materials for several consecutive years. As a result mutations accumulate in the material and selection of beneficial mutations is enhanced. Each year one seed from each X1 spike is pooled and irradiated. After 4 or 5 years the material is grown up in individual plant progenies which are carefully analysed.

The techniques of inducing somatic mutations or bud sports in fruit trees have been described by Granhall (1954). Dormant scions of apples, pears, plums and cherries have been exposed to X-rays, gamma rays and fast and thermal neutrons. They had been grafted on root stock. Radio active isotopes (P32) and (S 35) had been injected into stems, branches, twigs or buds of potted plants. During the leafing out period in early spring, the X-ray material which has been studied revealed mutation effect on colour, shape and time of ripening of fruits.

Beneficial Mutations : Induced Mutations in Certain Crop Plants

Wheat :

1. Increased straw strength combined with short straw.
2. Slightly earlier.
3. Increased backing quality and superior gluten content.
4. Increased resistance to stem rust.

Leaf rust resistance had been transferred to wheat through substitution of an *Aegilops umbelluata* chromosome. But undesirable growth characters are also conditioned by the substitution of this chromosome. But Sears, through X-raying had separated the desirable and undesirable effects.

Polyploidy as Method of Plant Improvement

Polyploidy is a condition in which individuals have more than two chromosome sets, or genomes, in their somatic cells. In contrast to the normal diploid (2n), they may be triploid (3n), tetraploid (4n) pentaploid (5n), hexaploid (6n) and so on. Polyploid plants may arise by duplication of the chromosome sets from a single species, autoploidy, or by combining chromosome sets from two or more speices, alloploidy. Alloploidy is the more common method of ploidy in nature. Autoploidy which are produced by doubling the individual chromosomes with a plant, also hold considerable interest for the plant breeder. Generally autoploids are more strong and vigorous than the diploid from which they were derived and tend to have larger leaves, dark green colour, larger flowers and seed and larger cells and nuclei. Autoploid tomatoes and yellow corn produce more vitamin C than the diploids from which they originate. The tendency of autoploids to have greater vegetative growth and reduced seed production suggests that autoploidy would be more useful in breeding crops harvested for their vegetative parts or for root crops. As a result autoploidy has been viewed with most interest by the breeders of forage crops, sugarbeets etc., The interest in polyploidy as a tool of the breeder was given a tremendous boost when it was discovered that Colchicine an alkaloid extracted from seeds or corms of the autumn crocus (*Colchicum autumnale*) could be used to double the chromosome number.

On the other hand, recessive mutations that are deleterious to the plant may be covered up by their dominant alleles to a greater extent in polyploids so that they are not expressed as frequently in the phenotype of the plant. Polyploidy has been an important factor in the evolution of plant species. A knowledge of its mechanics is important in understanding the relationships between certain plant species. The chromosome relationships in the various species of *Brassica* have been cited as an example. The genetic origins of common wheat, tobacco, cotton have been traced in a similar manner. A knowledge of these genetic relationships is useful to the breeder in planning crosses between species or between genera. Interspecific crosses of this nature have been used many times to add genes for disease resistance or other characters from a wild species to a closely related cultivated species. From a three way interspecific cross with cotton (*Gossypium arboreum* × *G. thurberi*) × *G. hirsutum* greater lint strength was obtained.

Cyto-taxonomical studies and correlation of polyploidy and geographical distribution of the species have revealed that the polyploids are more frequent in the Northern latitudes and higher altitudes than in the tropical zones. It has also shown that perenniality is more common in the polyploids. Darlington gave the following in regard to rhizome in relation to polyploidy.

Genus	2n	4n	6n	8n
	No rhizome	Little rhizome	Much Rhizome	Extensive rhizome
Chrysanthemum	*C. coronarum*	*C. indicum*	*C. morifolium*	*C. deciseanum*

Origin of Some Crop Plants

1. In banana *Musa paradisiaca* autotetraploids are inferior to triploids in that they have weaker leaves and increased fertility. But, they offer the only available chance of adding disease resistance to commercially successful varieties. In banana, autotetraploids are produced by chance fertilization of an unreduced triploid egg (AAA) by a haploid pollen from the disease resistant diploid parent. A large number of such tetraploids have been produced, but they have not yet gained any commercial success.
2. The tetraploid maize has 43 per cent more carotenoid pigment and vitamin A activity than the diploid. Collins (1912) formulated a hypothesis that maize ($2n = 20$) originated as a hybrid between teosinte (*Euchloena mexieana*, Schrad. $n = 10$) and an unknown grass belonging to the tribe Andropogoneae. Vijendra Das (1970) cytologically studied autotetraploid ($4n = 40$) Zea mays L., and found quadrivalent configuration in diploid and octovalent configuration in autotetraploid and supported the hypothesis of Collins (1912) about the possible origin of maize.
3. Certain distant crosses are not successful at the diploid level, but are relatively successful at the autotetraploid level. e.g., 4n *Brassica oleracea* x *B. chinensis* is successful, but when *B. oleraceae* is diploid it is unsuccessful.

4. The autotetraploid varieties of forage crops have been considerably successful. The most successful examples are, tetraploid red clover (*Trifolium pretense*) and rye grass (*Loliium perenne*), other examples are tetraploids of alsike clover (*Trifolium hybridum*) variety Tetra and berseem (*Trifolium alexandrium*), variety Pusa Giant Berseem. Auto tetraploid red clover and rye grass are more vigorous, digestible and palatable, and have greater resistance to nematodes as compared to the diploids.
5. The experimentally produced distant hybrids were highly sterile, occasional branches were fertile and had twice as many chromosomes as present in the F_1 hybrid, e.g., F_1 from the cross *Primula verticillata* x *P. floribunda* produced a fertile branch with twice the chromosome number. The progeny obtained from this fertile branch were allotetraploid and were named as *P. kewensis.*
6. Allotetraploids generally combine the morphological and physiological characteristics of the parent species. But it is difficult to predict the precise combination of characters that would appear in the allopolyploid species. For example, the aim in producing *Raphanobrassica* was to synthesis a crop species that would combine the root of radish (*Raphanus sativus*) and the leaves of cabbage (*Brassica oleraceae*). *Raphanobrassica* did combine the characteristic root and shoot systems of the two parental species, but in the opposite direction, that is, it had leaves like radish and roots like cabbage direction, that is, it had leaves like radish and roots like cabbage (Experiment of Russian Botanist, Karpechenko).
7. *Brassica juncea* (2n = 36) was synthesized by Frandsen during 1943 through crosses between *B. nigra* (2n = 16) x *B. campestris* (2n = 20). The synthesized amphidiploid closely resembled *B. juncea* and crossed readily. The offspring was fertile and meiosis was normal with regular bivalents.
8. Another feature of allopolyploids is that many of them are apomictic. Apomictic species are common in grasses, e.g., Poa. Apomicts are known in many other groups of plants, e.g., *Traxacum, Parthenium, Rubus, Fritillaria, Tulipa, Solanum.* Apomixis arises in response to near complete sterility of interspecific hybrids. Each sterile plants are able to reproduce asexually through apomixis.
9. Polyploid series of some of the crop plants

Scientific name	**Common name**	**Gametic chromosome number (n)**
Brassica nigra	Black mustard	8
B. oleraceae	Cauli flower or cabbage etc.,	9
B. campestris	Sarson	10
B. carinata	Abyssinian	17
B. juncea	Indian mustard	18
B. napus	Rape	19
Gossypium arboreum	Asiatic cotton (desi)	13 (A2)

Scientific name	Common name	Gametic chromosome number (n)
G. herbaceum	Asiatic cotton	13 (A1)
G. thurberi	Wild American cotton	13 (D1)
G. barbadense	Sea Island cotton	26 (A2D2)
G. hirsutum	American upland cotton	26 (A1D1)
Nicotiana sylvestris	Wild tobacco	12
N. tomentosa	Wild tobacco	12
N. tabacum	Cultivated tobacco	24
N. rustica	Cultivated tobacco	24
N. bigelovii	Wild tobacco	24
Saccharum officinarum	Noble canes	40
S. barberi	Indian canes	41, 45, 46, 58, 62
S. sinense	Indian canes	58, 59
S. spontaneum	(Kans) wild canes	20 to 64
S. robustum	Wild cane	30 to 74
Sorghum versicolor	Wild sorghum	5
S. bicolor	Jowar	10
S. halepense	Johnson grass	20
Triticum boeticum	Wild einkorn	7 (A)
T. monococcum	Einkorn wheat	7 (A)
T. dicoccoides	Wild emmer	14 (AB)
T. dicoccum	Emmer wheat	14 (AB)
T. turgidum	Solid stem wheat	14 (AB)
T. cartalicum	Persian wheat	14 (AB)

CHAPTER 13

Laboratory Exercises

(Contribution of Dr. M. Rizwana Banu, Centre for Plant Breeding and Genetics, Tamil Nadu Agricultural University, Coimbatore)

1. DETERMINATION OF CHROMOSOME COUNT FROM DIVIDING POLLEN MOTHER CELLS, ROOT TIPS AND POLLEN GRAINS

Killing and Fixing

The killing and fixing of a material is an important process. Reduction division goes on in pollen mother cells in anthers. Collect anthers of flower buds in a very young stage to study meiosis. Because the cell division is a continuous process and so it cannot be fixed at a definite stage. Careful and critical examination and experimentation have to be carried out to secure proper fixation of the material.

Killing

This may be defined as a sudden and permanent termination of life of cells.

Fixation

This is commonly said to be the preservation of different cellular structures or cellular components as nearly in living condition as possible. Chemicals are used as fixatives. Both process can be obtained by a single chemical fluid which is a mixture of different chemical agents. A good fixative is one that changes the cell chemistry the least but preserve the cell structure the best. Killing preceeds fixation. To study the chromosome morphology the chromosome should be stained and spread well. When a cell is treated with chemicals, it undergoes certain changes which are irreversible and it is called as destruction.

Primary requisites for study of chromosomes :

1. Coagulation of protein and precipitation of chromatids.
2. Chemical should have rapid penetration and kill the cell.
3. When cells are treated with chemicals, they should check the autolysis of protein.
4. The chemicals should maintain the cells in aseptic condition
5. It should be able to give clear and dark staining of chromosomes.
6. It should enable the hardening of tissues.

All the properties are not found in a single fixative and hence different compatible fluids which have these properties can be used. There are two types of fixatives.

1. Acid fixation image. It preserves chromatin, spindle fibres and nuclei and dissolves mitochondria and Nucleoplasm.

2. Basic fixation image. It dissolves chromatin, spindle fibre and preserves nucleoli, mitochondria and Nucleoplasm.

Nearly all fixative agents give acid fixation image. Fixing chemicals are lethal in action.

Main chemicals used as fixatives are classified as metalic and non-metalic fixatives.

1. Metalic fixatives. Chromic acid, Osmic acid, Platinum chloride, Mercuric chloride, Uranium nitrate etc.

2. Non-metalic fixatives. Ethyl alcohol, methyl alcohol, acetic acid, formaldehyde, chloroform, propionic acid.

Advantages of non-metalic chemicals over metallic chemicals are washing is not required whereas a thorough washing is required in metallic chemicals.

Preparation of stains

Acetocarmine

A mixture of 45 ml of glacial acetic acid and 55 ml of distilled water is gently boiled. A quantity of one gram of Carmine powder is slowly added stirring gently for a few minutes. After boiling, the contents are removed from the flame and allowed to cool at room temperature. Then the solution is filtered into a clean dropping bottle with Watman No.1 filter paper (double). The filterate is light reddish in colour and is 1%. Ferric chloride or Ferric acetate may be added one or two drops if necessary.

Acetoorcein

The procedure for the preparation of acetoorcein is the same as that of aceto carmine. But here instead of Carmine, Orcein of pure quality is used.

One gram of Orcein is dissolved in 45 ml of glacial acetic acid and 55 ml of distilled water after it is heated to boiling point. The contents are allowed to boil for one minute and then removed from the flame. The contents are allowed to cool to room temperature. The mixture is filtered through double Watman No.1 filter paper into a clean bottle.

McCallum's Propiono carmine

One gram of Carmine powder is dissolved 45 ml of Propionic acid and 55 ml of distilled water and prepared similar to Acetocarmine. Propiono carmine gives better fixation and staining properties than glacial acetic acid. Propionic acid dissolves more Carmine powder and gives a more intensive reaction together with greater clarity.

Feulgen or Nuclear reaction

The Feulgen reaction has proved to be particularly useful of its specificity to DNA.

Procedure

Dissolve 0.5 gram of basic fuchsin in 100 ml of boiling distilled water and shake thoroughly. Cool the contents to 50°C and filter. To the filtrate add about 10 ml of Normal Hcl and 0.5 gram of Potassium meta bisulphate. Shake thoroughly, stopper tightly and place in a dark place for about 18 hours. Store the reduced colourless solution in dark in a well stoppered coloured bottle. If the solution is coloured it has to be discarded, and prepared freshly again.

Precautions :

1. The water should be glass distilled.
2. The normal Hcl is prepared by adding 82.5 ml reagent Hcl to one litre of glass distilled water.

Crystal violet

Crystal violet is also referred to as Gentian violet. The disadvantage with this stain is over staining of the cytoplasm. 100 ml of distilled water is heated to boiling point in a clean dry beaker. One gram of crystal violet is added to the boiling water and cooled down to room temperature. Then the contents are filtered to give one per cent Crystal violet solution.

Heidenhain's Iron Hematoxylin

Hematoxylin is a chromogen dye derived from logwood, **Hematoxylin campechianum** L., and is one of the most important stains. The dye solution itself has little or no affinity for tissues, unless iron or aluminium is present in the latter, consequently mordanting in some form is necessary. The chemical most commonly used for mordanting is ferric ammonium sulphate, colloquially known as "iron alum" although other substances may also be used. The strength of the solution is usually 3%. Some technicians use a 2%, others a 4% solution, in water. The solution of the dye (Hematoxylin crystals) as used in Heidenhain's iron hematoxylin technique is simply a 0.5% solution in distilled water. It will be more satisfactory to prepare a 10% solution in absolute ethyl alcohol and to dilute a portion with distilled water to the 0.5% strength when required. As a rule, the slides should be left in the stain atleast as long as they remained in the mordant, but in most cases 24 hours is the optimum time.

Preparation of 1 Normal HCl

Concentrated HCl is 12 Normal of molecular weight of about 36.46. One CC of concentrated HCl of 12 Normal +11 CC of distilled water = 1 N.HCl. For hydrolysis, 9 drops of aceto-orcein stain + 1 drop of 1 N. HCl, gently heated on a watch glass and cooled at room temperature before squash.

Glass cleaning fluid

The following are the chemicals needed, viz., 1. Potassium dichromate—2 grams; 2. Distilled water—10 cc, 3. Concentrated sulphuric acid 10 cc. Dissolve potassium dichromate in distilled water and add very slowly concentrated sulphuric acid just to dissolve stirring with a glass rod. For cleaning slides and coverslips the total volume may be made to 1000 CC with distilled water. Leave the slides and coverslips in the solution for 24 hours. Rince thereafter in tap water by taking the slides with foreceps. The solution can be used again for further cleaning of slides. Use some soap water if necessary for final cleaning of the slides, and wipe dry with a clean muslin cloth.

Fixing chemicals

1. Ethyl alcohol (Ethanol) C_2H_5OH. Percentage suitable varies from 70% to 100% (100% absolute alcohol) . 98% alcohol is available as rectified spirit. Possess immediate penetrating and dehydrating properties. Causes irreversible denaturation of proteins - brings about desirable hardening of tissue.

2. Methyl alcohol (Methanol) CH_3 OH. Occassionally used in chromosome studies. Causes swelling of chromosomes—colourless and poisonous fluid.

3. Acetic acid CH_3 COOH. It can be combined with any other fixative. It has a pungent smell. Used from very low to very high concentration (1 to 100%). It freezes to form resembling ice at 0°C and hence known as glacial acetic acid. It possess remarkable penetrating properties even higher than alcohol. It is mainly desirable for the study of structures - in checking their shrinkage and to preserve chromosome structure.

4. Formaldehyde (HCHO). This is a gas in commercial form known as formalin. It contains 40% formaldehyde in water. For fixing, 10 to 40% formaldehyde in water (commercial formalin) is used. It has a very low precipitating action on proteins in high concentrations. It hardens the tissue to a remarkable extent and hence should not be considered as a good fixative for chromosomes. It has a capacity of preserving cell constituents in an excellent way so that the life structure is maintained. It is very difficult to smear due to excessive hardening of tissues placed in a fixative containing formaldehyde often show well scattered chromosomes after setting in paraffin blocks.

5. Propionic acid (C_2H_5 COOH). Of late this is used extensively. This is colourless with acrid odour. Mixable with alcohol, water and ether in all proportions and it is a good solvent for Analin dyes. It can be obtained from the oxidation of normal propyl alcohol. It can be used as a substitute for acetic acid.

6. Chloroform (CH Cl_3). This is a derivative of methane CH_4 and sparingly soluble in water. This has a narcotizing action and hence used as anaesthetic. It gets slowly converted into highly poisonable carbonyl chloride in the presence of light and air. It is a good solvent for fatty substances—a property used with advantage in fixatives. In study of plant chromosomes it is used as a fixative to dissolve fatty and waxy substances, secretions etc., from the upper surface facilitating the penetration. Chloroform has slight hardening action on all plant tissues.

Killing and fixing fluids

Mixture giving acid fixation images

Carnoy's fluids

Two different fluids bear Carnoy's name. The second fluid is the one more commonly employed.

1.	100% ethyl alcohol	—	15 CC	3:1
	Glacial acetic acid	—	5 CC	
2.	100% ethyl alcohol	—	30 CC	6:3:1
	Chloroform	—	15 CC	
	Glacial acetic acid	—	5 CC	

Action of both fluids is very rapid, because of the great penetrating power of each.

Formalin-Aceto alcohol

This fluid more familiarly known as FAA, might almost be called the standard preservative of botanical microtechnique, since it is used more extensively than anyother. The standard proportions are

50% or 70% ethyl alcohol	—	90 CC	18:1:1
Glacial acetic acid	—	5 CC	
Formalin	—	5 CC	

This reagent may be used with almost any plant material intended for anatomical or morphological study. It is unsuitable for chromosome studies. Material may be left in it almost indefinitely without appreciable damage; this property of nearly perfect preservation makes formalin—aceto—alcohol the ideal fluid to take on long collecting trips. The minimum time of fixation is 18 hours. Washing in two changes of 50% ethyl alcohol is all that is necessary.

Formalin-Propionic-Alcohol

Since Propionic acid appears to give more adequate fixation than acetic acid, it may be substituted for the latter, in the same proportions, in the formula given above. The time of fixation and manner of washing are identical.

Method for root tip squashes (onion)

Collect root tips between 9-10 AM, when there is bright sunshine in 3:1 Carnoy's fluid freshly prepared. After one hour wash the root tips in running water and place on a watch glass in 9 drops of aceto-orcein and add one drop of 1 N Hcl, gently heated, allowed to cool and squash on a clean slide with a few drops of the stain.

Method of root tip squashes (difficult material of root tips)

The root tips are collected for pre-treatment chemicals viz., saturated μ Bromonaphthalene (for leaf tip saturated para dichlorobenzene solution) and kept in the solution for 2-3 hours. Then the material is thoroughly washed in running water and transferred to 3:1 alcohol : acetic acid mixture for overnight. The next day wash the root tips and transfer to 9 drops of aceto-orcein and one drop of 1 N Hcl, gently heat, cool and squash with a drop of the stain of aceto-orcein.

Method of flower bud smears (Vijendra Das and Mensinkai, 1966a)

The very young flower buds are fixed in freshly prepared 6:3:1 alcohol : chloroform : acetic acid for 2 hours. Then the anthers are dissected out and smeared in a few drops of aceto-carmine stain, on a clean slide. For difficult materials like cotton the following procedure is adopted. A modified fixative of the 6:3:1 Carnoy's was used. The flower buds in which the corolla tip was slightly visible were fixed in 3 cc of absolute alcohol; 2 cc of chloroform; and 1 cc of propionic acid. The buds were kept in the fixative for 5 hours and then transferred to 0.25% of colchicine to destroy the spindle mechanism. After 15 hours in the aqueous solution, the buds were transferred to 70% alcohol for preservation. The stain was prepared in the same way as follows: 11½ cc propionic acid was added to 13½ cc of distilled water and heated to fuming. 200 mgm of carmine powder was added little by little and stirred constantly. The solution was boiled for a few minutes and the heater was switched off to reduce the temperature. When the boiling ceased, once again the temperature was raised to boiling and lowered again to stop boiling. This method of alternative boiling and cooling was done three times. The solution is cooled at room temperature. After cooling it was filtered with double filter paper of Watman No.1 into a dropping bottle and an attractive pinkish stain is obtained. The anthers are smeared in 0.5 per cent to 1.0 per cent of propiono carmine. For permanent preparation of the slides, they are passed through i) 45% glacial acetic acid for 5 minutes ii) 1:1 glacial acetic acid and N. butyl alcohol solution for 2 or 3 minutes iii) pure N. butyl alcohol for a few seconds. Then the slides are mounted in neutral Canada balsam.

Schedule for making the temporary slides permanent (Mc Clintock, 1929)

Butyl alcohol, acetic acid medium

Gently warm the sealed temporary slide to melt the wax and remove the wax with a needle. Invert the slide in a mixture of glacial acetic acid and a part of butyl alcohol 3:1. Keep until the coverslip has fallen down. It is necessary to have some marking to know as to which side of the coverslip was intact with the slides. The slide and cover slip may be kept for a minute or two in the following mixtures.

(a) Equal parts of glacial acetic acid and butyl alcohol (1:1)

(b) Three parts of N-butyl alcohol and one part of glacial acetic acid

(c) Pure N-butyl alcohol.

The slides are then removed and the coverslips fixed in position with a drop of Canada balsam. Keep the slides on the top of an oven or at room temperature without disturbance for about a week until the balsam has dried completely. Inscribe the particulars on the slide with a glass marking pencil and preserve in the slide cabinet.

Paraffin methods

This technique is mainly used to study the embryo sac and ovule structure, leaf sections, stem cuttings cross sections etc., some times for anther sac also. The preparation of permanent slides by the microtomy involves the following essential steps.

1. Killing and fixing
2. Washing and dissection of the material
3. Dehydration
4. Hardening
5. Infiltration

6. Embedding 7. Microtoming 8. Mounting
9. Down grades of alcohol 10. Staining
11. Upgrading of alcohol and making permanent.

1. Killing and fixing. Collect the spikelets or ovaries of flower buds in a freshly prepared mixture of F:A:A (1:1:18) Formalin : acetic acid : alcohol. Material may be left in the fixative almost indefinitely without appreciable damage. The minimum time of fixation is 18 hours. The amount of the fixing fluid should be always 10 times than the volume of the material.

2. Washing and dissection of the material. It is necessary to wash out the killing and fixing fluid thoroughly before proceeding further. If spikelets are fixed, the ovaries have to be dissected out, other materials like stem or leaf should be reduced to requirement and make it small.

3. Dehydration. Next step is to remove water from the tissues. Dehydration in water : alcohol mixture as follows :

50% alcohol	—	3 hours
70% alcohol	—	over night
80% alcohol	—	3 hours
90% alcohol	—	3 hours
100% alcohol (two changes)	—	over night

In the last solution, put a pinch of eosin (erythrosine) dye to give pink colour to the material to be visible in wax while embedding.

4. Hardening. Transference or hardening through the steps of alcohol - chloroform mixture.

3 part of alcohol : 1 part of chloroform	—	2 hours
1 part of alcohol : 1 part of chloroform	—	2 hours
1 part of alcohol : 3 part of chloroform	—	2 hours
Pure chloroform	—	over night

5. Infiltration. Add paraffin shavings to the material in the chloroform little by little until saturated. This can be done for two days. As soon as saturation is effected, the corks of the test tubes are removed and the test tubes are placed on the top of an oven (flat bottomed test tubes alone should be used), at about 30°C. Shake the contents. Add more was shavings until dissolved for 24 hours. Then the tubes are transferred inside the oven, kept at temperature of 60°C. Add more wax shavings until dissolved. Then take the test tubes out, one by one and pour out the molten chloroform wax and add fresh wax shavings and keep inside the oven, at 60°C. After 2 to 3 hours again pour the molten wax outside and add fresh wax shavings to the test tubes and keep inside the incubator. This process has to be followed for a week's time until the smell of chloroform is completely ceased from the material. The melting point of the paraffin wax should be about 58°C which will help for infiltration at 60°C.

6. Embedding. Paper boats have to be prepared. The material with molten wax have to be poured in the boat and further molten wax have to be added to make the volume and quickly arrange the material with hot needles in groups in horizontal or vertical manner suitable for cutting and making blocks. The paper boats should have been smeared with glycerine before pouring the material inside. The paper boats then allowed to float in a trough of water to solidification. Once solidified, the papers are torn and removed and the wax blocks are kept in water for overnight.

7. Microtoming. The next day the wax blocks with material are placed on the table and cut into pieces with the material with hot scalpel and mounted on the wooden blocks ready for cutting in a microtome. Set the scale in the microtome (for ovaries 10 to 12 microns). Move the wheel of the microtome with a steady speed so that the wax fixed on wooden blocks cut into the desired thickness and forms ribbons. The microtome knife has to be properly sharpened and gently heated with a spirit lamp so that the wax ribbons will not break into pieces. The wax ribbons should be straight and not curved, and this can be done by adjusting the screws on either side of the knife.

8. Mounting. The ribbons are arranged ¾ of the length of the slide. An adhesive is necessary for affixing the sections of ribbons. Place a drop of the adhesive on the slide, smear it evenly and immediately flood with 3% formalin in distilled water.

Adhesive – Egg albumin + equal volume of glycerine + 0.5 g of Sodium salicilate, stir well in a beaker and filter with a glass wool. Place the ribbon on the flood, then put the slide on a warm plate or gently heat with a spirit lamp and stretch the ribbons with needles and flatten the materials. By this the wrinkles of the ribbon will be removed. If a warm plate is used, the temperature should be about 48°C., and should not melt the paraffin. The ribbons can be stored properly in wooden trays, and kept inside drawers.

Staining Procedures

9. Down grades of alcohol. Before the sections can be stained, the paraffin must be removed. Place the paraffin slides in a jar of xylol for 2 hours or overnight. Then follow the following series.

1:1-xylol and absolute alcohol	—	5 minutes
100% alcohol	— —	5 minutes
90% alcohol	— —	5 minutes
80% alcohol	— —	5 minutes
70% alcohol	— —	5 minutes
50% alcohol	— —	5 minutes
30% alcohol	— —	5 minutes
10% alcohol	— —	5 minutes

10. Staining

Mordant in 4% alum (Iron alum)	—	20 to 30 minutes
Rince in water 2 or 3 times	—	5 minutes
Stain in 0.5% Hematoxylin	—	15 minutes
Rince in water 2 or 3 times	—	5 minutes

Destain 5 minutes in saturated aqueous Picric acid (if necessary) Rince in water 2 or 3 times to remove picric acid. Then pass through quickly in alcohol-xylol series.

Gentian violet or Crysal violet stain of 1% may also be used. The excess stain may be washed in water. Picric acid need not be used. Mordant with iron alum is also not necessary. Then quickly pass through upgrading of alcohol (dehydration) and mount in neutral Canada balsam or euparal.

11. Dehydration (upgrading of alcohol)

20% of alcohol	—	2 minutes
50% alcohol	—	2 minutes
70% alcohol	—	2 minutes
90% alcohol	—	2 minutes
100% alcohol	—	2 minutes
Pure xylol I	—	30 minutes
Pure xylol II	—	15 minutes
Pure xylol III	—	10 minutes

Mount in neutral Canada balsam or Euparal, to make permanent.

2. PREPARATION OF KARYOTYPES FROM DIVIDING ROOT TIP CELLS AND POLLEN GRAINS

The karyotype study is useful to study the morphology of chromosome since it is a constant factor for a species. When it is drawn it is called ideogram. In the root tip cells metaphase plate constitutes the best stage to study and measure the chromosome length. It will clearly show the position of centromers, length of chromosomal arms, secondary constrictions of satellites. For the students, *Allium* (onion 2n = 16) root tips are the best material to study the karyotype of a species. Normally it is not studied in pollen grains. In pollen grains diameter is measured.

Measurement with occular micrometer.

Stage micrometer has 100 divisions of 1 mm

Hence 1 stage division = 0.01 mm

The occular micrometer is fixed in the eye piece of the microscope.

For example : 10 occular divisions = 4 stage divisions

1 occular divisions = 4/10 = 0.4 m (microns)

Hence the length = 0.4 x 0.01 = 0.004 mm

For measurement of pollen grain (diameter)

For example : Occular divisions — 60

Its length — 60 × 0.004 = 0.24 mm

3. DETERMINATION OF INTRASPECIFIC VARIATION IN CHROMOSOME NUMBER FROM LOCALLY AVAILABLE TAXA

The root squash technique and pollen mother cells smear techniques have been described in the topic determination of chromosome count in the item (1). The students are expected to grow the root tips of different intra specific variable plants and also fix the material and take up the study. First the student should try with onion 2n = 16 which is cytologically easy material for chromosomal study. A through experience is necessary to study all the mitotic stages especially metaphase stage. For difficult materials, different techniques have to be used which has been explained in the first part of the laboratory exercises. Next the student should study the pollen mother cells of maize 2n = 20 especially diplotene stages to observe the chiasma. It is easy to observe the anaphase separation in maize PMCs. The next easy material is Sorghum 2n = 20 wherein full meiotic stages can be easily observed. As the chromosomes will be dot like, the student will develop his cytological technique. Paddy 2n = 24 is slightly difficult as the chromosomes are small dot like. Then the student should try cumbu (2n=14) *Pennisetum americanum* (old name *typhoides*). The cumbu-napier (*Pennisetum typhoides x Pennisetum purpureum*) 3n = 21 is a triploid with bivalent and univalent formation. The student can observe in the PMCs the univalents and bivalents with slight improved smear technique explained earlier. The wild sorghum *Sorghum halapense* (2n = 40) is also easy material to study the quadrivalents formation. At every stage student observes, he should make clear observation notes on the mitotic or meiotic stages. The very easy material for PMC study is *Aloe vera* (2n=14), Reho discolor (2n = 10).

4. STUDY OF SEX CHROMOSOMES AND THEIR BEHAVIOUR DURING MEIOSIS FROM GRASSHOPPER AND ANY APPROPRIATE DIOCECIOUS PLANT (e.g Coccinia)

In grasshoppers two types of sex chromosomal mechanisms are found. XO type and XX type The XO type is male grasshopper, which has only one X chromosome. The presence of one unpaired X chromosome determine the male sex. During meiosis, the male being heterogametic sex produces two types of gametes, half with X chromosome and other half without X chromosome. The females are XX types, and one from male combines with one X from female produces female sex.

XX and XY types

While in dioecious plant it is similar to *Drosophila*, the female possesses two homomorphic X chromosomes (XX) and they produce one kind of eggs and each with one kind of X chromosome. The males possess one X chromosome and one Y chromosome (XY). The males since possess two heteromorphic sex chromosomes, produces two kinds of gametes, half with X chromosome and other half with Y chromosome. If an egg is fertilized X bearing gamete, produces a female. If fettilized Y bearing gamete, a male is produced. The *Drosophila* is the easy material to study the sex chromosomes. A banana is kept outside which will attract *Drosophila* (fruit fly) and the infected banana is placed inside a milk bottle and covered with a cotton wool with some aeration. The larvae will be crawling inside the milk bottle which can be taken and with the help of pointed needles and place on a clean slide and the mouth portion is pulled to bring out the salivary glands. With a drop of aceto-carmine stain observed under a microscope for the banded chromosomes. By repeating the procedure, clear pictures of salivary gland chromosomes can be observed with details.

5. DETECTION OF ANOMALIES IN CHROMOSOME PAIRING AND DISJUNCITON CAUSED BY MUTANT GENES AND STRUCTURAL ALTERATIONS OF CHROMOSOMES

The pollen mother cells have to be fixed in proper fixative and a thorough study is necessary with keen observation. For this Colchicine treatment to the flower buds is necessary to destroy the spindle mechanism so that the chromosomes will spread evenly in the cell. By proper staining technique the chromosomal pairing and disjunction can be studied.

6. PREPARATION OF CHROMOSOME MAPS FROM 3-POINT TEST CROSS DATA

Double cross over usually do not occur between genes less than 5 map units apart. For genes further apart, it is advisable to use a third marker between the other two in order to detect any double crossovers. Suppose we test cross trihybrid individuals of genotype ABC/abc and find in the progeny the following.

36% ABC/abc	9% Abc/abc	4% Abc/abc	1% AbC/abc
36% abc/abc	9% aBC/abc	4% abC/abc	1% aBc/abc
72% parental type	18% single cross over between A & B (region I)	8% single cross over between B & C (region II)	2% double crossover

To find out the distance A – B we must count all crossovers (both single and double) that occurred in region I = 18% + 2% = 20% or 20 map units between the loci A and B. To find out the distance between B – C we must again count all the crossovers (both single and doubles) that occurred in region II = 8% + 2% = 10% or 10 map units in between the loci B and C. The AC distance is therefore 30 map units. When double cross overs are detected in a three-point linkage experiment and 26 map units when double crossovers are undetected in the two-point linkage.

7. IDENTIFICATION OF MUTANT GENOTYPES IN DROSOPHILA AND ARABIDOPSIS STOCKS MAINTAINED BY THE DEPARTMENT

Mutations are haphazard not only in the nature of their effects, but also in point of origin, a fact clearly shown in Drosophila. When these flies are reared in the laboratory, they are often kept in milk bottles with banana as food. The condition within a given bottle is probably pretty much the same for all the flies in the culture; still a mutation might take place in just one fly in the bottle, and not in any of the rest. Moreover, even in the one fly, the mutation is confined when it first occurs to a single cell and to a single gene within that cell. Not even the allelic gene changes, as a rule, though it is the same kind of gene as the one that mutated and is located within the microscopic confines of the same cell. The factors which cause mutations are themselves microscopic in dimension and so might influence one minute part of a cell such as a gene without influencing the rest of the cell. New mutations of any particular kind are very infrequent. In *Drosophila*, for example, only one germ cell in several hundred thousand contains a new mutation such as red eyes to white. In *Drosophila*, according to some estimates, at least one sperm or egg cell in every ten contains a new mutation at one locus or another. However, the vast majority of these do not cause any change in the outward appearance of the fly, and they would often escape detection. This fact makes it difficult to determine the precise over-all mutation rate.

8. CORRELATION OF FLORAL STRUCTURE WITH POLLINATION SYSTEM (e.g. SESAMUM, PEA, LATHYRUS, WHEAT, RICE, MAIZE, RICINUS)

Rice (Oryza sativa L.)

The inflorescence is a panicle which bears single flowered spikelets. The rice flower differs from that of other cereals in having 6 stamens. The flower is surrounded by a lemma and palea structure which form the hull or husk that encloses the threshed grain or paddy. The outer glumes are usually obscure, being only about one-fourth the length of the lemma and palea. The blooming of rice normally between 8 and 11 a.m. The flowers in a single panicle bloom over a period of 7 to 10 days, but most of the flowers bloom between two to four days after emergence of the panicle from the boot-leaf. Pollen is generally shed at the time of opening of flowers. Blooming of the spikelets starts at the apex of the panicle and proceeds downwards. The rice flower is normally self-pollinated. The extent of natural crossing varies from 0 to about 3% with an average of about 0.5% depends on the variety, season and environment.

Maize (Zea mays L.)

Maize bears monoecious flowers with staminate flowers produced in the tassel and pistillate flowers on the shoot. Pollination is accomplished by the transfer of pollen from the tassel to the silk. About 95% of the ovules on a shoot are cross pollinated and about 5% are self-pollinated. The main stem of the maize plant terminates in a tassel bearing two-flowered staminate spikelets, each flower having three stamens. As the tassel flowers open, the anthers are pushed out by the elongating filaments and pollen grains are emptied from the entruded anthers. Pollen shedding begins one to three days before the silk have emerged from the husks of the same plant, and usually continues for a period of several days after the silks are really to be pollinated.

Gingelly (Sesamum indicum L.)

The flower of sesamum has a two-lipped, tubular corolla, with five lobes which are united at the base. The corolla varies from white to purple in colour and is covered with short hairs. Four stamens are grouped in two pairs with one pair of stamens shorter than the other. Flowering occurs in early morning with the anthers beginning to dehisce around 3 a.m., shortly before the flowers open. The stigma becomes receptive with the dehiscence of anthers and remains so until 7 or 8 a.m. The stigma is usually covered with pollen by day break when insects visitation begins, so self pollination is the rule. However, pollen carried by insects may result in some cross pollination, usually around 5 per cent. The corolla withers and falls off towards the afternoon.

Castor (Ricinus Communis L.)

Castor plant is usually monoecious with about 50 to 70% male flowers occurring towards the base of a racemose inflorescence and 30 to 50 per cent pistillate flowers in the upper portions. Variation in this pattern of flowering include.

(a) racemes with pistillate and staminate flowers throughout

(b) racemes with 70 to 90 per cent pistillate flowers

(c) racemes with 100 per cent pistillate flowers

(d) racemes with a few hermaphrodite flowers

The flowers are wind and insect pollinated and vary from 5 to 46 per cent natural cross pollination.

Pulses (Chickpea) *Cicer arietinum L.*

Chickpea is having typical papilionaceous flowers consisting of 5 sepals, five petals comprised of one standard, two wings and two keels, ten stamens, nine fused to form a staminal column and one free. In chickpea, anthesis starts between 9 and 10 a.m., and may continue upto 3 p.m. The flowers remain open for 2 days, the flowering process being over early on the second day. The plant is primarily self-pollinated as anthers dehisce one day earlier prior to opening of flowers. A very small percentage of cross-pollination may result from visitation by insects after the flowers open.

Wheat (Triticum aestivum L.)

Wheat is a self-pollinating crop. Blooming starts several days after the wheat spike emerges. The main culm flowers bloom first and the tillers later, in the order of their formation. Flowering continues throughout the day, with two to three days required for a spike to finish blooming. The glumes normally open during the flowering process, the anthers protrude from the glumes, and part of the pollen is shed outside the flowers. Entry of foreign pollen while the flower is open may result in a small amount of cross-pollination. Normally, cross pollination is less than 1 per cent. If conditions are unfavourable for the opening of the glumes, the anthers may shed their pollen without being extruded. Wheat flowers are emasculated by clipping the glumes and removing the anthers with fine pointed tweezers. Pollinations are made one or two days later by breaking a ripe anther over the stigma.

9. FIELD EXPLORATION FOR DETECTION OF MALE STERILE PLANTS AND ESTIMATION OF THEIR POLLEN FERTILITY IN LOCALLY GROWN CROP PLANTS SORGHUM, TOMATO AND LINUM

Pollen grains are very minute in size varying from 10 to 200 (microns). Each pollen grain consists of a single microscopic cell and possess two coats, the exine and intine. The exine is tough, cutinized which is often provided with spiny outgrowth or reticulations of different pattern and in some cases it is smooth. The intine is however smooth, soft, delicate. When the pollen grain germinates, the intine grows out into a tube called the pollen tube through the pores.

Castor Pollen grains

The pollen grain is somewhat triangular shaped and the exine is hard and intine is soft. There are three germ pores on three angles of the triangular shape structure and medium sized.

Cotton (Gossypium L.)

The exine of cotton pollen grain is round with small spiny projections, on the surface. These are bigger in size than any other pollens.

Sorghum (L.) Moench.

The exine is smooth, round. There is a single germ pore and smaller in size.

Assessment of pollen fertility

The fertility of pollen grain is assessed by treating with 1:1 aceto carmine and glycerine solution.

Procedure

The pollen grains are dusted on a slide from anthers of the flowers and a drop of acetocarmine : glycerine 1:1 mixture is added and covered with a coverslip. The fully developed pollen grains take the darker stain and swell showing the fertility when observed under the low power of the microscope. The sterile grains do not take any stain and shrivelled. Based on the stainability, the fertility of pollen is assessed for unit area.

Fertility of pollen grains in a hybrid derivative of *Sorghum durra* x *Sorghum halapense*

Replication	Fertile	Sterile	Total	Calculation
1	18	3	21	
2	19	5	24	
3	23	2	25	Fertility = 218/260 x 100 = 83.84%
4	27	5	32	Sterility = 16.16%
5	31	7	38	
6	25	5	30	
7	18	2	20	
8	36	7	43	
9	21	6	27	
Total	**218**	**42**	**260**	

Pollen germination

For pollen germination the following medium is used.

A. Boric acid 0.1 g
Calcium Nitrate 0.3 g
Magnesium sulphate 0.2 g
Potasium Nitrate 0.1 g
Distilled water 100 ml

B. Sucrose solution : 26 to 32%

Culturing

9 ml of B is mixed with 1 ml of A. Cellophane bits or pieces of filter papers are floated over the solution and pollen grains dusted on it. After a few hours the bits are carefully taken with fine pointed foreceps placed on a slide and 1 part of aceto carmine and 1 part of glycerine is added and observed under the low power of the microscope.

Pollen squash technique

A pollen squash technique applicable to sugarcane and allied genera is given here. Anthers containing actively dividing pollen are treated for one hour at 28°C with 0.5% solution of Colchicine and another one hour in 0.002 M 8-Hydroxy quinoline, washed in water and fixed in a solution containing methanol 60 ml; chloroform 30 ml; distilled water 20 ml; picric acid 1 g and mercuric chloride 1 g for 24 hours. Washed in alcohol (70%, 30% methyl alcohol) and brought down to water, hydrolysed in 1N Hcl at 60°C for 15 minutes, stained in leuco-basic fushion for 30 minutes and smeared on a slide in aceto-carmine. The slides are made permanent by separating cover glass and slide, placing in a 1:1 mixture of acetic acid and normal butanol; passing through n-butanol, and mounting in Canada balsm.

Rapid temporary preparations are made by treating anthers in saturated para dichlorobenzene for one hour, fixed in acetic alcohol (1:3) for 3 hours, washed in water, hydrolysed in 1N Hcl for 15 to 20 minutes at 60°C; stained in leuco-basic fushion for one hour and smeared in a drop of aceto-carmine.

Study of stomata

For preparation of leaf epidermal peels for the observation on the nature of stomata, the leaf bits from middle lamina of fully grown leaf are taken and transferred to a test tube containing concentrated Hcl. After boiling for 3-5 minutes, the contents of the test tube are immediately poured into a petry dish containing water. By careful observation the peels of epidermal layer can be found out floating on the surface of water. Good epidermal peels are gently taken and put on a clean slide with a drop of water. The leaf peel is gently spread with the help of needle and water is drained off. Then a drop of analin blue or cotton blue stain is added and a coverslip to be placed. Then observed under low power of the microscope. The stomata are minute openings which occur in epidermis of leaf in plants. The stomata are surrounded by two specialized epidermal cells look like kidney or bean shape in appearance. The guard cells contain green plastids. The number of stomata in an unit area varies with kind of plant as well as with the conditions under which it grows. It also varies from leaf to leaf of the same plant and even in different parts of the same leaf. The size of the stomatal pore also varies with species.

Bougainvillae — 15 per unit area, size : length 9.5 and Breadth 6.8 m
Solanum species — 20 per unit area, size : 7.2 × 5.2 m
Sorghum — 10.5 per unit area, size 10.5 × 6.0 m

10. ESTIMATION OF POLLEN OVULE RATIO AND ITS BEARING ON POLLINATION SYSTEM

Self pollination is the transfer of pollen from an anther to a stigma within the same flower or to a stigma of another flower on the same plant. Cereal crops, such as wheat or barley in which the flower is enclosed by floral bracts are seldom pollinated except from pollen originating in an anther within the same flower. This is in contrast to the pollination in the cross-pollinated plant, in which pollen is carried by the wind, e.g. Corn where the pollen is carried by the wind to silks, of other plants. The pollination of red clover flowers by insects which carry pollen from one plant to the stigmas in flowers on the other plants. Fertilization resulting from the union of a sperm and an egg (gametes) produced on the same plant is self fertilization. The union of sperm and egg gametes from different plants is cross fertilization.

From the breeding standpoint, field crop plants which reproduce by sexual means may be grouped according to their usual method of pollination as normally self pollinated, normally cross pollinated or both self and cross pollinated. The amount of natural crossing or natural selfing within these crops will vary with (a) the variety or strain of the crop (b) the seasonal conditions (c) the velocity and direction of the wind, and (d) the insect population. Some common crop plants that are normally self pollinated are: wheat, barley, oats, soybeans, cowpeas, field peas, burr clover, crotalaria, Korean lespedeza, rice, tobacco, flax, tomatoes, potatoes, peanuts, common vetch, hairy vetch, black medic, hop clover, common lespedeza, slender wheatgrass, annual bromegrass, subterranean clover, foxtail millet. The amount of natural cross-pollination that may occur within crops of this group may vary from none up to 4 or 5 per cent. The plant breeder working with self-pollinated crops will need to determine the extent of the natural crossing under particular conditions. The percentage of natural crossing in a crop may be estimated by a simple procedure. Two varieties are selected that are pure for different forms of a character which is easily recognized and simply inherited. The varieties are planted in such a manner that individual plants recessive for the character are completely surrounded by plants dominant for the character. Then the progenies of the recessive plants are grown to determine the percentage of plants which exhibits the dominant character.

11. EMASCULATION AND BAGGING OF FLOWERS OF BRASSICACEAE, POACEAE, PAPLIONACEAE, MALVACEAE AND LINACEAE, POLLINATING THEM MANUALLY AND ESTIMATING FRUIT AND SEED SET

Brassicaceae* : (mustard) *Brassica juncea* (L.) *Czern. et Coss.

Flowers in this family have 4 petals, the petals bifureated with varying degrees of incision and deep yellow to pale yellow or cream in colour. There are 6 stamens, the anthers are lower than the stigma at bud stage, but prior to flower opening, the filaments elongate and carry the anthers upward, so that they are as high as or above the stigma. Flowers are emasculated in the evening and pollinated the next morning. Flower buds that will open the next day are selected and the remainder of the buds and flowers on the flowering branch are removed. The petals as well as the stamens of the selected buds are removed with a pair of tweezers and the emasculated flowers are bagged. Ripe anthers are collected the next morning and pollinations are made by dusting pollen from the ripened anthers over the stigma. After pollination the flowers are again bagged. Selfed seed is set freely only in self-fertile species. Flowers to be selfed should be bagged before they open to avoid natural cross pollination. The bag should be such size as to allow lengthening of the inflorescence. Cross-pollination of a group of varieties of the brown sarson or toria can be obtained by growing them under cages in which are included honey bees, or by installing the varieties at a safe distance from other plantings of the same crop in the field.

***Poaceae* (*Oryza sativa L.*)**

Several methods have been used for emasculating rice flowers. Before emasculation immature spikelets or those previously flowered are removed from the panicle keeping only flowers that will bloom the following day. The conventional method, still used by some rice breeders, is to separate the glumes, with a pair of forceps, in the early morning about 1 to 2 hours before pollination and remove the 6 stamens. Emasculation and pollination may be facilitated however, by cutting off the end of the florets in a manner similar to that used with other cereals viz., wheat and barley. Clipping across the anthers just above the attachment of the filament will remove most of the anthers at the same time and expose the remnants of the anthers so that they can be removed with fine pointed tweezers. Emasculations are done in the afternoon and pollination the following morning. After emasculation, flowers are covered with a butter paper bag.

Hot water may be used to open the florets of rice before crossing. In the hot water treatment the rice panicle is immersed in water contained in a thermos flask at 42 to 44°C for a period of 5 to

10 minutes. Panicle in the 2nd or 3rd day of blooming are chosen as female parents and spikelets have already flowered and immature florets are removed. An hour or so before blooming normally begins, the tiller is bent over carefully to avoid breaking and inserted into the hot water. The thermos bottle may be supported on a trough like holder at an angle of about 35 degrees to prevent loss of water. This treatment causes the florets to open in a normal manner and the stamens may be removed without injury to the stigma. Pollinations must be made within 30 minutes period before the glumes close naturally. Successful pollinations require mature and healthy anthers. Florets nearing the blooming stage are opened and the turgid anthers are taken with forceps and broken over the stigma. Pollen may also be dusted over clipped florets by shaking a shedding panicle over them. Pollen of rice does not normally remain viable for more than a few minutes.

***Paplionaceae* (Blackgram) Phaseolus mungo L.**

Flowers are generally emasculated in the evening and pollinated in the next morning. For emasculation, flowers that will open one or two days later are selected, and the rest of the flowers and buds in a branch are removed. The stamens of the selected buds are removed with a pair of fine pointed forceps by gently pushing the keels apart. The emasculated floral branch is then bagged. To ensure selfing the flowers need to be bagged, also as insects may sometimes carry pollen to the stigma and bring about cross pollination.

***Malvaceae* e.g. Cotton Gossypium L.**

The base of the cotton flower is surrounded by three leaf like triangular bracts to form a square. On the day preceding pollination, the twisted corolla emerges from the square. In American upland varieties, the corolla is cream coloured when the flower first opens, but it turns red the second day and later falls from the plant. The pistil is formed normally from three to four carpels in Sea Island or Egyptian cotton and four to five carpels in American upland cotton, corresponding to the number of locks in the boll. Normally, 7 to 9 seeds are set within each lock. The stamens form a tube which encloses the styles and is united with the inside base of the corolla. Pollen is shed directly on the stigma when the anthers open, or it may be carried to the stigma by insects. Pollen is wind-borne only to a very slight extent, if at all, on account of its heavy sticky nature. Pollen carried by insects may result in considerable cross pollination. The amount of cross-pollination normally ranges from 5 to 25 per cent although amounts in excess of 50 per cent have been reported in Tennessee and North Carolina. The amount of cross pollination depend more upon the relative abundance of pollen carrying insects than any other factor. Most natural crossing occurs between plants growing only a few yards apart; however, cross pollination has been reported at distances up to one fourth mile. Different amounts of cross pollination have also been observed with different varieties.

A simple technique has been developed for artificially cross pollinating cotton flowers. Crosses are made on the day preceding that in which the flowers would open normally. First the corolla is cut away with small scissors or a curved scalpel and the anthers are removed. Pollen is collected from the male flower and brushed on the receptive stigma. Some breeders recommend emasculating one day and pollinating the next day, but this procedure may result in unnecessary drying of stigma. Seed set about 75 per cent may be obtained after artificial crossing.

***Linaceae* (Linseed) Linum usitatissimum L.**

The linseed flowers begin to open and the anthers shed pollen shortly after sunrise. The linseed flower has 5 petals which may be white, blue, violet, purple or pink. The flower fully open by 7 a.m., and petals fall before noon. The linseed flowers are emasculated late in the afternoon or early morning. The petal is removed by pulling gently with the thumb and the index finger. Then one or two sepals are rolled back and held down while the five anthers are removed with the point of a pencil or fine pointed tweezers. Care must be exercised to prevent injury to the stigma or it will dry out rapidly. It is unnecessary to cover emasculated flowers since insects are not attracted to the flower after the petals have been removed. The emasculated flowers may be marked by typing tags or small pieces of coloured string around the petiole. Different colours may be used on different days or to mark different crosses. Pollinations are made in the morning after emasculation usually before 8 a.m. Delay in pollination may result in a poor seed set because the pollen dries out and deteriorates rapidly. Petals are removed from the flowers selected to supply the pollen. The anthers are brushed lightly over the stigmas of the emasculated flowers. One male flower will pollinate two or three emasculated flowers.

Suggested Readings

Anderson, E. (1949). Introgressive hybridization. John Wiley and Sons, New York.

Ashby, E. (1930). Studies in the inheritance of physiological characters I. A physiological investigation of the nature of hybrid vigour in maize. Ann. Bot. 46 : 1007–1032.

Auerbach, C. (1976). Mutation Research. Chapman and Hall, London.

Auerbach, C. (1967). Changes in the concept of mutation and the aims of mutation research. In heritage from Mendel. Ed. Brink, R.A., Univ. Wisconsin Press Madison.

Bateson, W. (1913). Mendel's Principles of Heredity. Cambridge, England. Includes accounts of the classical experiments of Bateson on sweet peas and poultry.

Beale, G. and Knowles, J. (1978). Extranuclear genetics. Edward Arnold, London.

Blakeslee, A.F. and Belling, J. (1924). Chromosomal mutations in the Jimson weed, Dature stramonium. J. Heredity 15: 195.

Brown, S.W. (1966). Heterochromatin. Science, 151: 417.

Bridge, C.B. (1916). Non-disjunction as proof of the chromosome theory of heredity. Genetics L:1.

Bridges, C.B. (1925). Sex in relation to chromosomes and genes. Amer. Nat. 59: 127.

Bridges, C.B. and Olbrycht, T.M. (1926). The multiple stick "Xple" and its use. Genetics 11: 41.

Bruce A.B. (1910) The Mendelian theory of heredity and augmentation of vigour. Science, 32: 627–628.

Carlson, E. (1959). The comparative genetics of complex loci. Quart. Rev. Biol. 34: 33.

Clausen, R.E. and Goodspeed, T.H. (1925). Interspecific hybridization in Nicotiana II. A tetraploid glutinosa tabacum hybrid. An experimental verification of Winge's hypothesis. Genetics 10: 278.

Collins, G.N. (1912). The origin of maize. J. Washington. Acad. Sci. 2: 520–530.

Collins, G.N. (1921). Teosinte in Mexico. Jour. Hered. 12.

Du Praw, E.J. (1970). DNA and chromosomes. Holt, Reinhart and Winston, New York.

East, E.M. (1910). A Mendelian interpretation of variation that is apparently continuous. Amer. Nat. 44: 65.

East, E.M. (1916). Studies in size inheritance in Nicotiana. Genetics 1: 164–176.

East, E.M. (1936). Heterosis, Genetics 21: 375–397.

Engler, J.A. and Van Bree, M.P. (1981). The nucleotids sequence and protein coding capability of the transposable element. Gene. 14: 155.

Falconer, D.S. (1961). Introduction to quantitative genetics. Ronald Press, New York.

Fisher, R.A. (1918). The correlation between relatives on the supposition of Mendelian inhertance. Trans. Roy. Soc. Edinb. 52: 399

Fisher, R.A. (1930). The genetical theory of natural selection. Clarendon Press Oxford (Reprint, Dover Publ. 1960). Interpretations of natural selection.

Fisher, R.A. (1932). The genetical interpretation of statistics of the third degree in the study of quantitative inheritance. Genetics 17: 107.

Fisher, R.A. (1947). Statistical methods for research workers. Edinburgh: Oliver and Boyd, Ltd., 10 ed.

Fisher, R.A., and Yates, F. (1963). Statistical tables for biological, agricultural and medical research. 6th ed. Oliver and Boyd, Edinburgh.

Goldschmidt, R.D. (1955). Theoretical genetics. Univ. California Press, Berkeley.

Green, M.M. (1980). Transposable elements in Dorsophila and other Diptera. Annul. Rev. Genetics, 14: 109.

Henderson, S.A. (1970). The time and place of meiotic crossing over. Annul. Rev. Genetics, 4: 295.

Henking, H. (1891). Z wiss zool. 51: 685.

Hutchinson, J.B. (1934). J. Genet. 28: 437 (Abstr. Biblio. of cotton Breed. and Genet. 98, 1954).

Jenkins, J.B. (1979). Genetics. Houghton Mifflin Co., Boston.

Johannsen, W. (1909). Elements der exacten Erblichkeitslehree. G. Fisher, Jena, East Germany.

Jones, D.F. (1917). Dominance of linked factors as a means of accounting for heterosis. Genetics, 2: 466–479.

Karpechenko, G.D. (1935). Die Theorie der Krcuzung entfernter Formen. In vavilov, N.I. (ed). "Theoretische Grundlagen der Pflanzenzuchtung". Bd I, Kap. 14.

Leaderberg, J. (1947). Gene recombination and linked segregation in Escherichia coli. Genetics, 32: 505, 525.

Levine, L. (1971). Papers on Genetics, A book of Readings, Toppan, Japan.

Lewin, B. (1976). DNA sequences coding for more than one protein. Nature 264: 11.

Lewis, E.B. (1948). Pseudoallelism in Drosophila melanogaster. Genetics, 33: 113.

Lewis, D. (1979). Sexual incompatibility in plants. Edward Arnold, London.

Love, A. and Love, D., (1943). The significance of differences in the distribution of diploids and polyploids. Hereditas, 29: 145.

Mather, K. (1951). Measurement of Linkage in Heredity. Wiley, NewYork.

Mather, K. and Jinks, J.L. (1971). Biometrical Genetics. Chapman and Hall, London.

Mc Clintock, B. (1929). A method for making aceto-carmine smears permanent. Stain Tech. 4: 53–56.

Mc Clintock, B. (1942). The fusion of broken ends of chromosomes following nuclear fussion. Proc. Natl. A cad. U.S. 21: 458.

Mendel, G. (1865–66). "Versuche uber Pflanzen hybriden. Verh. d. Naturf. Vereins in Brunn: 4: 3–47. (Mendel's Original paper). Reprinted in original German; J. Hered. Vol. 42: No. 1.

Morghan, T.H. (1910a). Amer. Nat. 44: 449.

Morgan, T.H. (1910b). Sex limited inheritance in Drosophila. Science 32: 120.

Morghan, T.H. (1911). Random segregation versus coupling in Mendelian inheritance. Science 34: 384.

Morghan, T.H., Sturtevant, A.H., Huller, H.J., and Bridges, C.B. (1923). The mechanism of Mendelian heredity, New York, Henry Holt & Co.

Muller, H.J. (1912). Some genetics aspects of sex. Amer. Naturalist. 66 : 118–138.

Muller, H.J. (1914). The bearing of the selection experiments of Castle and Philips on the variability of genes. Amer. Naturalist. 48 : 567-576.

Muller, H.J. (1916). The mechanism of crossing over. Amer. Nat. 50 : 193.

Muller, H.J., and Altenburg, E. (1919). The rate of change of hereditary factors in Drosophila. Proc. Soc.. Expr. Biol. and Med. 17: 10–14.

Muller, H.J. (1927). V. Int. Kongr. Vererbgsl. 234.

Muller, H.J., and Painter, T.S. (1932). The differentiation of the sex chromosomes of Drosophila into genetically active and inert regions. Zeitschrift fur Inductive abstammungs und Vererbungslehre, 62: 316–265.

Muller, H.J. (1947). The dance of genes. Genetics, Medicine and Man. Ithaca, New York. Cornell Univ. Press. 35–65.

Muller, H.J. (1948). The production of mutations (Nobel lecture delivered at Stockholm, Dec. 12, 1946). Stockholm: P.A. Norstedt und soner.

Muller, H.J. (1950). Radiation damage to genetic material. I. Effects manifested mainly in the descendants. Amer. Scientist, 38: 33–59.

Muller. H.J. (1956). On the relation between chromosome changes and gene mutations. Brook haven Symp. Biol. 8 : 126.

Muntzing, A. (1936). The evolutionary singnificance of autopolyploidy. Hereditas, 21: 263–378.

Nilsson–Ehle, H. (1909) Kreuzungsuntersuchungen an hafter und Weizen. Lunds Univ. Aarskr. 2, 5, 2: 122.

Punnett, R.C. (1912). Inheritance of coat colour in rabbits. J. Genetics, 2 : 221.

Punnett, R.C. (1923). Heredity in poultry. Mac Millan, New York.

Reinhart, R.R. and Ratty, F.J. (1965). Mutation in Drosophila melanogaster cultured on irradiated food. Genetics 52: 119.

Rhoades, M.M. (1933). The cytoplasmic inheritance of male sterility in Zea mays. J. Genetics, 27: 71–93.

Rhoades, M.M. (1941). The genetic control of mutability in maize. Cold spring harbour symposia on quant. biol. 9: 138–144.

Rhoades, M.M. (1946). Plastid mutations. Cold spring harbour symposia on quant. biol. 11: 202–207.

Sears, E.R. (1976). Genetic control of chromosome pairing in wheat. Annul. Rev. Genetics 10: 31.

Shull, G.H. (1908). The composition of a field of maize. Annual report of American breeders association. 4 : 296–301.

Shull, G.H. (1914). Z indukt Abstamm-u-Vererblehre 12 : 97.

Shull, G.H. (1952). Beginning of the heterosis concept. In Heterosis, Iowa State Colleges Press, Ames, Iowa.

Sonneborn, T.M. (1949). Beyond the gene. Amer. Scientist, 37: 33–39.

Stadler, L.J., (1928). Mutations in barley. Induced by X-rays and Radium. Science 68: 186–187.

Stadler, L.J. (1930). Some genetic effects of X-rays in plants. Jour. Hered 21: 3–19

Stadler, L.J. (1951). Spontaneous mutations in maize. Cold spring harbor symposia on quant. biol. 16: 49–64.

Stebbins, G.L. (1950). Variation and evolution in plants. Columbia Univ. Press. New York.

Stern, C. (1950). The birth of genetics. Genetics 35, supplement.

Sturtevant, A.H. (1913a). The linear arrangement of sex linked factors in Drosophila as shown by their mode of association. J. Exptl. Zool. 14: 43.

Sturtevant, A.H. (1913*b*). The Himalayan rabbit case with some consideration on multiple allelomorphs. Amer. Nat. 47: 234.

Sturtevant, A.H. (1925). The effects of unequal crossing over at the Bar locus in Drosophila. Genetics, 10: 117–147.

Sturtevant, A.H. (1945). A gene in Drosophila that transforms females into males. Genetics, 30: 297.

Sutton, W.S. (1902). Chromosomes in heredity. Biol. Bull. 4: 231–251.

Swanson, C.P. (1957). Cytology and Cytogenetics. Prentice-Hall, New York.

Tatum, E.L. and Lederberg, J. (1947). Gene recombinations in Escherichia coli. J. Bact. 53: 673–684.

Vijendra Das, L.D. and Mensinkai, S.W. (1966a). A simple smear technique for **Gossypium arboreum** Linn. Madras Agric. J. 53: 83.

Vijendra Das. L.D. and Mensinkai, S.W. (1966b) Pleiotropism in Gossypium arboreum Linn. Science and Culture, 32: 375.

Vijendra Das, L.D. (1970). Chromosome associations in diploid and autotetraploid Zea mays Linn. Cytologia, 35(2): 259–261.

Vijendra Das, L.D. and Sree Rangasamy, S.R. (1973). Effect of irradiation with gamma rays on embryo sac in M_1 generation of rice. Current Sci. 42: 431–438.

Vijendra Das, L.D. (1978). Effects of radiations on the R_1 and R_2 progenies of Pennisetum typhoides S. & H. Environmental and Experimential Botany, 18: 121–125.

Whitehouse, H.L.K. (1973a). Towards an understanding of the mechanism of heredity. St. Martin's Press, New York.

Whitehouse, H.L.K. (1973b). Towards and understanding of the Mechanism of Heredity. 3rd ed. St. Martin's Press, New York.

Wilson, E.B. (1928). The cell in heredity and development, New York. The Mac Millan Co., 3rd ed. (The standard text of cytology and chromosome theory of heredity in English language).

Wilkie, D. (1964). The cytoplasm in heredity. Methuen, London.

Winge, O. (1917). The chromosome–their numbers and general importance. Comp. Rend. Trav. Lab. Carlsberg, 13: 131.

Young, M.W. and Judd, B.H. (1978). Nonessential sequences, genes, and the polytene chromosome bands of Drosophila melanogaster. Genetics, 88: 723–742.

Zinder, N.D. (ed.) (1975). RNA phages, Cold Spring Harbor, N.Y. Cold Spring Harbor, Laboratory.

PLANT BREEDING

CHAPTER 1

Plant Type Concept

PLANT IDEOTYPES

Plant ideotype refers to an ideal plant type to get maximum yield.

Introduction

The main function of the plant is to produce complex carbon compounds from CO_2 and H_2O, utilizing solar energy. It is difficult at present, to replace this process by artificial chemical synthesis in a laboratory. Only higher plants are capable of synthesizing complex compounds from atmospheric carbon for human consumption.

$$CO_2 + H_2O \xrightarrow[\text{Chlorophyll}]{\text{Light}} \overset{\text{Sugar}}{(CH_2O)_n} + O_2$$

n = No. of sugar molecules.

This is carried out in chloroplasts which absorb solar radiation. The irradiated chlorophyll transfer their energy to water, producing hydrogen and oxygen. This H+ reacts with CO_2 producing $(CH_2O)_n$. As this is a reduction reaction, energy is needed to the formation of sugar compounds. In plant breeding it is considered that an ideal plant type should make full use of solar energy.

When nutrient supply is a limiting factor for crop growth, the problem of plant type is not so important. However, when the nutrient supply is sufficient, or the price of fertilizer is low compared with the price of crops produced, ideal disposition of leaves for maximum sun harvesting becomes important, and plant breeders are expected to breed new varieties which have a greater potential for high yields.

Phenotypes reflect both genotype and environment. Since environment influences physiological processes, for a plant to be successful, it must be accurately matched to its surroundings.

Farrer (1898) the first Australian wheat breeder, was able to describe in phenological terms the features that made introduced wheats poorly adapted to his environment. "They were too late in ripening for our climate, and their ears were blasted, and their grains pinched by hot winds and heat of our summers". He visualised an ideotype with narrow leaves, few tillers and early maturity, coupled with an ability to thrive on dry soils. These types of characters, together with disease and pest resistance have constituted the main goal of genetic improvement of crop and pasture species. Breeders were usually not trained in plant physiology and environmental influence and frequently were concerned with developing cultivars for a particular situation.

Stakman et al. (1967) defined ideal rice type as one which could mature in a short growing season, photoperiod insensitivity, short stiff straw to provide lodging resistance and tolerance to serious pest and diseases.

Much has been researched and written about plant ideotype (shoot type only), architecture, determinate (vs) indeterminate habit, dwarfness, earliness, leaf angle, green area retention etc., but very little attention has been paid on similar aspects of root growth on which the whole architecture depends mainly because of difficulties in study due to their underground position. Improvement in different aspect of shoot growth have resulted in yield increases ; similar improvements in roots are bound to result in further yield increases, and under critical stress conditions it may even determine the survival and production or failure of the crop. There is very little study on root variability, selection, heritability, heterosis and breeding and their (1) architecture-deep penetrating roots for sub-soil water and nutrient absorption, root branching behaviour and ranification diameter (surface-volume ratio), nodulation and N fixation, regeneration capacity, etc. (2) efficiency in nutrient absorption and uptake and (3) stress tolerance, for example, the stresses of toxicions, soil acidity, salinity and sodicity, water logging and anoxia, cold etc. The little work done is also diffuse and not easily accessible.

Vedrov (1974) presented a discussion on the importance of root type in breeding new forms of plants. In breeding wheat for resistance to drought, selection for the vigour of the primary root was considered important. In their monograph "Root systems and yield in crop plants", Ustimenko et al. (1975) have discussed the anatomy, morphology and biology of the root, taking account of genetic, specific and varietal features in the development of root systems and differences in hybrids in this respect. Attention has been drawn to the inheritance of the growth and activity of the root system and the value of selection for this character. Correlations are indicated between the development of the root systems and various economic characters in such crops as wheat, barley, maize, panicum, millet, sunflower, peas and perennial herbage.

Passioura (1972) emphasized the effect of root geometry on the yield of wheat growing on measured quantity of water. In an experiment involving 40 varieties of spring wheat, determined the number of embryonic roots ; the varieties having the greatest number of root at germination were most productive in both rainfed and irrigated conditions. The number of primary roots was positively correlated with the number of grains per ear and with 1000 grain weight (Alekseeva and Malina, 1976). In a study, four drought resistant sorghum lines showed deeper root penetration and produced heavier and more numeraries, primary and secondary roots than the four susceptible lines, bhanetal. (1974). In barley, the degree of development of the root system is positively correlated with the productive tillering and grain weight per plant (Oleinik, 1977). The correlation between yield and number and length of embryonic roots was high, while that between yield and number of crown roots was low (Oleinik, 1977).

Phaseolus beans selection with tap root morphology outyielded others possessing fibrous roots (Guadamuz, 1980). In soybean, cultivars were selected with fast tap root elongation rate and slow tap root elongation rate. The cultivars with fast elongation rate had deeper maximum rooting depths and greater number of roots at 100, 150 and 300 cm depths than those with slow elongation rate (Kaspar, 1982). The cultivars with fast elongation rates obtained a larger percentage of their water from below 120 cm while others with slow elongation rate absorbed water from above this zone. Wheat varieties with long roots are more resistant to Aluminium toxicity than those with short roots (Collins et al. 1981).

Plant Type Concept in Rice

The term plant type refers to a set of morphological characters associated with the yielding ability of rice varieties. The plant type concept in rice, similar to the wheat ideotype emerged gradually from many observations and comparisons between low and high yielding varieties in Japan.

Better varieties of crop species must have been selected for higher yields in the past. In practice, the high yields have been achieved by increased inputs, particularly nitrogen fertilizers. Hence, varieties must have been selected on the basis of their response to nitrogen fertilizers. Tsunoda, a Japanese worker compared low and high yielding varieties of rice, sweet potato, and soybean, and summarized their morphological characteristics.

- Low nitrogen responders have long, broad, thin drooping, pale-green leaves, and tall, weak stems.
- High nitrogen responders have erect, short, narrow, thick, dark-green leaves, and short, sturdy stems.

Tsunoda postulated from physiological knowledge of crop photosynthesis that the thick, dark-green leaves loose less light through reflection. Reduced leaf size and erect habit permit uniform light distribution to all leaves and reduce respiration. Consequently, dry matter and yield increases even under low light conditions. The uniform light penetrations and the short, sturdy culms minimize lodging ; thus there is little or no yield loss once panicles are well developed.

The following morphological characters deserve particular attention in rice :

Short, stiff culms,
Erect leaves,
High tillering

Short, stiff culms. The increased resistance of improved varieties to lodging appears to be the single character most responsible for high yields. A short, stiff culm makes the rice plant more lodging resistance. Although height is the most important plant character associated with lodging, a short variety is not always lodging resistant. There are differences in the stiffness of short culms.

The importance of lodging resistance has long been recognized, but only in recent years has the semidwarf gene been effectively introduced into tropical rice varieties. In southern Japan, Hoyoku and its sister high yielding varieties are largely characterized by increased lodging resistance.

Erect leaves. Erect leaves allow the deeper penetration and more even distribution of light, which results in increased crop photosynthesis. In one model, the crop photosynthesis of an erect-leaved canopy is about 20% higher than that of the droopy-leaved canopy when the LAI (leaf area index) is extremely high.

This model assumes that all the leaves are uniformly oriented at angles of 0° or 90° with respect to the horizontal plane (Fig. 1).

When we consider leaf angles and the light environment of individual leaves within a canopy, almost intuitively we visualize that a combination of short, erect upper leaves grading to droopy, longer lower leaves is ideal for maximum crop photosynthesis. In fact, theory and experimental evidence shows that this plant architecture leads to the maximization of incident solar radiation.

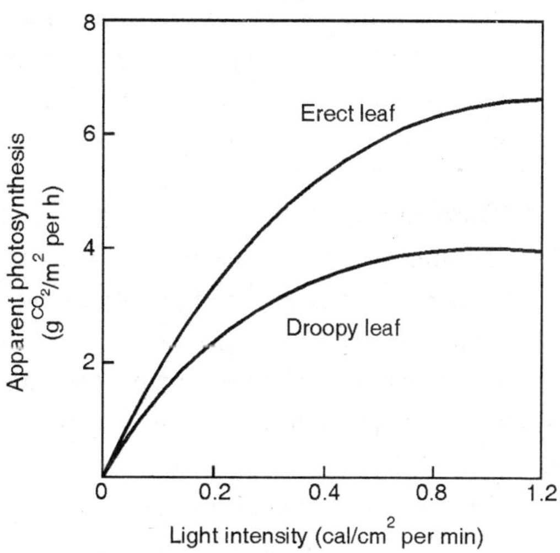

Fig. 1. Difference in light photosynthesis in rice with erect and droopy leaves.

There is only one rice experiment indicating that the droopy-leaved canopy has a lower crop photosynthesis than the erect-leaved canopy at high light intensities and accordingly, the droopy-leaved canopy was lower than the erect-leaved canopy in post flowering dry matter production by 34% and in grain yield by about 33%. In this experiment, Nihonbare, an erect-leaved variety, was field grown up to the beginning of heading, when the field was divided into two plots. In one plot, the leaves were drooped by attaching lead beads about the size of rice grains to the leaf tips. In the other plot, the leaves were kept erect. The LAI (leaf area index) of the crop at heading was 7.1. The crop photosynthesis was measured 2 days after the treatment. The results of this experiment support predictions from the crop photsynthesis model. However, it is possible that the droopy-leaved canopy was damaged by attaching lead beads to the leaf tips ; hence, the results might have been exaggerated in favour of the erect-leaved canopy.

High tillering capacity. Within a range of spacing from 50 × 50 cm to 10 × 10 cm, tillering capacity affects the grain yield of a variety. The grain yield of IRI54-45-1, a low-tillering selection, increased with decreased spacing upto 10 × 10 cm. A high tillering variety, IR8, reached maximum yield at 20 × 20 cm spacing and averaged at 10 × 10 cm. IR8 is less affected by spacing changes than IR 154-45-1. In transplanted rice, conventional spacing ranges from 20 × 20 cm (25 hills/m^2) to 30 × 30 cm (11.1 hills/m^2). High and early tillering habits are advantageous under such conditions. Under direct seeding, however, tillering capacity rarely affects grain yield within conventional seeding rates (Table 1) because total panicle number per square meter depends more on the main culm than on tillers.

Table 1. Response of a low-tillering variety (IR154-45-1) and a high-tillering variety (IR8) to different seed rates in direct seeding, IRRI, 1968 wet and dry seasons.

Seed rate (kg/ha)	*Plant (no./m^2) (dry season)*		*Planicles (no./m^2)*				*Yield (t/ha)*			
	IR154-45-	*IR8*	*IR154-45-1*		*IR8*		*IR154-45-1*		*IR8*	
			Wet	*Dry*	*Wet*	*Dry*	*Wet*	*Dry*	*Wet*	*Dry*
50	205	129	585	530	595	555	4.32	7.27	5.20	7.57
100	416	241	610	662	680	645	4.54	7.34	5.33	7.44
200	800	503	740	667	670	644	4.42	6.81	5.04	7.68

Yoshida (1972).

Medium Tillering capacity was once considered desirable for high yielding rice varieties. Low yields were believed caused by fast growth and a LAI (leaf area index) beyond the optimum, which, in turn, is closely related to high tillering capacity. This conclusion was based on the comparison of unimproved indica and improved japonica varieties. With an improved, lodging-resistant indica variety, their is no optimum LAI for dry matter production, although a critical LAI does exist. A LAI as large as 12 at heading is not detrimental to grain yield unless the crop lodges.

Missing plants due to poor and levelling, insects, diseases, and other stresses may occur in the field. A high-tillering variety is able to compensate for such missing plants.

Grain yield increased with increasing plant densities upto 182-242 plants/m^2. The number of panicles per unit of land area increased with increasing plant density but the increase was accompanied by a decrease in paddy weight per panicle. In another experiment, all the plants produced panicles at a plant density as high as 1,212 plants/m^2. Thus, of the three crops, rice is highly tolerant of high plant densities, wheat is less tolerant, and corn is the least tolerant. In both wheat and corn, grain yield decreases when the plant density exceeds a certain value (Yoshida, 1972).

The above discussion leads to the conclusion that exceptional tolerance for high densities makes the rice plant flexible in its response to changes in plant density, and high tillering capacity is desirable for achieving maximum yields in transplanted rice cultivation.

Wheat. The newly engineered wheat varieties in America and Mexico are as successful in giving higher yields as the IRRI (International Rice Research Institute) rice varieties. This success is referred as "Green Revolution". In 1952, a Japanese breeder named Inazuka bred a new cultivar, crossing Turkey Red × Fultz Daruma. He obtained some semidwarf breeding lines in F_6 which in 1935 was named Norin-10. The characteristics of Norin-10, are short stature, stiff straw, high tillering, resistance to lodging, resistance to yellow rust, and a good response to nitrogen. Norin-10 was then sent to America and Mexico as breeding material, and Vogel in America and Borloug at CIMMYT (Centro Internaceonal de Mejoramiento de Maiz Y Trigo—International Centre for Maize and Wheat Improvement) in Mexico, produced a series of high yielding varieties. These were particularly successful in America, where a new variety named Gaines (Norin 10 × Brevor) gave yield of 14t/ha and captivated agronomists. The contribution of Mexican wheat bred by Borloug is very well known, and has done much in changing the world food situation.

Maize. The success in rice and wheat breeding described above made clear that the efficiency of crops in absorbing solar energy is the key factor in obtaining higher yields. This concept is now fully accepted by plant physiologists who have produced many outstanding research reports. However, in the case of maize the concept of ideal performance or plant type is not as simple as with rice or wheat. The success of hgih yielding hybrid maize is a convincing factor that there is not urgent need to engineer a different plant type for meeting this objective. Rather, more attention should be paid to disease resistance, especially to virus diseases, and to the improvement of cultural practices much research has also been done involving the analysis of the phenomenon found in hybridization.

Of course, there is also much research analysing geometrically the ideal type of maize from the stand point of efficient use of solar energy. Japanese workers have calculated the density of leaf area every 10 cm and attempted to analyse the ideal type of photosynthetic system giving criteria for screening the most suitable plant type amongst hybrids.

Plant Ideotyes in Grain legumes

The legume grain yield is dependent on number of pods per unit area, number of seeds per pod and seed weight, the product of these three determining net yield. Number of pods per unit area is most influenced by different environmental factors, followed by number of grains per pod. Seed weight is the most stable character. Encouraging too many lateral branch formation delays synchronous maturity of the crops. Number of pods is dependent on number of nodes and the leaf area. Ishag (1973) reported a positive correlation in **Vicia faba** between leaf area of a node and the number of pods it retained.

Adams (1975) described three generalized plant types already existent in grain legumes as follows :

Type A. The vegetative structure is formed by the onset of flowering. The flowers are produced in a flush over a 5 to 8 day period, followed by both pod formation and some flower and pod abscission.

Growth phases (vegetative and reproductive) are non-overlapping leaves and generally restrictive in number but not in size. Pods may be many or few, seeds small to large, maturity early to late and plant form is generally determined. Some cowpeas and gardenpeas approach this type.

Type B. The vegetative structure develops upto onset of flowering and continuous over a prolonged period of time (20-30 days), during which vegetative structure continuous to form and leaves expand. Vegetative and reproductive growth stages are overlapping. Leaves are unrestricted in number or size. Pods usually are numerous may be few to many seeded, maturity generally late, plants intermediate. Examples include some south east Asian soybean and cowpea and the vining pea.

Type C. After the initial seedling stage, there is a long vegetative phase, at the end of which there is a long floral and pod setting stage. The vegetative and reproductive growth phases usually overlap. Plants may be tall, slightly to fully branched, and intermediate though not viny. The north American soybean and their descendants are the closest examples.

Adams (1975) tried to predict the ideotype of dry beans assuming that it is to be grown in monoculture, in as non-restrictive environment with respect to soil conditions, climate factors, length of growing season and crop cultural practices, with the following features :

1. Plant height-70-75 cm.
2. Determinate growth or short indeterminate vine.
3. Numerous nodes with upper nodes becoming gradually longer than lower ones.
4. None or only 2-4 branches arising at basal nodes.
5. Profile-narrow, erect, tall, of uniform width from base to top, or slightly tapering.
6. Leaf size-Medium to small, leaflets more narrow than broad. Leaf orientation only critical at high levels of Leaf Area Index (LAI), leaves with numerous small mesophyll cell and high stomatal index.
7. Long racemes at each node from base to top with eight florests per raceme.
8. Pods-long with 8-10 seeds per pod.
9. Seeds-medium sized, 25-30 mg. a seed.
10. After initial seedling and basal branching stages, vegetative and reproductive phases should overlap to a great extent with development of full plant structure consisting of the sequential formation from base to top. This should commence early and continue over the entire length of growth season.
11. The partitional factors should favour a high ratio of reproductive to vegetative tissue, consistent with demands of sink upon source organs and or activity to give high harvest index of 0.6 to 0.7.
12. High rates of N-fixation, mineral uptake, photosynthesis and transport are required.

A group of scientists (convened by FAO) meeting in India, suggested the following plant type for chick pea.

(*a*) High pod number potential.

(*b*) Greater number of seeds/pod.

(*c*) Seed size should be as large as in acceptable to consumers.

(*d*) A plant and canopy profile designed and structured for maximum light interception. This is perhaps the most important requirement. It follows that the plant must be erect, less leafy and less spreading.

(*e*) Node number, Internode length and branching pattern in keeping with the requirement set above.

(*f*) Leaf size and orientation consistent with the above requirements set above.

(*g*) Leaf area duration to fit the needs of the cropping system in use. It should be as long as possible in the reproductive period. Present varieties have an unduly long vegetative phase and relatively short reproductive phase. Photosynthesis becomes a limiting factor in pod formation.

(*h*) A root and stem morphology capable of rendering the crop efficient in nutrient uptake at all levels of nutrient availability.

(*i*) High harvest index.

(*j*) A programmed phase development and reproductive growth to provide the optimum balance of time and environmental resources for these two phases.

(*k*) Medium statured plant for mixed cropping tall statured plants for mixed harvesting.

(*l*) Short duration plants (approximately 120 days) for comparatively warmer climate and long duration plants (140-150 days) for cooler climates.

In addition to the above, chickpea breeders have suggested that the vareities should have multiple resistance to diseases and pests, drought tolerance and better quality and quantity of protein.

Soybean. There is a general agreement among agronomists that the main objectives in breeding soybean are short stature (lodging resistant) high yield, high protein content, resistance to diseases and nematodes, and suitability for mechanical harvesting (non-shattering). However, the improvement of soybean varieties is not as simple as that of rice and wheat. To increase the yield, it is necessary to increase (1) number of pods per plant (2) number of seeds per pods and (3) weight of each seed. When the pod number is increased, the number of seeds per pod inevitably decreases. Similarly, when we try to increase the number of seeds in each pod, the size and weight of seeds decreases. Therefore, although it is not difficult to engineer increase a variety which fulfills one or two of the above conditions, it is extremely difficult to fulfill all three. We have to compensate this by improving cultural practices, on the basis of the physiological traits concentrating on canopy structure.

The tendency now-a-days is not to breed varieties with extremely high yield, but to keep potential yield at a moderate level of 3.5 to 4.0 t/ha leaving any additional yield to the improvement of cultivation practices. Plant breeders are aiming at non-lodging, disease resistance and suitability for mechanical harvesting (non-shattering).

Brassica. Rape cultivars differ in their responses to vernalization when grown under different temperatures and photoperiods in a controlled environment. Given and optimal time of flowering, selection for early initiation and a long stem elongation phase will result in yield increases with the formation of more numbers of floral primordia (Thurling and Vijendra Das, 1979).

REFERENCES

Adams,M.W., (1975). In: Report of the TAC working group on the biology of yield of grain legumes DDDRL:IAR/77/2 FAO of the United Nations, Rome, 1-16.

Alekseeva, E.S. and M.M. Malina. (1976). (Features of breeding buckwheat, taking into account the development of the root system). Doklady Vsesoyuznoi Ordena Lenina Akademii Seleskohozyaistevennyhkh Nauk Imeni Lenina No. 1:15-17.

Bhan,S., H.G. Singh and A. Singh (1974). Note on root development as an index of drought resistance in sorghum (*Sorghum bicolor* (L) Moench). Indian J.agric. Sci.43:828-830.

Collins,F.C., D.A.Hinkle and R.B.Baser. (1981). Tolerance among wheat varieties to low soil pH. Arkansas Farm. Farm Research.30 (4):15.

Farrer,W. (1898). The making and improvement of wheats for Australian conditions. Agric. Gazette of New South Wales 9:131-168.

Guadamuz, M.E. (1980). Relationship between root morphology and yield component in French bean, (*Phaseolus Vulgaris L.*) Thesis, Universidad de costal Rica, San Jose, Costrarica, 69pp.

Ishag,H.M. (1973), J.Agric.Sci.Cambridge, 80:181-189.

Ishag,H.M. (1973), J.Agric.Sci.Cambridge, 80:191-199.

Kaspar, T.C. (1982). Evaluation of the tap root elongation rates of soyabeans cultivars (Abstr.) Int.B.43:1323 B.

Oleinik, A.A.(1977). Efficacy of selecting plants of spring barley for number of embryonic roots. Seskokhozyaistvennaya Biologiya 12:42-44.

Passioura,J.B. (1972). The effect of root geometry on the yield of wheat growing in stored water. Aust. J.Agric. Res.23:745-752.

Stachman, E.C., R.Bradfield and P.C. Mangelsdorf. (1967) Campaigns against hunger. Cambridge, Belknap, Harvard.

Thurling, N. and L.D. Vijendra Das (1979). Variation in the pre-anthesis development of Spring rape (Brassica napus L.). Aust. J. Agric. Res. 28 : 597-607.

Ustimenko, A.S., P.P. Danialchuk and A.T. Grodzdikovskaya. (1975). Root systems and yield in crop plants. Kiev, Ukrainian SSR, Vrozhai, 368 pp.

Vedrov, N.G. (1974). Root system of cereal crops as an objective in breeding in : Sorti Vdobrenie, Inkutsk, USSR:139-145.

Yoshida,S. (1972). Physiological aspects of grain yield. Annu.Rev.pl.physiol.23:437-464.

CHAPTER 2

Genotype X Environment Interaction and Stability Nature

GENOTYPE X ENVIRONMENT INTERACTION

Cultivars of a crop are grown under a wide range of conditions. They are exposed to different soil types, soil fertility levels, moisture levels, temperatures, and cultural practices. All of the variables encountered in producing a crop can be described as the environment.

When cultivars are compared in different environments, their performance relative to each other may not be the same. One cultivar may have the highest yield in some environment and a second cultivar may excel in others. Changes in the relative performance of genotypes across different environments are referred to as genotpe *x* environment interaction.

Types of Interactions

Every factor that is a part of the plant has the potential to cause differential performance that is associated with genotype *x* environment inteaction. Environmental variables can be classified as either predictable or unpredictable factors (Allard and Bradshaw, 1964). Predictable factors are those that occur in a systematic manner or are under human control, such as soil type, planting date, row spacing, plant population, and rates of nutrient application. Unpredictable factors are those that fluctuate inconsistently, including rainfall, temperature and relative humidity.

Predictable factors can be evaluated individually and collectively for their interaction with genotypes. Studies have been made of genotype *x* soil type, genotype *x* row spacing, genotype *x* planting date, and genotype *x* plant population interactions.

Unpredictable factors contribute to the interactions of genotypes with locations and years. Genotype *x* location, genotype *x* year, and genotype *x* location year interactions have been evaluated in many crop species.

The relative performance of genotypes across environments determines the importance of an interaction. There is no genotype *x* environment interaction when the relative performance among genotypes remains constant across environments. In figure 2, cultivar I has the same yield superiority over cultivar 2 across two environments. No genotype *x* environment interaction is present because the yield differential between the cultivar is 50 units in both environments.

Genotype x environment interaction can occur in two ways.

1. The difference among genotypes can vary without any alteration in their rank. In Fig. 2a genotype *x* environment interaction is present because cultivar 1 yield 20 units more than cultivar 2 in environment A and 50 units more in environment B.

Figure 2. The relative performance of two cultivars in two environments. (*a*) No genotype *x* environment interaction is present. (*b*) Genotype *x* environment interaction is present but does not alter genotype ranking. (*c*) Genotype *x* environment interaction is present and alters genotypic ranking.

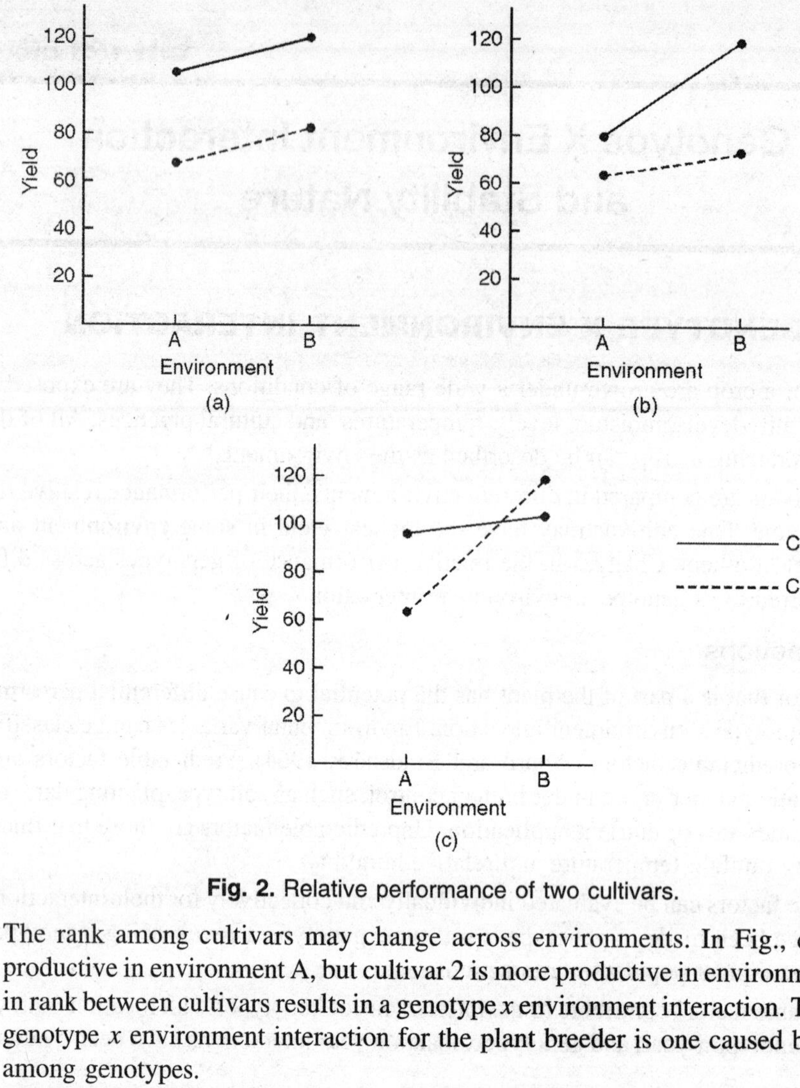

Fig. 2. Relative performance of two cultivars.

2. The rank among cultivars may change across environments. In Fig., cultivar 1 is more productive in environment A, but cultivar 2 is more productive in environment C. The change in rank between cultivars results in a genotype *x* environment interaction. The most important genotype *x* environment interaction for the plant breeder is one caused by changes in rank among genotypes.

Genotype *x* environment interactions are of interest of breeders for several reasons.

1. The need to develop cultivars for specific purposes is determined by an understanding of the interaction of genotypes with predictable environmental factors. Unique cultivars may be required for different row spacings, soil types, or planting dates.
2. The potential need for unique cultivars in different geographical areas requires an understanding of genotype *x* location interactions. The importance of the interaction can determine if division of a large geographical area into sub areas is needed for testing new genotype and obtaining data on cultivar performance for crop producers.

3. Effective allocation of resources for testing genotypes across locations and years is based on the relative importance of the genotype *x* location, genotype *x* year, and genotype *x* location *x* year interactions.
4. The response of genotypes to variable productivity levels among environments provides an understanding of the stability of performance. An understanding of the environmental stability of genotypes helps in determination of their suitability for the fluctuation in growing conditions that are likely to be encountered.

Assessment of Genotype X Environment Interactions

Determining the importance of genotype *x* environment interaction requires appropriate experimental procedures. An understanding of the steps involved in the design, conduct, analysis, and interpretation of such an experiment can be useful.

Genotypes for Evaluation. The genotypes chosen for an assessment of possible interactions are an important consideration in design of the experiment. Some analyses of genotype *x* environment interaction are not based on an experiment specifically designed for that purpose, particularly the assessment of the importance of interactions with locations and years. Instead, breeders utilize data from cultivars and experimental lines that have been evaluated over locations and years as a part of normal testing programmes. The main disadvantage of such an approach is that the cultivars and experimental lines may not be a random sample of available genotypes. Estimates of genotype *x* environment interaction obtained with selected genotypes may be higher or lower than those would be obtained with random individuals. The preferred procedure is to use a random sample of genotypes from those that are available for testing.

Test must be conducted at two or more locations and years to obtain estimates of genotype *x* location, genotype *x* year, and genotype *x* location *x* year interactions. The locations of testing generally are those routinely used for the breeder. Locations may be considered a fixed effect when they are not randomly chosen from all possible sites in an area. Some breeders consider them a random effect., however, because the breeder has no control over the climatic conditions that will occur at a location in any year. For the same reason, years of testing are considered random effects.

At least two replications are needed in each location and year to obtain an estimate of experimental error with which to test the significance of the interactions of interest. Any additional replications will provide a more reliable estimate of the experimental error.

Data Analysis

Data analysis includes the calculation of mean values, determination of the statistical significance of the source of variation, and calculation of estimates of appropriate variance components (Snedecor and Cochran, 1980 ; Steel and Torrie, 1980).

The sources of variation in an experiment are partitioned into main effects and their interactions. The mean squares for the sources of variation are determined, and appropriate F-tests are made to assess the probability that a source of variation is significant. Components or variance can be calculated for the main effect of genotype and its interactions with years and locations, standard errors can be computed for each component of variance.

Data Interpretation

Data interpretation includes consideration of the statistical significance of sources of variation and an assessment of the practical importance of variation observed among mean values. The genotype *x* location interaction measures the consistency of performance among genotypes at different locations. The consistency of performance of genotypes in different years is indicated by the genotype *x* year interaction. The genotype *x* location *x* year interaction measures the consistency of performance among genotype for each combination of year and location. An experiment conducted at two locations in 2 year has four year-location combination : year 1-location-1, year 1-location 2, year 2-location 1, and year 2-location 2. A significant genotype *x* location *x* year interaction indicates that the relative performance among genotype was not the same for each of the year location combinations. For all of the just mentioned interactions, an examination of mean values is necessary to determine if a significant interaction is due to a change in rank among genotype or to changes in the differences among genotypes without variation in rank.

The lack of any statistically significant interactions involving genotypes simplifies the nature of the testing programme required for cultivar development and simplifies cultivar selection by the producer. Theoretically, the lack of a significant interaction of genotype with location, year, or location *x* year indicates that a test at one location during one year would be sufficient to identify genotypes with superior genetic potential. Cultivars with the best performance at one location in one year would also be superior at other location in other years.

The practical implications of statistically significant genotype *x* environment interactions depend on the cause of the interaction. Genotype *x* environment interactions are not a problem for the breeder or producer if they are not due to changes in rank of performance among genotypes. Under these circumstances, a test at one location in 1 year could be used to identify superior genotype, if genetic differences among lines were adequately expressed. The same cultivars would be superior in all locations and years, although the amount of superiority would vary. Significant genotype *x* environment interaction that involve changes in rank of performance among genotypes are common. In determining the practical implication of the interactions, the breeder must consider the extent of the changes in rank and their potential impact on genetic improvement. Subjective judgements often must be made ; therefore, two breeders evaluating on the same data may adopt different courses of action. The options available to the breeder are different for each type of interaction.

Genotype X Location. Wide fluctuations in the rank performance of genotypes at test locations suggest that it may be desirable to develop genotypes for different locations through independent selection and testing programmes. The cost of establishing independent programmes for different geographical areas is substantial ; therefore, the decision can be difficult. Before establishing independent breeding programmes, the breeder should make for the genotype *x* location interaction. If the differences among locations are due to soil type or other factors that are consistent from year to year, independent programmes may be appropriate. Temporary differences among locations associated with unusual climatic conditions would not justify independent programmes.

Another consideration in determining the implications of genotype *x* location interaction is that fluctuations in rank may preclude selection of superior genotypes for multiple locations. Assume that a group of genotypes was divided into three classes; good, intermediate, and poor. A genotype *x* location interaction could be caused by fluctuations in rank among genotypes within the three groups, but not among groups. Such an interaction would be unlikely to justify the establishment of breeding programmes for independent locations, at least for the initial stages of testing.

Genotype X Year. An inconsistent ranking among genotypes grown in different years is in some regards more difficult to deal with than a genotype *x* location interaction. A breeder does not have the option of establishing independent breeding programmes for different years. The primary option available is to identify genotypes that exhibit superior performance on the average across years. This involves the genotypes in several years before selection of one for release as a cultivar. To reduce the length of time for genetic improvement, multiple locations in 1 year often are used as a substitute for years. The substitution is only effective when the divergence in climatic conditions among locations is comparable to differences among years.

Genotype X Year X Location. When there are fluctuations in the ranking of genotypes associated with individual location-year combinations, the breeder must identify genotypes with superior average performance over locations and years. This can be accomplished by testing over multiple locations and years. For example, an analysis of genotype *x* environment interaction for tobacco yield in North Carolina indicated that the mean squares for the genotype *x* year and genotype *x* location interactions were not significant (Jones et al., 1960). The ranking among cultivars were similar each year when averaged over locations. Ranking of cultivars were also similar at each location when averaged over years. But the genotype *x* year *x* location interaction was significant in the experiment. The interaction seemed to be associated with specific conditions, such as rainfall pattern and disease infestation, that caused the ranking of cultivars to vary among certain year location combinations. If the cultivar with the highest average performance over years, is chosen, it would be expected to have acceptable performance the next year, but it may not be the best in that particular season. Producers often reduce the effect of fluctuations caused by genotype *x* year interaction by growing more than one cultivar each season.

Stability of Genotype Performance

The reliability of cultivar performance across locations and years can be an important consideration in plant breeding. Some cultivars are adapted to a broad range of environmental conditions. While others are more limited in their potential distribution. There are cultivars that perform similarly regardless of the productivity level of the environment, and others whose performances is directly related to the productivity potential of the environment.

The stability of cultivar performance across environments is influenced by the genotype of individual plants and the genetic relationship among plants of the cultivar. The term homeostasis and buffering have been used to describe the stability in performance of individual plants or groups of plants over different environments.

The terms developmental homeostasis and individual buffering have been used to describe the stability of individual plants (Allard and Bradshaw, 1964 ; Briggs and Knowles, 1967). It has been shown that heterozygous individuals, such as F_1 hybrids, are more stable than their homozygous parents. The stability of heterozygous individuals seems to be related to their ability to perform better under stress conditions than homozygous plants.

The terms genetic homeostasis and population buffering have been used to describe the stability of a group of plants that exceeds that of its individual members. (Allard and Bradshaw, 1964 ; Lerner, 1954). Heterogeneous cultivars generally have more stability on the average than do homogeneous ones.

Methods of Stability Analysis

A number of statistical procedures have been developed to enhance our understanding of genotype *x* environment interaction and its relationship to stability.

Analysis of Variance. The environmental stability of a group of genotypes has been evaluated with standard analysis of variance procedures. The significance of interactions involving genotypes is determined with an F-test. The relative magnitude of the genotype *x* location, genotype *x* year, and genotype *x* locations *x* year variance components can be used to determine the effect of locations and years on the stability of a group of genotype.

The relative environmental stability of different group of genotype has been compared with use of the analysis of variance procedure. It is found that genotype *x* location and genotype *x* year interactions are to be greater significance in maize single-cross hybrids than in double-cross hybrids.

Pairwise Analysis of Variance. The standard analysis of variance procedure for a group of genotypes does not provide information on the environmental stability of individual genotypes. Information on individual genotypes can be obtained by conducting a combined analysis of variance for every pairwise combination of genotypes at all locations in a given year (Plastid and Peterson, 1959). For each genotype, the mean of estimates derived from its combination with all other genotype can be calculated. These means provide a measure of the contribution of each genotype to the genotype *x* location interaction.

Regression Analysis. The environment stability of individual genotype has been estimated by the use of regression analysis (Finlay and Wilkinson, 1963 ; Eberhart and Russel, 1966). A group of genotypes is grown over a range of environments. The mean performance of the genotypes at each environment is referred to as the environmental index. The performance of each genotype is regressed on the environmental index to obtain its mean performance over all environments, its linear response to varying environments, and an estimate of deviations from linear regression at individual environments. A desirable genotype was described by Eberhart and Russell (1966) as one with a high mean, a regression coefficient of 1.0, and deviations from regression of O. Such a genotype would have increased performance as the productivity of the environment improves.

Geometric Analysis. Hanson (1970) has proposed a measure of genotypic stability based on deviations from expected yield over environments. These deviations define the coordinates of a genotype within a stability space having a number of dimensions equal to the number of environments. Genotype stability is expressed as a euclidean distance, either from a stable genotype (relative stability) or between any two genotype (comparative stability).

Cluster Analysis. Cluster analysis also has been used to classify genotypic stability. On the basis of similarities in phenotypic responses in 16 environments, Ghaderi and colleagues (1980) arbitrarily grouped winter wheat genotypes into 10 clusters. They concluded that this method was effective in identifying groups of genotypes with various combinations of means and stabilities.

REFERENCES

Allard, R.W., and A.D.Bradshaw (1946). Implications of genotype-environmental interactions in applied plant breeding. Crop Sci.4:503-507.

Brigs, R.N., and P.P.Knowles.(1967). Introduction to plant breeding, Reinhold Books, New York.

Eberhart, S.A., and W.A.Russell. (1966). Stability parameters for comparing varieties. Crop.Sci.6:36-40.

Finlay, K.W., and G.N.Wilkinson. (1963). The analysis of adaptation in a Plant-breeding programme Australian J.Agr. Res.14:742-754.

Ghaderi, A.E.H.Everson and C.E.Cress (1980). Classification of Environments and Genotypes in wheat-crop. Sci.20:707-710.

Hanson, W.D. (1970). Genotype stability. Theor. Appl. Genet. 40:226-231.

Jones, G.L. D.F.Matzinger, and W.K. Collins (1960). A comparison of Flue-cured tobacco varieties over locations and years with implications on optimum plot allocation Agron.J.52:195-199.

Lerner, I.M. (1954). Genetic homeostasis. Oliver and Boyd, London.

Plaisted, R.L., and L.C.Peterson. (1959). A technique for evaluating the ability of selections to yield consistently in different locations or seasons. Am.Pot.J.36:381-385.

Snedecor, G.W., and W.G.Cochran. (1980). Statistical methods. 7th Edition. The lowa state university press, Ames.

Steel, R.G.D., and J.H. Torrie. (1980). Principles and Procedures of Statistics. McGraw Hill, New York.

CHAPTER 3

Response of Genotypes to Different Stress Conditions

Introduction

The plant cannot alter an environmental stress external to itself, but it can effect the stress only by preventing or decreasing the penetration of the stress into its tissues. This kind of resistance is called *stress avoidance.* Stress avoidance is stress resistance by avoiding thermodynamic equilibrium with the stress. The plant with avoidance is able to exclude the stress, either partially or completely, either by means of a physical barrier which insulates its living cells from the stress or by a steady state exclusion of the stress (a chemical or metabolic barrier). By avoiding the stress, it also avoids the strain. The plant may be resistant in spite of permitting the stress to enter its tissues provided that it decreases or eliminates the strain. This kind of resistance is called stress tolerance. Stress tolerance is stress resistance by an ability to come to thermodynamic equilibrium with the stress without suffering injury. The plant with stress tolerance is able to prevent, decrease or repair the injurious strain by the stress. Though tolerating the stress, it may either avoid or tolerate the strain.

Types of Environmental Stresses

Environmental stresses are of two main types *viz biotic* and *physio chemical* ones. The former belongs to the field of pathology and ecology. Among the physio chemical stresses, to which a plant may be exposed, some that have been unimportant until now may become more important due to man's explorations of outer space. Some of the stresses have been given below :

1. Plant Response to Low Temperature

The inability of the plant to control its own temperature is shown by radiation cooling of plants below the air temperature, at night and by the radiation heating of the cambium on the sunny side of stress to as high as 30 ºC above the shady side in winter. Plants thus poikilotherms, they tend to assume the temperature of their environment. This means, that they cannot develop low temperature avoidance. Exceptions are the fleshy inflorescences of Arecare and so called snow plants that develop temperatures above that of their environment due to rapid growth and the consequent rapid release of respiratory heat.

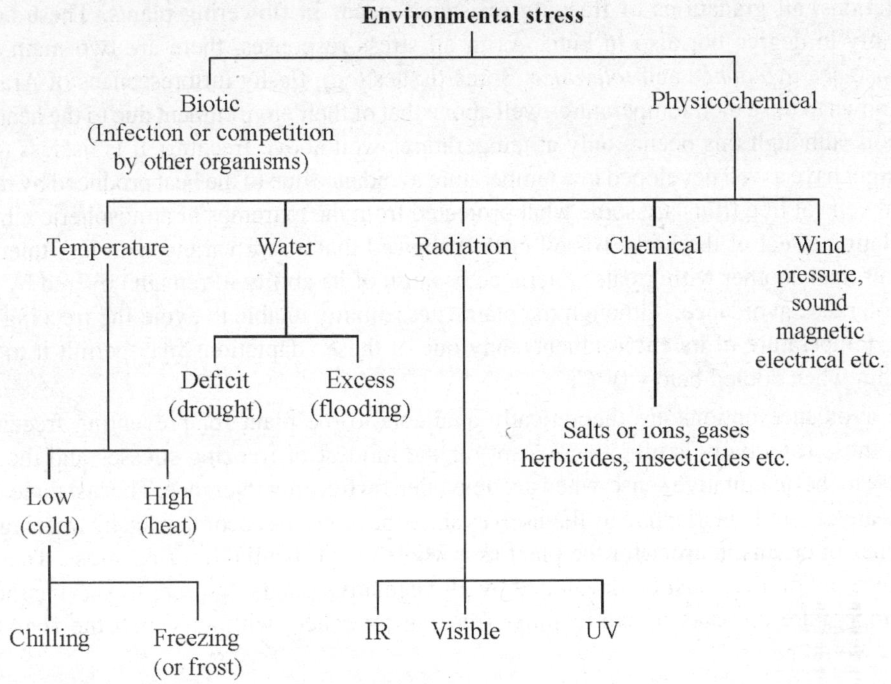

2. Plant Response to High Temperature

The temperatures of leaves at high air temperatures in the full sun have long been a matter of controversy. A moderately high temperatures (up to 40 °C) is or it now appears established, both from experimental evidence and theory that the leaf temperature cannot be more than a very few degrees below the air temperature. At very high temperatures (40-50 °C) this is not true, and leaf temperature may be considerably (as much as 15 °C) below the air temperature in this way exhibiting pronounced high temperature avoidance. The leaves of most plants however survive inspite of temperature above that of the air and they must therefore possess high temperature avoidance.

3. Response of Genotypes to Chilling Injury

Plants from tropical and semitropical climates show varying degrees of resistance to chilling injury. Since the plants are poikilotherms chilling resistance must be due to tolerance, and no avoidance related structural change is expected. Early evidence indicated that plants which are susceptible to chilling injury may harden in the field, becoming more resistant and therefore capable of surviving chilling temperatures (Sellaschop and Salmon, 1928)

4. Response to Radiation

Ionizing radiations are highly penetrating and highly absorbed by the plant. Plants that survive such radiations must possess stress tolerance. Ultra violet radiation may be largely reflected and transmitted by the leaf of a whole or absorbed by the leaf surface. Survival of these radiation may be due to avoidance.

5. Response of Plants to Freezing

The freezing stress may induce a fatal strain when the tissues of a plant are frozen below a threshold freezing temperature specific for that tissue. In order to survive, the plant must prevent such a fatal strain. Many have developed adaptation which protect them to varying degrees. As a result of these adaptations, all gradations of freezing resistance occur in flowering plants. These adaptations differ not only in degree but also in kind. As in all stress responses, there are two main classes of freezing resistance-*avoidance* and *tolerance*. Some tissues (eg), fleshy inflorescences of Araceae have long been known to raise their temperatures well above that of their environment due to the heat produced by respiration, although this occurs only at temperatures well above freezing. It is useless to look for plant that might have a well developed low temperature avoidance due to the heat produced by respiration. The cambial cells of tree fruits are some what protected from the extremes of atmospheric temperatures by the insulating effect of the bark. Torsell (1959) showed that a rare variety with less tolerance may survive better than another with greater tolerance because of its ability to remain covered by snow *i.e.,* its greater tolerance avoidance. Although the plant is essentially unable to avoid the freezing stress *i.e.* the freezing temperature of its environments, any one of these adaptations may permit it to avoid the freezing strain when cooled below 0 °C.

Five avoidance options are theoretically available to the plant for preventing freezing injury. The first of these essentially is effective except for the mildest of freezing stresses and the second is effective only in the quantitative sense-when accompanied by freezing tolerance. The last three avoidance mechanisms are generally ineffective by themselves since they must be accompanied by freezing tolerance in other tissues of organs in order for the plant as a whole to survive the freezing stress. Therefore, the only resistance option that must be developed by all vegetative plants, in order to survive the freezing stress of temperature climates is extracellular freezing tolerance, without which the five avoidance mechanisms would be useless.

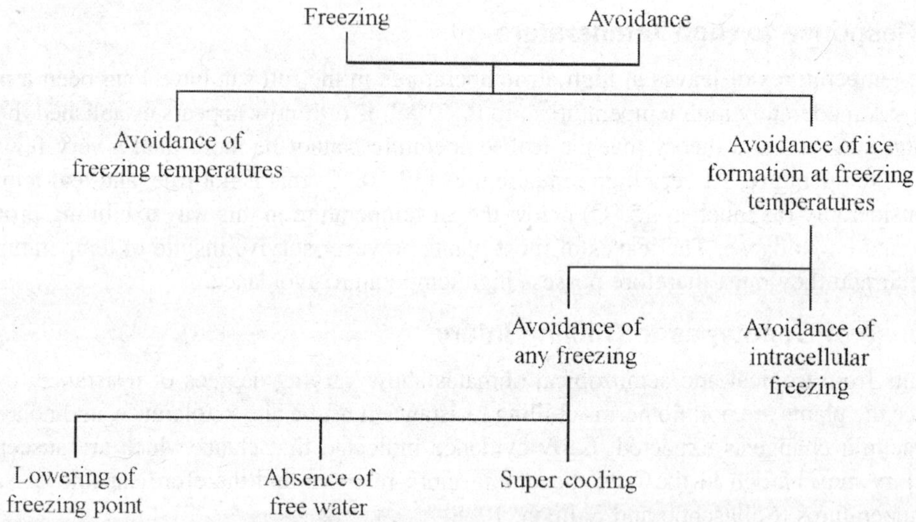

The different kinds of freezing avoidance

6. Plant Responses Associated with Drought

Resistance. Breeders instinctively look for new sources of variation when attempting to improve plants, but such empirical selection in the case of drought resistance has been difficult. The reasons are that drought resistance is not a simple response, but is mostly conditioned by a number of component responses which interact and may differ for different crops and in response to the different types ; intensity and duration of water deficit. Also selection under controlled environments rarely correlates with performance in the field where it is difficult to obtain consistent ranking of response to drought stress because of year to year variability in the environment.

Despite these difficulties, physiologists have described a number of individual responses in crop plants. Although it may be simplistic to claim a direct relationship, there is some evidence that some of these traits may contribute to particular forms of drought resistance.

A sample of these plant responses is shown in the following table. All appear to be under some form of genetic control.

One plant response not mentioned in the following table is called "early warning system" recently described by Zhang et al., 1987 (Table 2). This relates to the abscicic acid (ABA) produced in the root

Table 2. Plant responses to water deficits which are considered to be associated with drought resistance when exposed to intermittent and/or terminal drought.

*Response**	*Resistance to type of drought (intermittent or terminal)*
Drought escape	
Phenology (early flowering)	Terminal
Photoperiod sensitivity	Terminal
Developmental plasticity	Intermittent
Remobilisation of assimilates	Terminal
Drought tolerance	
Dehydration avoidance (Postponement)	
Root depth and density	Both
Low hydraulic resistance in roots	Terminal
High hydraulic resistance in roots	Terminal
Early seedling vigour	Both
Osmotic adjustment (Osmoregulation)	Both
Reduced stomatal and cuticular conductance	Intermittent
Leaf movements (rolling parahelitropic)	Intermittent
Increased leaf reflectance (glanconsness)	Both
High transpiration efficiency	Both
Desiccation tolerance	
Low lethal leaf water status	Intermittent
Maintenance of leaf area	Intermittent
High temperature tolerance of leaves	Intermittent
High relative leaf and stem elongation rates	Intermittent
Synchronisation of male and female flowering (maize)	Intermittent

*In a number of cases these responses are interrelated. For example, Osmo regulation, which enables the continued growth of roots at depth, may also assist the survival of vegetative parts under conditions of increasing water stress.

tips of a range of plants (including wheat) in response to the initial drying of soil profile. Abscicic acid (ABA) is subsequently transferred to the shoots where it includes stomatal closure. This response which occurs before the plant is subjected to any significant water stress, may be important in postponing dehydration.

If plants vary in the sensitivity with which they 'anticipate' declining soil moisture, it may be possible to select for this response by assaying ABA production and thereby increase water use efficiency in situations of terminal drought.

Except in maize, yield can be analysed in terms of three yield components. The number of ears (or panicle), number of seeds per panicle and the mean weight per grain. These components are "determined' at different stages of growth and so are affected in different ways by drought, depending on when the drought occurs.

Tillering occurs during early growth and the rate and duration of tiller emergence is affected by crop spacing, temperature, photoperiod, nutritional status (particularly nitrogen and phosphorus). There is also much genetic variation in tillering habit. Many late formed tillers die depending especially on plant spacing and the availability of water and nitrogen. In wheat and barley, ear initiation occurs during the time that stem elongation and the death of tiller is occurring. The number of grains per ear is thus subject to environmental influences; including drought during this time and it is often the yield component which varies most in response to season to season variation in water supply. The sensitivity of this yield component to drought is enhanced because at ear emergence, the stomatal control of transpiration is less than at other times (eg. Henson et at., 1984). Drought during grain filling, especially if accompanied by high temperature, as is common, hastens leaf senescence, reduces the duration of grain filling and reduces the mean weight of the grains.

Except for rice, cereals are remarkably well adapted to adjust their growth patterns to maximize yield in the face of drought which develops slowly. Some of the adaptive responses to drought is shown in the following Table 3. Some yield is obtained when the sole source of water is that present in the profiled sowing. This adaptation occurs because water deficit restricts leaf expansion, tiller growth and accelerates tiller death so giving small plants which set few grains. These responses to drought cause the plant to exploit soil water reserves less rapidly so ensuring that some is available for grain filling. All these responses can be manipulated genetically by breeders to produce varieties which make optimum use of available water. Breeders need to ensure that their varieties have sufficient phenotypic flexibility to be able to utilize fully the available water in years of above average rainfall, while not being too adversely affected by drought in years of below average rain fall.

Table 3. Development–dependent adaptive Responses of small grain cereals to drought on a time scale from seed to maturity.

Stage of development	*Effect of drought*	*Consequences for yield*
Seed	Germination reduced and delayed	Altered yield components reduced yield if less than ca. 1000 plants per m^2

Seedling	Increased tiller death Reduced N uptake	Reduced ears/m^2 and yield
Tillering, ear initiation	Death of young florets	Accelerated senescence Reduced grain set and yield.
Stem elongation ear growth	Reduced stem elongation	Greater accumulation of soluble carbohydrates in stems, partially buffers plant against loss of photosynthetic capacity during grain filling
Anthesis Maturity	Accelerated senescence	Reduced mean grain mass

Drought Responses as Selection Criteria

It is operationally, and conceptually, much easier to use as selection criteria whole plant or crop responses to drought which are associated with resistance to stress. Such responses are the result of the operation of several resistance to stress. Such responses are the result of the operation of several mechanisms but have a number of advantages over the mechanisms themselves.

- — Responses to stress are a much more highly integrated measurement of crop performance than individual traits, both in a hierarchical and a temporal sense.
- — In many cases they are easier to assess as they are integrators of events over a period of time and are usually measurable by simple means (balance, ruler or calendar).
- — As whole plant or crop processes (usually growth or development processes) they were much easier to relate to final yield than are individual mechanisms but.

As more highly integrated traits, their inheritance is not likely to be simple and their heritability to be low.

Crop Responses to Stress as Selection Criteria

Basic mechanisms	*Crop response*	*Consequence*
Deeper rooting	1. No delay in silking (maize)	Maintenance of grain number
Water conservation assimilate transfer	2. Better grain setting and/or filling (millet)	Maintenance of grain mass per panicle

Response of Genotypes to Soil Alkalinity/Salinity

Plant resistance to abiotic stress factors (such as salinity, drought and temperature) denotes a relative rating of the plant's response to a given level of stress factor under defined soil and weather conditions. Plant resistance in the context of salt affected soil is thus not a simple reaction but it is a complex interaction between the plant and the operating soil stress factor and its expression is affected considerably by numerous variables relating to the plant itself (genotype, ploidy level, growth stage, maturity period etc.) the soil status (texture, topography, fertility, ground water table, irrigation schedule, management level etc., and the atmospheric conditions (temperature, humidity, light, rainfall, wind pollution etc.) terms like "salt resistant" or "salt sensitive" crops or varieties do not convey any useful information to the breeder unless the stress as well as response parameters are also stated.

The salt stress resistance (SSR) denotes a plant's ability to prevent, reduce or overcome the possible injurious effects caused directly or indirectly by the excessive presence of soluble salts/toxic ions in its root zone.

Different species or species groups appear to have developed their own strategies on certain anatomical, morphological and developmental features to cope with environmental demands while genotypes within a species, often show notable variations of the specific scheme extending the range of adaptation of that species. Thus, despite the preponderance of unwanted or even potentially toxic ions in the soil liquid phase surrounding the root cells, a resistant type of plant tries to maintain its own characteristic ionic composition in its tissues so as to survive, grow and reproduce (yield).

Halophytes like some species of the family chenopodiaceae respond to salinity by taking up sodium and chloride at thigh rates and then accumulating these ions and isolating them in vacuoles of leaf cells. This compartmentation has great significance for the performance of halophytes in a saline environment (Flowers et al., 1977).

Plants have also evolved a variety of mechanisms for adaptation to osmotic stress, one of which osmo regulation that is cellular adaptation.

Inter Specific Differences

A survey of reported data showed on the response of crop plants to increasing salinity levels about eight fold differences among them regarding tolerance to salinity stress, threshold salinity level (SY 100) beyond which yield begins to decline. An interesting point that has emerged from the work done at the CSSRI (Central Soil Salinity Research Institute) is that some allopolyploid crop species turned out to be remarkably more tolerant to both alkali soil as well as saline soil conditions as compared to their putative diploid progenitors and relatives of lower ploidy levels (Rana 1978; Rana et al., 1980). Data based on response evaluation of a large germplasm 'collection' showed that bread wheats ($2n = 6x = 42$) were found to be more tolerant compared to the durums ($2n = 4x = 28$) and einkorn wheat ($2n = 2x = 14$). Among ten species of *Triticum*, whose response to soil salinity and alkalinity were studied critically under comparable conditions, hexaploid species *aestivum* (including *spharococcum*) was characterised by a distinctly high tolerance rating and was superior to tetraploid relatives, namely durum, turgidum, polonicum and carthlicum. Diploid species *monococcum* showed maximum growth reduction and failed to flower in alkali soil (pH. 9.1) as well as in saline soil of sandy loam type.

Likewise allopolyploid species *Brassica juncea* ($2n = 4x = 36$) was found to be superior in salt tolerance compared to its diploid progenitors B.campestris ($2n = 2x = 20$) and **B. nigra** ($2n = 2x = 16$). Allopolyploid species **Medicago sativa** ($2n = 3x = 32$), **Sesbania grandiflora** ($2n = 4x = 24$). **Oryza coartata** ($2n = 4x = 48$), **Chloris barbata** ($2n = 4x = 40$) were also characterised by a high level of salinity or alkalinity stress tolerance. In **Gossypium,** however there was a reverse situation since diploid asiatic species **arboreum** and **herbaceum** turned out to be more tolerant compared to the new world tetraploid species **hirsutum** and **barbadense.** Besides pointing to the significant role of genomic interactions in enhancing adaptation of some species to edaphic stress environments, these make it evident that salinity/alkalinity tolerance rating of crop species is largely influenced by its previous history of cultivation and management.

Intra Specific Variation

Early studies on plant response to salinity stress, conducted on materials not particularly selected for genetic diversity, showed no significant differences within a species and created a widely shared impression that there was no appreciable genetic variability within crops to justify breeding efforts for improvement of salt resistance.

Beginning with the systemic study of salt tolerance of 60 strains of **Agropyron desertorum** by Dewey (1962), voluminous and convincing evidence has accumulated over the years showing that considerable exploitable intra specific variability, exists regarding tolerance to salinity/alkalinity and mineral deficiency/toxicity (Singh and Rana, 1984; Nelson, 1983). Varietal differences have been shown in bread wheat regarding the response of genotypes. When grown in reclaimed (non-stress) and partially reclaimed alkali soils as judged by the alkalinity caused reductions in grain yield several reviews have appeared in recent years citing numerous reports on considerable intraspecific variation for tolerance to edaphic stresses (Downton, 1964). The CSSRI has made sizable collections of indigenous cultivars of rice, wheat, and barley materials that have for a long time been grown traditionally in different situations of salt affected soils under little management are evaluation of these materials, along with large stocks of germplasm collections of these crops built up over the years regarding their responses to soil salinity/alkalinity. Parameters used for monitoring varietal response to edaphic stress included seed germination, rate of seedling emergence, seedling growth rate, plant height, dry matter production and grain yield.

Salinity interacts strongly with other environmental factors and this often diffuses or mates the genotypic differences in tolerance making inheritance studies uncertain.

Genes and chromosomes with Major Effect

Genes which have a major effect on varietal responses to edaphic stress factors have been reported in many crops. Mostly controlling uptake/utilization of nutrients or exclusions of toxic ions (Devine, 1982).

Genes with major effects influencing ion uptake/use

Gene	*Reported effect*	*Reference*
fe	Controls efficiency of Fe utilization in soybeans	Weiss (1943)
	Recessive homozygotes lack ability to reduce Fe^{+3} to Fe^{+2} at the root surface	Genetics 28: 253-268
Al_p	Confers resistance to Al toxicity in winter barely	Reid (1970) Barely genetics II 409-413
fer	Recessive homozygotes show Fe in efficient response in tomato	Brown et al., (1971) Physiol. Plant: 25: 48-53
NCl	Controls chloride exclusion from plant tops in soybeans.	Abel (1969) Crop Sci: 9: 687-698

Chromosomal Regulation of Plant Response

Author	*Reported conclusion*	*Reference*
Shootmater (1974)	Resistance to soil acidity conferred by D genome in wheats	Euphytica 23-505-513
Prestes et al., (1975)	Al tolerance factor carried on chromosome 5 D in wheats	Agron. Abs. 67-60.

Rana et al., (1980)	Resistance to alkali and saline soil conditions increased with ploidy level in Triticum and Brassica spp. wide adaptation of bread wheats to salt affected soils owing largely to D genome (2D, 3D, 5D)	Inter natl. symp. CSSRI, Karnal pp. 487-493

Conclusion

From the above discussions of plant response to different stress conditions, we can conclude that plant is trying its level best to survive in the stress environment and reproduce. We can also conclude that, the plant response to the different stress conditions is not only controlled by the physiological and biological factors but also controlled by genetic factors like intraspecific variations, inter specific variations and chromosomes.

REFERENCES

Baker, F.W.G. (1989). Drought Resistance in cereals CAB. International Walling ford, United Kingdom.

Chopra, V.L. (1986). Approaches for incorporating drought and salinity resistance in crop plants. Oxford-IBH Publishing Co. Pvt. Ltd., New Delhi.

Devine, T.E. (1982). Genetic fitting of crops to problem soils. In: Breeding plants for less favourable environments Eds. M.N. Christiansen and C.F.Lewig. John Wiley, New York. pp. 143-173.

Dewey, D.R. (1962). Breeding crested wheatgrass for salt tolerance. Crop sci: 2:403-407.

Downton, W.I.S. (1984). Salt tolerance of food crops: Perspectives for improvement CRC critical reviews in plant sciences 1:183-201.

Flowers, T.J., P.F. Troke and A.R. Yeo (1977). The mechanism of salt tolerance in halophytes. Ann. Rev. Plant. Physiol. 28:89-121.

Nelson, L.E. (1983). Tolerance of 20 rice cultivates to excess Al and Mn. Agron. J. 75:134-138.

Henson, I.E., V. Mahalakshmi, G. Alagarswamy, and F.R. Bidiger, (1984). The effect of flowering on stomatal response to water stress in pearl millet (Pennisetum americanum). Journal of Experimental Botany 35:219-226.

Rana, R.S. (1978). Wheat variability for tolerance to salt affected soils. In: Genetics and Wheat improvement. Ed. A.K. Gupta, New Delhi, Oxford and IBH Publishing Co., New Delhi, pp. 180-184.

Rana, K.S., Kn.N. Singh and P.S. Ahuja (1980). Chromosomal variation and plant tolerance to sodic and saline soils : Symposium papers-International Symposium on salt affected soil, CSSRI, Karnal-1.

Sellaschop, A.W., and J. Salmon. (1928). "In Temperature and life" Springer Publ., New York.

Singh, K.N. and R.S. Rana (1984). Genetic variability and path analysis in Triticales grown in alkali soils. Indian J. Agric. Sci. 54: 950-954.

Torsell, H. (1959). The hardiness of plants. Academic Press, New York.

Zhang, J., U. Shurr, W.J. Devies, (1987). Control of stomatal behaviour by abscisic acid which apparently originates in the roots. Journal of Experimental Botany 38:1174-81.

CHAPTER 4

Stresses Due to Drought, Cold, Salinity and Alkalinity

Introduction

In its widest context, the stress environment is one of the major factors responsible for the current global map of crop distribution patterns. The initial stages of plant domestication from native vegetation must have involved genetic adaptation of the prevailing environment. Even at the early stages of modern agriculture the importance of the environment in effecting crop distribution and adaptation was recognized. For the many factors that affect crop evolution and distribution (solar radiation, atmospheric composition, soil, water, temperature, wind, fire, man, animal and other plant organisms), the majority may be considered as components of environmental stress in the narrow sense.

Undoubtedly, current world crop production is limited largely by environment stress. It was estimated that only 10% of the world's arable land may be categorized as free of stress. About 20% of the land is under some kind of mineral stress, 26% is affected by drought stress, and 15% by freezing stress, where long-range crop production in farmers fields could be investigated. It was concluded that environmental stress was the main factor responsible for the disparity between potential and actual yield.

Thus, genetic improvement of stress resistance is an economically viable solution, even when research and development costs are considered, as long as the improved seed can reach the farmer at a reasonable cost. On the other hand, stress resistant varieties should not be necessarily associated with situation of poor or marginal crop performance, and their importance should not be limited to conditions of low-input and high-risk farming.

Physiological Parameters of Drought Stress

Water stress develops in the plant as the demand exceeds the supply of water. The disparity may occur due to atmospheric or soil conditions. It is reflected in a gradient of potentials developed between the soil or the soil-root interface and the transpiring organ. Since a gradient is involved, the immediate implication is that drought stress may develop to a different rate in different plant organs along the gradient (Begg and Turner, 1976). Since the leaf is directly involved with the production of assimilate for growth and yield, leaf water potential is the common immediate measure of plant water stress.

As water is being removed from the cell, osmotic potential is reduced due to the simple effect of solute concentration. However, if during the course of cellular water loss solutes are actively accumulated, osmotic potential would be reduced beyond the rate dictated by the mere effect of concentration. Such accumulation of solutes during the development of water deficit is termed *osmotic adjustment* or *osmoregulation*. Osmotic adjustment therefore allows maintenance of a higher turgor potential, in an osmotically adjusted tissue. As soil moisture is reduced so is leaf-water potential. Leaf-water potential displays the typical diurnal change in accordance with the change in evapotranspirational demand. Plant water stress is maximised at or just after midday, when leaf-water potential is at its lowest level. In a very generalised manner, the cellular implications of water deficit can be summarised in

(1) the spatial and structural effects on membranes and organelles.

(2) changes in hydration and structure of macro molecules such as proteins and nucleic acids.

(3) changes in the pressure differential across the membrane, cell wall complex, which affect cell expansion and

(4) involvement with plant hormones.

Cell expansion is the most sensitive process, while translocation is probably one of the most resistance processes. Cell division followed by photosynthesis is moderately susceptible in this generalised but useful framework.

Plant production under conditions of water stress will be affected by the integral effects of water stress at the various levels of plant organization. A simplified scenario of the development of drought stress in the field is as follows :

Plant growth can be inhibited by the moderate levels of stress that induce reduced cell enlargement. When leaf-cell enlargement is arrested, leaf-growth becomes unimportant as a sink for photosynthesis. Since photosynthesis per unit leaf area is not yet seriously affected at this level of stress, the photosynthetic product is used, at least partly for osmotic adjustment, or it may be partitioned to other organs, such as the roots the resulting increase in root/shoot ratio and the limitation of leaf-area expansion allow avoidance of the development of larger root-to-leaf gradient of potentials. Osmotic adjustment protects the cells from extreme desiccation and allows a continued gas exchange. As water deficit increases older leaves senescence to various degrees, and leaf area thus decreases. Crop water use may now be reduced in proportion to the reduced leaf area and the increasing hydraulic resistances in the plant. As long as bulk leaf turgor is maintained (in the live leaves) due to osmotic adjustment or because of the reduced evapotranspiration (ET), stomata do not close completely and carbon is being fixed. The rate of carbon fixation and its relation to transpiration (transpiration ratio) during this stage is determined partly by the stomatal resistance and partly by mesophyll resistance to CO_2. The relative importance of the two resistance is conversial.

If a moderate level of water deficit that may cause a reduction in cell growth and division takes place in plant meristems, a direct effect will take place on the differentiate organs that constitute yield components. It is however, fortunate and perhaps not a coincidence that reproductive meristems are often situated remotely from the sites of the lowest water potential in the plant. In the cereals, the differentiating apex is situated at or just above the solid surface. Meristems also often have a relatively greater capacity for osmotic adjustment. Thus, the level of water stress experienced by the reproductive meristems is less than that experienced by the transpiring leaves on a given day.

With a further increase in water deficit, turgor is lost, stomata fully close, plant growth ceases, the remaining live leaves (of cereals) roll up, gas exchange is nil, and the plant enters the prelethal, non-productive stage of survival. During this stage, carbon is lost by respiration, and tissue-water-status continues to decrease slowly with no transpiratoin, the temperature of the remaining live leaves increases to lethal levels, and all leaves die. The only remaining viable tissues are those of the protected meristems. The plant would be considered dead when its meristems are unable to recover and regrow upon rehydration.

It can therefore be seen that in terms of plant introduction the water-stressed plant undergoes two different and important stages. During the first stage, plants maintain their capacity for gas exchange and are therefore gaining carbon. In the second stage, plants only survive while losing carbon, plant breeding for drought resistance has to recognize and address each of the two stages in the proper context of the crops environment.

Physiological Parameters of Cold Stress

Levitt (1980) categorized plants into the following six response classes with respect to low temperatures:

1. "Chilling Sensitive"-Sensitive to low temperatures, above 0 °C.
2. "Tender"-Sensitive to light frost or temperatures near 0 °C.
3. "Slightly hardy"-will survive freezing temperatures to 5 °C.
4. "Moderably hardy"-will survive freezing temperatures within 0.5 °C to – 10 °C.
5. "Very hardy"-will survive freezing temperatures within 10° to 20 °C.
6. "Extremely hardy"-the most freeze-tolerant species, characterised largely by the ability for super cooling.

In the agricultural domain, six classes can be aggregated into two groups based largely on the nature of stress, its physiological repercussions, the nature of tolerance, and the agronomic implications involved. One group includes chilling sensitive and tender plants. While the other includes all the hardier species. The two groups may overlap in terms of stress and the response to stress at 0 °C, when water in plants may freeze. However, by and large, the two groups differ in that one is normally subjected to above-freezing (chilling) temperatures and the other subjected to freezing temperatures. Breeding for chilling tolerance and breeding for freezing tolerance carry different meanings in theory, practice and implications.

I. Chilling Stress

Chilling stress can be measured in various plant processes, such as germination, growth, photosynthesis, fruitset, yield and fruit quality. As is the case for heat stress, the primary transducer of chilling stress is at the cellular and organelle level, with the resulting repercussion at higher levels of plant organization. There is an almost general agreement that the primary lessions of chilling stress are in the cellular and organelle membranes. (Lyons et al. 1979, and Christiansen 1982).

The examination of cellular responses to chilling temperatures reveals various disorders, such as the loss of turgor, vacuolization, disruption of the cytoplasmic reticulum, vesiculation, reduced protoplasmic streaming, and general disorder of organelles. Studies show that membrane disruption precedes other cellular disorders which ensue as stress intensity or duration increases.

The effect of chilling stress on photosynthesis and respiration are of primary interest with respect to economic plants. The range of temperature optimums among various species does not fully reflect their maximum range of temperature adaptations (Berry and Raisions, 1981). Since chilling-sensitive species are typically warm-seasons (tropical) plants, the initial reduction of net photosynthesis with the decline in temperature may already be observed at the range of 25 to 30 °C. This initial reduction is not associated with chilling injury proper. This sub-optimal range of temperature for photosynthesis is undoubtedly above the range where membrane disruptions are observed in chilling-sensitive plants. With a further decline in temperature, net photosynthesis is reduced until the "critical" temperature is reached, at about 10 to 12 °C (Larcher, 1981). As temperatures reduce below the optimum all photosynthetic reactions decline according to their individual temperature coefficients, (Berry and Raison, 1981). A limit to photosynthesis will be set by the sensitivity of any given process at a specific temperature range. Disturbance of chlorophyll production and impairment of chloroplast function were shown to result from chilling stress (Mellerd and Mc William 1968; Kabak et al., 1983). This is expressed in the loss of normal green colour in leaves developed under stress. Electron transport may be an important limitation at chilling temperatures, and its temperature dependence within this range of temperatures has been associated with the onset of phase separation of chloroplast membrane lipids. RUBP (ribulose-1, 5-biphosphate) Carboxylate activity was also considered among the more chilling-sensitive components of the photosynthetic system (Moller et al. 1984). Because of the different role of RUBP Carboxylase in C_4, and C_3 species, it was considered to be more important in chilling sensitivity of C_4 plants.

Respiration is more tolerant of temperature changes than photosynthesis. At the higher levels of plant organization, chilling injury results in several well-known effects such as poor seedling establishment, stunted growth, wilting, chlorosis, necrosis, poor fruit set, etc., water stress is often observed in chill-affected plants. This effect is largely localized at the root, as chilling of the intact root (but not the shoot) results in wilting (Berry and Raisons 1981; Mc William 1983; Christiansen and John, 1981).

Another important growth stage that carries a great practical significance under chilling stress is flowering. Interest in chilling tolerance at flowering is largely associated with either the abnormal development of flowers or the failure of set seed or fruit at chilling temperatures. Pollen sterility is a common cause for low fruit or seed set under chilling temperatures at flowering in plants such as rice, sorghum strawberry, and tomato (Simon et al. 1976; Lin and Peterson 1975; Suzuki, 1982). However, studies on pre-anthesis development amongst rapeseed cultivars (Brassica napus) conducted in controlled environments indicated that Japanese cultivars which evidently had a marked vernalization (low temperature) requirement for floral initiation (Thurling and Vijendra Das, 1977).

II. Freezing Stress

Plant evolution is apparently responsible for the fact that most plants that are adapted to survive freezing enter the cold period in a relatively dormant state, with a minimum amount of active tissues. Thus hardy, deciduous trees shed all their leaves, while annuals and grasses enter the cold period in the form of a dormant crown after most leaves are killed by frost. The lesson to be learned is that a dormant state is conductive to resistance and that resistance in actively growing tissues is rare. This situation is different in principle from the other stresses, such as drought, heat or chilling where resistance in actively growing plants is usually required and achieved. For the most part, freezing resistance involve the conservation of life to the extent that survival and regrowth after freezing will be made possible.

As water in plants (at this point, without a reference to tissues and cells) cool below 0 °C ; it may super cool and not necessarily freeze, the formation of ice requires sources of nucleation. Plant water may supercool to very low temperatures in some hardy trees, apparently due to lack of an immediate source of nucleation. The phenomenon of supercooling to very low temperature in some plants on one land, and the understanding of the repercussion of ice formation in plants on the other, are at the basis of the agreement that ice formation rather than temperature per se is the most common stress factor in plants adapted to grow at sub zero temperatures.

Physiological Parameters of Salinity and Alkalinity Stress

In terms of their adaption to saline environments higher plants are classified as halophytes and glycophytes which are considered as essentially adapted and non-adapted to saline habitats, respectively. Claims were made that thee is no real justification for this classification as a continum of sensitivity or resistance to salinity experts across all higher plants (Wynn Jones 1981). However, Greenway and Munns (1980); Flowers et al. (1977) claim some unique responses for one group or the other. For the purpose of discussing practical breeding problems in crop plants, most information discussed here will be derived from work with glycophytes. It is, however, recognized that the study of halophytes may open the way for future developments in breeding for salinity resistance in crop plants. The exact nature of stress involved and how it integrate towards an effect on growth and yield has not yet been understood clearly. Empirical association between a given soil salinity level and the effect on growth are important (van Genuchten and Hoffman, 1984). However, the problem essentially lies in our inability to predict the rate of plant stress and its effect at high levels of plant organization from the physiological perturbations cause by salinity. The reason partly lies in the fact that when plants grow in saline conditions, they are subjected to three types of stress.

1. Water stress caused by osmoticum
2. Mineral toxicity stress caused by the salt and
3. Disturbances in the mineral nutrition of the plant

The soil solution of a wet saline soil may be at a water potential of 24 bar (which is the value for sea water) and up. Plant challenged by this potential developed a large soil- to-leaf gradient of potentials. Symptoms of water deficit can be observed in salinized plants, especially under conditions of high evaporative demand and ultimately may wilt and dessicae. Water stress is therefore considered a component of salt injury. However, there are many cases where water-stress symptoms are not in evidence and plants appear to maintain turgor. The lack of symptom in salinized plants led to the belief that water stress is relatively unimportant in causing injury by salinity. Hanks (1984) showed a linear relationship between yield and evapo transpiration (ET), when variable ET was induced by different salinity treatments. Hence, the effect of salinity on ET reduction may be taken as an expression of water stress. Hanks recognized the possible effect of ion toxicity towards ET reduction, say, by reducing root or leaf growth.

Plants adjust osmotically in response to salinity, and therefore may not show symptoms of wilting. It can be seen that osmotic adjustment may occur in response to initial plant dehydration caused by salinity-induced water stress. Various components were identified in the increase of cellular osmoticum under salinity stress, including ions (K^+, Na^+, Cl^+) and metabolites (proline, glycine-betaine, and various organic acids). The amount of Na^+ and Cl^- found in such cases can only be accounted for by the massive uptake of these ions from the saline medium.

Osmotic adjustment as a means for combating water stress under saline conditions entails a heavy cost. There is no doubt that osmotic adjustment by the uptake of salt will subsequently involve severe problem of Na^+ and Cl^- toxicity.

Another problem raised by osmotic adjustment and turgor maintenance under salinity stress is transpiration. The transport of potentially toxic ion, such as Na^+, may vary with water flow rates. Thus, the maintenance of turgor enables greater tanspiration which in turn, may cause a further increase in plant toxification (Wynn Jones 1981). It was recently proved that shoot growth reduction under salinity stress is not mediated by turgor loss in wheat and barley on a short-term basis. This would provide further support to factors other than water relations in mediating plant injury by salinity.

Several specific effects on plants of Na^+ and Cl^- of NaCl were identified. However, experimental work on the metabolical and physiological effect of salt or ions did not always offer a critical evidence on the direct ionic effects because the effect of water stress could not always be overruled.

NaCl affects the activity of enzymes of glycolytic pathway in a way that can be considered as damage. Salinity affects oxidative phosphorylation and the ATP/ADP (Adenine Tri Phosphate/Adenine Di Phosphate) ratio in the root tissue or in isolated root mitochondria. The interference of salinity with the normal respiratory metabolic pathways can bring about a reduction in the supply of energy. Salinity interferes with protein synthesis as evidenced by the inhibition of amino acid incorporation into protein.

The involvement of plant hormones in mediating plant response to salinity is an attractive proposition. An effect in this respect has been ascribed to cytokinins and abscisic acid (ABA). Cytokinins are synthesised in the root and are transported to the shoot. ABA is most likely synthesised in the leaves and transported to the root. As salinity stress increases, less cytokinin is found in the xylem exudate, while ABA content of the shoot increases. Both hormones are similarly affected by water stress.

REFERENCES

Abraham, B. (1986). Plant Breeding for stress environments, CRC Press, Inc. Boca Raton, Florida.

Begg, J.E. and N.C. Turner., (1976). Crop water deficit, *Ady. Agron.*, 28:161.

Berry, J.A. and J.K. Raison (1981). Response of macrophytes to temperature, in physiological plant ecology. I. Responses to the physical Environment, Large, U.L., Nobel, P.S., Osmond, C.B., and Ziegier, H., Eds., Springer-Verlag, Berlin pp 278.

Christiansen M.N., (1982). Temperature stress and membrane lipid modifications, *Adv. Phytochem.*, pp 177.

Christiansen, M.N. and J.B.St.John, (1981). The nature of chilling injury and its resistance in plants, in *Analysis and Improvement of Plants Cold Hardiness*, Olien, C.R. and Smith, M.N., Eds., CRC press, Boca Raton, Fla., Vol-I.

Flowers, T.J., P.F. Troke and A.R.Yeo (1977). The mechanism of salt tolerance in halophytes, *Ann. Rev. Plant physiol.*, 28:89.

Greenway, H. and R. Munn (1980). Mechanisms of salt tolerance in nonhalophytes, *Ann. Rev. Plant Physiol.*, 31:149.

Hanks, R.J. (1984). Prediction of crop yield and water consumption under saline conditions. Salt tolerance data, in Soil Salinity under Irrigation. Processes and Management, Ecological Studies No. 50, Shainberg. I. and Shalhevet, J., Eds., Springer-verlag, Berlin, 272.

Kabaki,N., K.Tajima,and A.Amemiya,(1983). Physiological growth inhibition in rice as affected by low temprature.I. Physiological mechanism of growth retardation caused by cool temperature, Bull Nath Inst. Agric. Sc jpn, D.34:1.

Larcher,W., (1981). Effects of low temperature stress and frost injury on plant productivity, in *Environmental Factors Limiting Plant Productivity*, Johnson, C.B., Ed., Butterworths, London, p.p. 253.

Levitt,J., (1980). Responses of plants to environmental stresses chilling, Freezing and High temperature stresses, I.Academic Press , New York , pp. 487.

Lin, S.S. and M.L. Peterson, (1975). Low temperature induced floret sterility in rice, Crop Sci., 15:657.

Lyons, J.M., D.Graham, and J.K. Raison, Eds., (1979). Low temperature stress in crop plants: the role of the Membrane, Academic Press, New York, pp 565.

Mc William, J.R. (1983). Physiological basis for chilling stress and the consequences for crop production in *Crop Reaction to Water and temperature stresses in Humid, Temperature Climates*, Raper, D., Jr. and Kramer, P.J., eds., Westriew Press, Boulder, Colo, pp. 113.

Mellerd, A. and J.R. Mc William, (1967) and (1968). Studies on a maize mutant sensitive to low temperature, *Plant physiol,* 43.

Moller, G., G. Giesier, and P. Stamp, (1984). A serial assay procedure for simultaneous spectrophotomeric tests of four photosynthetic enzymes; adaptation of maize inbred lines and crosses to extreme temperatures for active growth, *Z. Acker Pflanzenbau*, pp. 153, 167.

Simon, E.W., A. Minchin, M.M. Mc Menamin, and J.M. Smith, (1976). The low temperature limit for seed germination, *New Phytol 301.*

Suzuki, S., (1982). Cold tolerance in rice plants with special reference to floral characters. II. Relations between floral characters and the degree and cold tolerance in segregating generations, *Jpn. J. Breed.*, Vol 32:9.

Thurling, N. and L.D. Vijendra Das, (1977). Variation in the pre-anthesis development of rape (Brassica napus L.). 3rd International Congress of the Society for the Advancement of Breeding Researches in Asia and Oceania (SABRAO), Plant Breeding papers: 2; 3a, 48-51.

VanGenuchten, M. Th. and G.J. Hoffman, (1977). Analysis of crop salt tolerance data in *Soil salinity under Irrigation. Processes and Management,* Ecological Studies No. 51, Shainberg, I. and shalhevet, J., Eds., Springer-Verlag, Berlin, 258.

Wynn Jones, R.G., (1981). Salt tolerance, in *Environmental factors Limiting plant productivity*, Johnson, C.B., Ed., Butterworths, London, pp 213.

CHAPTER 5

Development of Resistant Varieties to Drought

Introduction

Shortage of water is more limiting to crop production in the world than any other single factor. Thirty-six per cent of the crop land receives less than 30 inches of precipitation per year and much of the 64% suffers drought sometimes during the growing season. Drought received second highest rating of all biological/environmental stresses, ranking second only to diseases and insects. Drought is a real concern and a serious yield limiting factor on almost half of the worlds' 70 million hectares of rainfed rice. The ultimate goal is to improve the crops genetically through the breeding varieties that are able to produce more than our present varieties under stress conditions. Hence, the need for developing better crop adapted to semiarid growing conditions gain importance now-a-days.

Three parameters of drought resistance *i.e.* morphological, physiological and biochemical are considered to be important while developing varieties resistance for drought.

Morphological

1. Earliness 2. Reduced leaf area 3. Leaf rolling 4. Wax content 5. Colour of leaves 6. Root system 7. Awns 8. Hairs 9. Increased grain weight.

Physiological

1. Photosynthesis 2. Reduced transpiration 3. Stomatal closure 4. Osmotic adjustment 5. Reduced xylem vessel in roots 6. Reduced respiration losses 7. Desiccation tolerance 8. Canopy temperature.

Biochemical

1. Proline accumulation 2. Betaine accumulation 3. Protein synthesis.

Breeding for Drought Resistance

It falls under 2 categories viz.:

(*a*) Breeding for escape mechanism.

(*b*) Breeding for drought avoidance

(*a*) Breeding for escape mechanism

One of the simplest mechanisms by which plants deal with drought is drought escape. This is accomplished by rapid phenological development or developmental plasticity. Drought escape is the most dramatic in the "ephermeral" plants of desert regions which complete life cycles in a very short span. Drought escape plays an important role in crop species both through yield and their early maturity and developmental plasticity. Earliness, a highly heritable character is negatively correlated with yield. For this reason whenever breeding efforts were directed towards earliness, yield had to be sacrificed to certain extent. But escape proved to be very useful mechanism for coping with drought stress. Through adaption of biparental approach, undesirable linkages in the repulsion phase can be broken and it is possible to obtain recombinants that are both early and high yielding (Yunus and Paroda, 1982). The same is useful for developmental plasticity even for tiller numbers and their association with yield. The genetic architecture attributes such as earliness, tillering and determinate growth habit can be effectively used in breeding for drought mechanisms by employing breeding methods such as pedigree method and biparental approach, especially when negative linkages are to be broken.

(b) Breeding for drought Avoidance

Drought avoidance might permit a longer period of crop growth with reduced water use or increased water uptake, drought avoidance mechanisms often operate at the expense of photosynthesis. They reduce top growth at the expense of increased root development. Tolerance would be more desirable since the crop produce more yield at lower water potential. Levitt (1972) stated that drought avoidance is more important than drought tolerance in higher plants. Armenta-Soto et al., (1983) studied root system genetics using an aeroponic culture in rice and found that long roots and high root number are primarily controlled by dominant allels and these characters are highly heritable. Hence, efforts could be made to breed for this trait. Towards drought avoidance attributes such as rooting pattern, waxiness, high kernel weight, leaf rolling, leaf orientation play a considerable role and are under genetic control. For better genotype the following breeding approach is used.

1. Identification of promising and stable genotypes for yield performance under stress conditions by using multilocation tests.
2. Identification of source materials that posses desired physiological and morphological attributes related to drought avoidance mechanism.
3. Attempting crosses to incorporate the desired attributes one by one in the high yielding stable genotypes by using such breeding approaches as pedigree method, back cross method etc.
4. Multiple cross approach with a view to assembling different desired attributes simultaneously after the source genotypes have been identified. Breeding method such as recurrent selection could be used for this purpose.

Proposed Scheme for Breeding Drought Resistant Genotypes

Step I

1. Multilocation testing under stress conditions
2. Identification of stable genotypes
3. Crosses involving stable genotypes in stress and agronomically superior cultivar
4. Selection in F_3 generation under non-stress conditions for yield performance.
5. Testing of F_4 selected lines under stress conditions as well as non-stress conditions (For seed multiplication)
6. Same as 5 using F_5 selected lines
7. Multilocation tests and release

Step II

1. Crosses using selected F_5 lines under stress and identified genetic sou..ces for desired morphological and physiological traits.
2. Same as 4-7 of step I.

Criteria or Attributes for Drought Resistance

I. Morphological

1. **Root Growth.** The extent and pattern of root development are closely related to the ability of the plant to absorb water and hence are of great significance in drought resistance. Plant water stress decreases as the extent of root system increases. Plants with deep root systems are great importance in breeding for drought avoidance. Hurd (1976) found that as width, depth and branching of root systems increases the plant water stress decreases. Root thickness has also been positively related to drought resistance. The thick root character is common among prices from upland or hill-rice habitats, which are drought tolerant. It was shown that dominant alleles controlled the length of roots in upland genotypes of rice and that this feature could be easily incorporated in breeding for drought resistance.
2. **Earliness.** Early maturing varieties are known to escape the effect of a late drought and this confers a yield advantage. In sorghum, early maturing varieties was associated with reduced water use. Evapotranspiration is reduced in early maturing genotypes in proportion of their lower leaf area and their root density. Earliness in sorghum was negatively correlated with grain yield. But for a given stress environment there may be a balance between yield loss and water use economy as associated with earliness.
3. **Reduced leaf area.** Reduction in leaf growth and premature leaf senescence all occur under drought. The resultant reduction in leaf area increases plant survival but may reduce further cop growth and yield because of reduction in photosynthetic capacity.
4. **Leaf rolling.** Leaf rolling as a most important symptoms of drought. Rolling of leaves indicates a decrease in turgor or pressure potential of the specialized tissues giving the leaf lamina its lateral hydraulic extensibility. The result is a decrease in leaf area exposed to atmosphere and incident solar radiation.
5. **Wax content.** Deposition of epicuticular wax on the leaf lamina was found to promote drought avoidance in *Sorghum* (Blum, 1975). Greater deposition of epicuticular wax caused a reduction in net radiation of the canopy, reduction in transpiration and an improvement of stomatal control over transpiration. Water use efficiency in terms of the ratio between net carbon exchange and transpiration was also improved by greater wax deposition (Chatterton et al., 1975). High cuticular resistance is often mentioned as an attribute of drought resistant species. Clarke (1987) reports studies in wheat on excised leaf water retention, which is a measure of epidermal conductance using a technique suitable for screening large numbers of lines.
6. **Colour of leaves.** Light green and golden leaves reflect more light than dark green leaves and remains cooler. Light coloured canopies have lower leaf water potentials and dark coloured canopies have a higher-leaf water potentials.
7. **Awns.** Kaul (1971) concluded that drought resistant cultivar depend on longevity of its heavily awned spike; for additional photosynthetic production. Grundbacher (1963) concluded that awns of wheat and barely were associated with high yield under moisture stress. Sunsen and Ramage (1982) reported that awns increased yield of wheat by 7% on the average and larger

increases were observed when the flag leaf was inactivated. It was found that clipping of awns reduced yield by 20.8%. Awns increase the photosynthetic area of the spike which is important factor in drought condition.

8. **Hairs.** Leaf pubescence reduces radiation load and can reduce leaf temperature or transpiration or both in crops that have no epicuticular hairs, it would worthwhile to include this trait in selection.
9. **Increased grain weight.** The per cent of grain filling on the panicle and the 100 grain weight are also useful criteria in ascertaining reactions when drought occurs during the reproductive phases. Differences in grain yield reflect genetic variation in yield ability, which are confounded with plant reactions to water deficit occurring at different growth stages.

II. Physiological

1. **Photosynthesis.** The capacity of plants to photosynthesize during and following moisture stress is an important index of drought resistance. It was concluded that decreased photosynthesis with increased water stress was primarily due to stomatal closure.
2. **Stomatal closure.** Permanent damage to the plant processes caused by severe moisture stress may be directly or indirectly attributable to stomatal behaviour. For selection to be effective the degree of stomatal conductance needs to have a reasonably high heritability. Grover (1985) suggests that the superiority of sorghum crop in arid regions depends upon the capacity of its stomata to recover from severe drought. Rapid closure of stomata during the development of drought stress may account for the maintenance of higher water potentials in some cultivars and is an important index of drought tolerance. It was also pointed out the species that close their stomata earlier when stressed must have greater heat tolerance mechanisms.
3. **Osmotic adjustment.** Osmotic adjustment occurs in response to various environmental stress. Lowering of the osmotic potential by osmotic adjustment enables the plant to remain turgor at lower water potentials osmotic adjustment may postpone tissue death. Plants that can adjust osmotically will suffer less tissue death than others that cannot adjust osmotically at greater water deficit. Osmotic adjustment results in leaf elongation.
4. **Reduced xylem vessels in root.** Resistance to water flow can be increased by either reducing the number of root axes or reducing the diameter of the xylem vessels. Seminal roots may be more important than nodal roots because they grow deeper in profile. Drought resistant varieties must have seminal roots with larger vessels.
5. **Canopy temperature.** Canopy temperature at 50 percent flowering was linerarly related to relative spiklet sterility. There was a 0.2 per cent increase in relative sterility for every degree increase in canopy temperature. So the ability of a cultivar to satisfy evapotranspiration demand and maintain low canopy temperature and high plant water status could be attributed to its rooting behaviour and important in drought condition.

III. Biochemical

1. **Proline accumulation.** Parker (1968) suggested that proline may be simply a storage compound of nitrogen or it may result from protein breakdown. He also added that proline accumulates because moisture stress creates a blockage in the metabolic pathway by which some of the other aminoacids are synthesized. Proline content appears to be an indicator of degree of water stress.

2. **Betaine accumulation.** Betaine accumulation as a criterion for drought resistance is limited, and have importance in cellular functioning rather than the whole plant system.

Breeding for salt resistance

Plant mechanisms importing resistance to salinity and other soil stresses are not properly elucidated and reliable markers for such resistance are not so far available. Inter specific genetic diversity for resistance to salts and several other edaphic stress factors exists in several crops holding promise for genetic manipulations through breeding efforts to combine superior resistance to specific edaphic stress factor (s) with better yield ability. The knowledge of genes with jamor effects controlling uptake and utilization of nutrients and also exclusion of potentially toxic ions, reported in barely, wheat, rice soyabean and tomato.

The pivotal step in breeding for resistance to edaphic stresses is a reliable and efficient screen capable of isolating the resistant genotypes. Dowely (1962) involves isolating genotypes that survive at high salinities rather than picking those which are more productive at a lower salinity. Productivity of the surviving selections is then tested over a range of salinities and compared with that of other lines, ranging from salt resistant to salt sensitives. Narrowing down genetic base too much in the beginning ultimately leads to lines having outstanding survival and biomass production under conditions of high salinity.

While screening germplasm for tolerance ratings, criteria of seedling emergence index and early vigour are used for cutting down the number of entries to manageable numbers by selections. For advancement of generation are based primarily on spike weight which is strongly correlated with yielding ability in alkali/saline soils (Chopra and Paroda, 1986).

Roseiella and Hamblin (1981) concluded that selection should be made for increased mean productivity rather than for the tolerance stress.

New Approaches for incorporating salinity resistance

Use of cell culture techniques in developing salt tolerant crop plants is gaining importance. Protoplast culture technique is being employed to generate genetic diversity for salt tolerance at the cellular level by (i) passage through cell (ii) mutagenesis (iii) fusion with protoplasts and genetic engineering. Genetic engineering aims at producing an individual having a new combination of inherited properties.

Using the recombinant DNA technique a nitrogen fixing strain of *Rhizobium,* normally highly sensitive to osmotic stress was converted into a tolerant type with the help of broad host range plasmid by transferring DNA segment from *Escherichia coli* carrying the osm gene (s) that led to proline over production and consequent osmotic tolerance (Valentine, 1984).

Current research efforts are mainly focused upon critical assessments of inter varietal genetic variability and its exploitation for improving crop salt tolerance both by conventional breeding and by invitro methods that include tissue culture technique and recombinant DNA technology.

Distant hybridization offers another promising possibility for genetic improvement of crop plants through selective transfer or addition of desired adaptive gene complexes from relevant sources e.g., superior salt tolerance of wheat grass genes for enhancing adaption of wheat cultivar in diploid *T. elongatum* ($n = 2x = 14$) and diploid *Anisantha pontica* C. Koch, ($2n = 10x = 70$) have been reported to survive at salt concentrations as high as 1.5 times sea water. Somatic hybridization technique is expected to open up new vistas to this approach. (Table 4 and 5).

Table 4. Some Promising salt-resistant selections from locally adapted indigenous rice cultivars.

Sl No.	*Locally adapted parent material*	*Salt resistant Selection*	*Region of adaption*
1.	Johna	Johna 34a	Punjab/Haryana/ Western UP
2.	Budda Molagoluculu	MCM-2	Andhra Pradesh
3.	Kala Ratta	KRI-24	Maharashtra
4.	Arya	Arya 33	Karnataka
5.	Kalladachampava	MO.2	Kerala
6.	IRON Germplasm	AU-1	Tamil Nadu
7.	Patnai	Patnai 23	West Bengal

Table 5. Some salt tolerant rice varieties evolved in India through recombination breeding.

Sl No.	*Variety development*	*Original cross*	*Institution*
1.	PUR-1 'Kalar' ottu	SR 26B/MTU-1	State Agri Dept. Tamil Nadu
2.	Co 43	Dasal/IR 20	TNAU, Coimbatore (TN)
3.	MCM-1	CO.18/Kuthg	Rice Res. Station Machillipatnam (AP)
4.	SRI002	SR26B/MTU-1	State Agri Dept. Tamil Nadu
5.	SRI-2-1	Jaya/SR1002	State Agri Dept. Tamil Nadu
6.	MR-18	SR26B/Wannar-1	Rice Res Station Mandya Karnataka
7.	Usari	Jaya/Getu	CSA University, Kanpur (UP)

Conclusion

With reference to drought combination of some attributes (biochemical, physiological and morphological) could provide better index for resistance to water stress. So, there is an apparent need to intensify interdisciplinary research efforts involving plant breeders, crop physiologists, biochemists, soil physicists and meteorologists with reference to salt tolerance, the future strategy is to develop crop varieties that may withstand moderate increases in soil salinity without undergoing a significant yield loss.

REFERENCES

Armenta-Soto, J., T.T. Change, G.C. Loresto and J.C.O'Tolle (1983). Genetic analysis of root characters in rice. SABRO Journal 15:103-116.

Blum, A. (1975). SABRO J. 7:42-45.

Chatterton, N.J., W.W. Hanna, J.B., Powell, and D.R.Lee, (1975). Can. J. Pl. Sci., 55:641-643.

Chopra, V.L. and R.S. Paroda (1986). Approaches for incorporating drought and salinity Resistance in crop plants. Oxford & IBH publishing Co. Pvt. Ltd., New Delhi.

Clarke, J.M. (1987). Use of physical and morphological traits in breeding programmes to improve drought resistance of cereals. Wiley and Sons, New York, pp. 171-189.

Dowely, D.R. (1962). Breeding crested wheat grass for salt tolerance. Crop Sci. 2:403-407.

Ground Bacher, F.J. (1963). The Physiological function of the cereal awn. Biol, Rev. Cambridge Phil. Soc. 29:366-381.

Grover. A. (1985). Response of plants to water stress physical plant 63:87-92.

Hurd, E.A. (1976). In water deficits and plant Growth, Academic Press, New York 4:317-355.

Kaul, R. (1971). Zeitschrift for planzenzuchtung 65:233-243.

Levitt, J. (1972). Responses of plants to environmental stresses. Academic Press, New York.

Parker, J. (1968). In water deficits and plant growth Eds. T.T. Kozlowski Vol. I. Academic, Press New York, pp. 185-234.

Roseiella, A.A. and J. Hamblin (1981). Theoretical aspects of selection for yield in stress and non-stress environments Crop. Sci. 21:943-946.

Saghir, A.R., A.R. Khan and W.W. Worzella (1968). The effect of plant parts on grain yield, kernel weight and plant height of wheat and barely. *Agron.* J. 60:95-97.

Sunsen, C.A. and R.T. Ramage (1962). Competition between near isogenic genotypes. Crop Sci. 2:249-250.

Sunsen, C.A. and R.T. Ramage (1982). New variety yields 1.2. t/ha International Agril Dev. 2: 29.

Valentine, R.C. (1984). Genetic Engineering, California Agri 38(10):36-37.

Yunus, M. and R.S. Paroda (1982). Impact of biparental maturity on correlation coefficients in bread wheat. Theor. Apple. Genet. 62:337-3243.

CHAPTER 6

Plant Genetic Resources

"The theories of source material and origin of cultivated plants must be fundamental for scientific plant breeding".

N.I. VAVILOV

Introduction

N.I. Vavilov has often been considered as the father of plant genetic resource activities. He proposed that crop improvement should draw from wide genetic variation. For this he collected plants and their wild relatives from most parts of the world. These provided gene pool from which cultivars could be bread which could be suitably adapted to cultivation. Vavilov revealed that genetic variation in cultivated species was concentrated in certain regions of the world which he termed 'Centres of diversity'. He identified a number of these centres which have been the principal area from which breeders have collected raw material for their work. In recent years, however, the radical change which have taken place in modern agriculture have presented a formidable threat to the diverse resources held in the Vavilonean centres.

Survey and centres of origin of food plants

Life of N.I.Vavilov

Nikolai Ivandish Vavilov was born in Russia in 1887. He studied with William Batson at the John Innes Horticultural Institution. He became professor of Moscow University in 1914. He made expeditions to Persia and surrounding countries to collect cereals in 1916. Professor of agricultural botany and genetics at Saratov in 1917. He became the President of the Lenin Academy of Agricultural Sciences and Director of the institute of applied botany (Institute of Plant Industry) in 1921. His work involved establishment and direction of more than 400 research institutes and experiment stations with a staff of 20,000 during 1921-1934. Expeditions to Afghanistan, Abyssinia, China, Central America and South America to collect economic plants including 26,000 strains of wheat during 1923-1931. He undertook direction of comprehensive study of the world collection of plants and use of these in breeding. Similar work with livestock including horses, cattle and reindeer during 1931 to 1939. Elected academician of USSR 1929. He was Soviet delegate to International Congress on the history of science held in London during 1931. Invited to be President of International Congress of Genetics during 1939 and was elected as foreign member of the Royal Society of Great Britain in 1942. Died probably in the early months of 1942 (55 years) after prosecution by Lenin since he did not join politics and put him in concentration camp and died in the severe Russian winter without even warm cloths and sickness and not given any medical treatment.

Origin and distribution of food plants

According to Vavilov, the longer a population had been established in an area, the larger will be the number of species to be found there. He concluded that the wealth of forms and varieties of rice found in the southwest Himalayas which are closely alike to many Chinese varieties, points to this region as the centre of origin of rice.

According to Vavilov, at the present time the available data seem to establish the fact that there were atleast four chief centres in which our economic plants originated, and from which they were latter dispersed all over the world.

1. South western or Central Asia
2. The Mediterranean region
3. South eastern Asia
4. The highlands of tropical America

Central Asia

Native plants included Alfalfa, apple, barley, broad bean, cherry, flax, peas, garlic, hemp, mulberry, olive, onion, pomegranate, plum, radish, spinach etc.

Mediterranean Region

This was the home of asparagus, cabbage, cauliflower, cotton, fig, millets etc.

South Eastern Asia

Banana, bread fruit, peach, orange, soyabean, rice, sugarcane and yam were native.

American Area

Cocoa, American cotton, maize, potato, tobacco and tomato.

Vavilov also recognized addition of 5th centre of origin in Abyssinia and adjacent parts of northern Africa. He also suggested the possibility of a 6th centre of origin in Philippine islands where rice and coix had originated.

The History of Food Plants

Most of the food plants were domesticated from wild ancestors long before the beginning of the historical period and all available records indicates that they were as familiar to the people of the ancient world as they are today. Many investigations have been carried on in an attempt to determine their age and place of origin as well as their cultural history.

The work of De Candolle (Alphonse De Candolle, 1806-1893)

The classic work dealing with this phase of botany is De Candolle's "L origine des plants cultivees" which appeared in 1883. So careful and painstaking was his work that few of his conclusions regarding geographic distributions have had to be altered in the light of more recent studies. De Candolle based his conclusions on a great variety of evidences: the work of Theophrastus, Dioscorides and other old historians, Chinese writing, archeological and ethnological data such as monuments of Egypt, the ruins of Pompeii, the remains of the lake Dwellers of Europe and the ruins of south America, Phylological indications (evolution of language), involving the names of plants in Hebrew, Sanskrit and other ancient languages. His botanical conclusions based on distinguishing factors like number of varieties, presence or absence of wild types, length of cultivation such as over 4000 years and 2000 years etc., He arranged the useful plants in six classes. It will be interesting to give a few examples of each of these groups (De Candolle, 1864).

A. Old world species cultivated for over 4000 years. Example - apple, banana, barley, cabbage, cucumber, date, fig, grape, mango, millet, onion, rice sorghum etc.,

B. Old world species cultivated for over 2000 years. Example - alfalfa, carrot, celery, cotton, lemon, lime, mustard, orange, plum, radish, sugarcane, yam etc.

C. Old world species cultivated probably for less than 2000 years. Example - Coffee, buck wheat, goose berry, horse radish, okra, raspberry and strawberry.

D. New world species of ancient cultivation more than 2000 years. Example Cocoa, maize, sweet potato, tobacco, kidney beans etc.,

E. New world species cultivated before the time of Columbus. Antiquity not known. Ex.: Avacado, guava, pineapple, pumpkin, tomato, vanilla etc.,

F. New world species cultivated since the time of Columbus, (18th Century) Ex. Black berry, blue berry, cinchona, dew berry, goose berry, plum, rubber and straw berry. From these examples it can readily be seen that our most valuable economic plants including the cereals, most of the vegetables and fruits, tea, cocoa and the fibre plants were discovered, utilized and cultivated by man several thousands of years ago.

The Law of Homologous Series in the Inheritance of Variation (Vavilov, 1920)

The characteristic feature of studies of the earth vegetation from very olden days to our times has been the varied conception of systemic changes. A statistics of the plant kingdom deals for the present only with Linneons. Present day botanist estimate the number of species of higher flowering plants at approximately 1,60,000. Richest in species for the families of Compositae (about 14,000), Leguminosae (more than 13,000) and gramineae (about 6,000). At the beginning of 20th Century, the Italian botanist Comes found 700 botanical varieties in beans.

The rye-Scale cereale which was considered to be relatively uniform in comparison with wheat. There have been discovered hundreds of forms distinguished by morphological and physiological characters. Varieties of corn are distinguished by hundreds of morphological and physiological characters.

In studying the details of the race constitution of world vegetation, one can know certain regularities in the diversity of varieties and races which are contained within Linnean species, inspite of the great polymorphism. An hypothesis had been made to integrate the phenomena of intraspecific polymorphism and present the underlying regularity which had been observed in studying the forms of the plant kingdom, and have been called "The Law of homologous series in heritable variation".

The concept of parallel variation

In the origin of species, in the chapter on compensation and growth, Darwin (1859) emphasized that "various species show analogous (parallel) changes in such a way that one species acquires the characters of other related species or reverse to characters of an earlier ancestor".

As the beginning of the section of analogies or parallels in variations (variations of animals and plants under domestication), Darwin writes "by this term I mean that similar characters occasionally made their appearance in the several varieties or races descended from the same species, and more rarely in the off-spring of widely distant species".

The studying of racial composition of the world vegetation in certain related Linneon species within their limits of single genera, despite the sharp expression of polymorphism of many species, one can note a number of regularities in the diversity of varieties. The first regularity, seed is similar in the morphological and physiological characters which distinguished different varieties and races.

A detailed study of species in extensive materials collected in different geographic areas, shown a parallelism in phenotypic variation in Linnean species within the limits of a given genus.

Example : Wheat

As shown by numerous investigators, the species of wheat are grouped into the following three genetic and geographic types, distinguished according to the number of chromosomes (2n = 14, 28 and 48).

Group I 42 chromosomes

Triticum vulgare
Triticum compactum
Triticum spelta
Triticum sphaerococcum
Triticum macha
Triticum vavilovi

Group II 28 chromosomes

Triticum duram
Triticum turgidum
Triticum polonicum
Triticum dicoccum
Triticum persicum
Triticum timopheevi

Group III 14 chromosomes

Triticum monococcum
Triticum aegilopoides

Description : Group I

Triticum vulgare **soft wheats, with many varieties and races divided into following groups (42 chromosomes)**

1. awn, awnless and semi-awn forms with bent awns
2. white, red grey and black headed forms
3. forms with hairy glumes with smooth glumes
4. forms with white or red grains
5. winter and spring types etc.

The species which are closely related to soft wheats are ***Triticum compactum, Triticum spelta, Triticum sphaerococcum, Triticum macha, Triticum vavilovi*** exactly duplicate the variation of soft wheats.

Group II 28 chromosomes

In general they repeat the series of characters found in the first group. Here too we find forms with white, red or black head forms in glumes that are smooth or hairy with white or red grains and winter or spring types. Here also we find forms that are awn, short awn, awnless etc.

Group III 14 chromosomes

Triticum monococcum. This species duplicates the second group in its varietal constitution. Here we only fail to find awnless forms. But we do find varieties in long and short awns, as well as forms with white, red or black heads with smooth or hairy glumes and with red or white grains.

Triticum aegilopoides (14 chromosomes) a wild form of wheat consist large number of varieties basically similar to those in the other species of cultivated wheats.

Cotton

The same parallelism can be found in a study related species of cotton. Comparing Asiatic cotton (***Gossypium herbaceum***) which is widely cultivated in Iran, Afghanistan, China with the Indian cotton ***Gossypium arboreum*** and also with American species ***Gossypium hirsutum*** (upland cotton) and ***Gossypium barbadense*** (Egyptian cotton), one can note a striking similarities in the series of variations eventhough this species are separated by Continents, as has been brought out in the investigations of G.S.Zaitsev, S.E. Harland, F.M. Mayer and others. Varieties of these species resemble one another with respect to variation in fibre colour (white, brown, green), presence or absence of hairyness, colour of the seed, type of branching of the entire plant (monopodial or sympodial), forms of leaves, colour of stem etc., This is inspite of the fact that the American and Asiatic cotton are sharply distinct with respect to number of chromosomes ($4n = 52$ and $2n = 26$) and difficulty in crossing them with each other.

Variations in related genera

Rye and wheat (*Secale cereale* and *Triticum vulgare*). In comparing the racial constitution in related genera, we can observe the same regularity and polymorphism. We will compare wheat and rye. For example there were found in rye as well as in wheat, forms with the following contrasting characters.

1. Awned, half awned and almost awnless forms
2. Forms with hairy and others with smooth head
3. Forms having red, white and black ear heads
4. Forms with solid and others with hollow stem
5. Forms with brittle and others with tough rachis
6. Winter and spring types
7. Compact and loose headed forms
8. Forms with long and short heads
9. Forms with wide and others with narrow glumes
10. Forms with blackish and others with green seedlings

In short, the genus *Secale* in detail repeats the genus *Triticum* in its species composition. It was discovered in rye, those same characters which were known in wheat and existence of this or that form on the basis of the "Law of homologous series in variation" was usually verified by investigations. The parallelism of the series of variations in rye and wheat is understandable from the evolutionary point of view, since these genera are closely related to one another genetically. The genus *Aegilops* is also closely related to *Triticum* as is shown by the possibility of obtaining hybrids between members of the two genera.

Formulation of the law of homologous series

The regularities discussed above may be presented by symbols as follows. Different Linneons and different genera show great numbers of varying forms. At the same time the exchange of this variability is similar in related Linneons and genera. For brevity we will designate the different variable characters by letters *a, b, c, d, e, f, g, h, i* .... *k* etc. Different expressions of these characters may be designated by sub-scripts as a_1, a_2, a_3, a_4 .... b_1, b_2, b_3, b_4 ...... etc. For example the colour of glumes may be designated by the letter - a, white glumes will be a_1, yellow a_2, red a_3, gray a_4 ... etc.

Radicals (Fundamental or Origin)

Linneons and genera are distinguished not only by these characters, but also by specific complexes of morphological, physiological and genetic nature. These specific differences, we may call as radicals (fundamental or origin). There can be radicals of a species, genera and whole families. Thus for three closely related Linneons of the same genus, we may have the following expression of the morphological and physiological properties—L_1 $(a + b + c + d + e + f + g + h + ... k)$, L_2 $(a + b + c + d + e + f + g + h + ... k)$, L3 $(a + b + c + d + e + g + h +.. ... k)$.

The L_1, L_2 and L_3 are radicals distinguishing these Linneons from one another, *a, b, c* .... all the varying characters, such as colour and form of the glumes, leaves, stem etc., Each of these characters can be complete in itself and may be correspondingly divided into a greater or smaller number of morphological and physiological units, a_1, a_2, a_3 etc.

In example, if we compare three Linneons of wheat, *Triticum vulgare, Triticum compactum, Triticum spelta,* we may say that the radicals of these are distinguished morphologically according to the degree of compactness of heads. *Triticum spelta* is distinguished from other species also by the tightness in which the grain is enclosed in the spikelet.

The radicals of the species, *Triticum vulgare* and *Triticum durum* are distinguished primarily by the number of chromosomes (6n = 42 and 4n = 28). The same may be done with different genera. Let us consider rye (*Secale cereale*) and soft wheat (*Triticum vulgare*). These two are comparatively close in the similarity of directions of variability. Although at first glance it appears that there is no difficulty in distinguishing rye from wheat, there are actually very few characters specific for each of these genera. In other words, there are very few morphological characters in this case it can be represented. To represent the radicals of different genera by letters G_1, G_2, G_3 etc, if we indicate the formulae for the consideration and constitution of rye and wheat, we obtain the following :

G_1 $(a + b + c + d + e + ... k)$

G_2 $(a + b + c + d + e + ... k)$

The principle differences from the morphological point of view, in this case are differences in the glumes of rye and wheat and also in the characters of the grain. Rye has 14 chromosomes, soft wheat with which we are comparing rye has 42 chromosomes.

In many plants, for example in species of wheat, oats, barley, millet, cotton and tobacco, the conception of radical must include a consideration of the chromosome number in the species. Many species of plants are sharply distinguished by the presence of specific chemical components like oils, acids and alkaloids.

In general parallelism involves organs have the same function. "The Law of homologous series" appeared in 1920. Finally not in all cases, particularly in different genera, variations produced are by identical homologous genes. Phenotypic variations can be the result of different expressions of genes.

"The Law of homologous series" discloses the great possibilities of natural variation, it only states that with regard to the entire system, thorough investigation of all the variations characteristics of various species, reveals not a dis-organized process, but a definite regularity that follows from the nature of the evolutionary process.

Importance of Genetic Diversity in Crops

Now-a-days every major crop that we grow has a very narrow genetic base, and it is this which has contributed to their genetic vulnerability.

The problem was highlighted in the United States in 1970 with the southern corn leaf blight epidemic. Massive devastation of crops came about because most of the hybrid varieties in production at that time had a common form of a specific cytoplasmic gene which conferred susceptibility to a particular race of a fungal disease. Another example is the brown spot disease of rice, which caused the great Bengal famine in 1940.

The value of local land races of long-established cultivars and the gene which they may hold, often remain unappreciated until new foreign cultivars are introduced into a region. The semi-dwarf wheat of the "Green Revolution" when first grown in Mexico was overcome by the fungal disease black stem rot and stripe rust, while the tried and tested varieties resisted attack. Hurried crosses had then to be made using genetic resource available locally as one parent and these gave rise to successful high yielding and disease resistant crop. Thus, the value of local gene pool for the development of regionally adapted cultivar cannot be underestimated. When attempts have been made in West Africa to cultivate pearlmillets previously bred in India, they have suffered so much from mildew disease that they could not compete with the less highly bred traditional varieties of local origin.

History of Plant Genetic Resources Conservation

Food and Agricultural Organization (FAO) of the United Nations Organization conducted the first international technical meeting on plant exploration and introduction in 1961. One recommendation of the meeting was to establish a panel of experts to advice and assist FAO. This was placed on a formal footing in 1965, and another panel on Forest Gene Resources was established in 1968. At the same time FAO formed its crop Ecology and Genetic Resources Unit. During this early period of activity, two further international technical conferences were held under the auspices of FAO and the international Biological Programme (IBP). The first of these, held in 1967, paved the way for subsequent concerted international activity by making certain important recommendations.

The real work began with establishment of International Board of Plant Genetic Resources (IBPGR) by Consultative Group on International Agricultural Research (CGIAR). Now IBPGR works in close association with FAO, and its primary function lies in the development and administration of the International plant genetic resources conservation network. Amongst the Board's major tasks are the promotion, collection, conservation and evaluation of plant genetic resources of species of major economic importance and their wild relatives. It has defined over 50 priority crops needing urgent action and priority regions requiring survey (Table 6). The board has aimed at the establishment of standards, methods and procedures for exploration, evaluation and for conservation of stocks for both seeds and vegetative, material, to be applied particularly in the network of replicated storage centres of gene banks.

Table 6. Some of the Priority Crops needing urgent action (From IBPGR, 1993)

Crop	Global Priority-1	Global Priority-2	High Regional Priority
Cereals	Wheat	*Sorghum	Maize
		*Finger millet	quinoa

		*Barley *Pearlmillet *Foxtail millet *Rice	
Food legumes	Phaseolus beans	*Groundnut *Soyabean *Cowpea *Yard longbean *Winged bean *Chickpea *Vigna radiata V. mungo V. aconitifolia V. umbellata	Vicia faba Lentil Lupin
Roots and tubers	Cassava Sweet-potato	Potato	Yam Tato & Aroids Minor S. American tubers
Oil Crops		Oil Plam *Coconut *Oil seed brassicas	
Fibre		Cotton	
Sugar Crops	*Sugar-cane	*Beet	

*=A priority in at least one region.

Although having a lower global priority, these crops all have a first priority in at least one region.

The Genetic Variability of Crop Plants

The pattern of genetic diversity seen in crop plants results from the interaction of five important factors: namely gene mutation, migration, recombination, selection and genetic drift. The first three of these factors all enhance variation in natural populations, whereas selection and drift tend to act to reduce variation. By mutation new characters are found in populations. Migration can increase variation in populations by adding allied genetic material, but this is hard to demonstrate in populations of wild plants. However, this phenomenon is very important for evolution under domestication.

Recombination is the mechanism by which variation is arranged in different way. Here the reproductive biology of the plants, whether cross-pollinating or inbreeding should be considered. Variation in outbreeders is greater because of heterozygosity at different loci, but in inbreeders homozygosity reaches a high level.

Selection is the process by which undesirable genes or gene combinations are eliminated from populations. A successful gene combination will reproduce more frequently. So, there is some degree of differential fertility in populations between different genotypes.

The last factor to consider is genetic drift and it covers two important concepts : firstly the random change in genetic equilibrium in plant populations, especially small populations and secondly the founder principle which is an important feature of crop evolution.

Plant Genetic Resources Exploration (Sources of Germplasm)

There are different kinds of sources from which the entire range of germplasm for any major crop can be assembled (Fig. 3). Gene pools may be indigenous, introduced or locally developed. The various resources that need to be conserved may be categorized as follows :

1. **Wild relatives.** Wild relatives of crops share common ancestors with crops but have remained in the wild as products of nature.

Fig. 3. The spectrum of genetic resources : the five categories of germplasm. These five categories indicate first an evolutionary continum linking ancient wild forms with modern cultivated varieties (cultivars) and, second, an ecological continum linking wild with domesticated crops.

2. **Land-races & Primitive cultivars.** Local crop varieties developed in primitive agricultural systems, rather than being deliberately bred, farmers selected them over many generations.
3. **Obsolete cultivars.** Obsolete cultivars, left over from the early days of plant breeding, are now mainly found in germplam collections.
4. **Advanced breeding line, Mutations & other product of plant breeding programmes.** Advanced breeding lines and stocks are plants which have been developed by breeders for use in modern scientific plant breeding. They include cultivars not yet ready for release to farmers.
5. **Modern cultivars.** The high-yielding elite cultivars have been developed by scientific plant breeding for modern intensive agriculture.

They are the raw material of the plant breeder. All except 1, the wild relatives, are termed domesticates, as they cannot survive in nature.

Genetic Resource Exploration

It is now realised that the exploration and collection of plant genetic resources must be based on the application of sound scientific principles. The objective of the botanical collector and the plant genetic resources collector are not the same. *Whereas the botanical collector may look for uniformity or trueness to type, the field sampling procedure in plant genetic resources exploration are aimed at the fullest possible recovery of genetic variation of species irrespective of the relative frequency of rarity of any genes or linked genetic complexes.* An optimum sampling strategy should enable the collector to obtain, with 95% certainty, all the alleles occurring in the population at a frequency greater than 5%. In order to achieve this, it is desirable to understand something of the population structure of a species and its eco-geographical distribution.

Sampling Strategies

The theory of sampling strategies must be founded upon an extensive knowledge of the pattern of genetic variability of population, based on the extensive geographical variation exhibited by most species the pattern of extensive variation with population lies. Ecological factors are a major determinant of genetic diversity, and many species display differentiation into ecotypes in response to their habits (*Gregor, 1933*). This is also characteristic of the agricultural environment, and agro-ecotypes are most clearly distinguished in primitive cultivar. Climatic factors such as maximum and minimum average temperature, precipitation and the seasons of dormancy and growth, light intensity and day length are all reflected in corresponding developmental characteristics (*Bennett, 1970*). Such factors generally lead to clinal variation patterns, whereas typographically or edaphically determined differentiation may lead to either clinal (as in the case of altitude) or a mosaic distribution. Consequently the distribution of variation as affected by such factors should influence the frequency of sampling.

Any collection can only be a small sample of the total variation. Geographical variation pattern include, such characteristics as disease resistance morphological features and other conspicuous differences as well as variation in quantitative characters. The relevance of this to plant breeding is obvious, for although varieties or strains may look alike, they may differ greatly in factor useful to the plant breeder, especially in differences in physiological characters. The variation within populations, especially at the local ievel, will be controlled by the interaction of the breeding system of the species and the forces by which variation is maintained, the maintenance of heterozygosity in outbreeding species is clear. Populations of inbreeding species are commonly supposed to have a simple genetic structure consisting of a number of inbred line each of which maintains itself as a constant, genetically homozygous entity for a number of generations. But as *Allard (1970)* has indicated, there is now a considerable body of evidence which shows that such population contain much genetic variation and heterozygostity. Sampling method must be used to ensure the collection of representative within-population variation as well as that associated with geographical patterns of variation.

A number of studies have been undertaken on the nature and distribution of variation in wild species population and from these it is possible to formulate recommendations for the overall strategy. *Further questions relating to number of plant to sample per site, the total number of sites to sample and the distribution of sample sites within each area*, it is also convenient to discuss for different types of crops separately.

Seed Crops

Hawkes (1980) has given precise details of the number of plants to sample at each site, and Marshall and Brown have even stated that the relatively large samples recommended by *Bennett and Allard (200-300 plants per population) represent extreme wastefulness.* They indicate that, in most circumstances, sample size should not *exceed 50 plants per population and in no circumstances it is desirable to collect more than 100 plants per population.* Although collecting in the field becomes somewhat of a compromise between resources, it is recommended that the sample size should be as large as possible. In situation where populations are not so large as in the case of cereal, it may not be possible to collect a large sample. The collection of samples which are too small carries with it the great problem, of genetic drift upon consequent immediate regeneration of the seed. *Larger samples enable at least a proportion to be put into long-term storage right away, avoiding the need for immediate regeneration.*

There are two types of Sampling: Non-Selective and selective sampling. Non-selective sampling may only be applicable to wild vegetation, as used in ecological studies, but could not be utilized in a field of primitive wheat, for instance, without causing much damage. The collector enters the field and works his or her way back and forth across the field until the whole area has been sampled. Samples of seed are collected at every few places, depending on the size of the sample to be collected, and placed together in a bag, without any prior knowledge about population structure, one can recommend that only non-selective sampling will allow the collector to capture most of the common alleles in the population, and this is the primary objective of collecting genetic resources.

In addition to non-selective sampling, selective sampling is also recommended, whereby phenotypes which are distinguishable, where in terms of a morphological character, or the manifestation of physiological character, such a drought tolerance or disease resistance are collected separately. Only a bulk sample should be collected from each population. While this is acceptable for the non-selective sample, the selective samples should be retained as separate collections. Otherwise, there is the risk of quickly losing such rare allele with hope of retrieving them at a later date for breeding purposes.

Vegetatively-Propagated Crops

Often these crops are not sexually fertile and do not produce seed; consequently vegetative propagules (rhizomes, corms, bulbs, tubers and roots) are the only plant parts available. It cannot be assumed that the population structure of a vegetatively propagated crop conforms to the same pattern as a seed crop. Each plant provides only a singly type, and as has been shown in the case of potatoes for example (*Jackson et al. 1980*), a field may contain several genotypes of which one or two comprise the majority of the plant.

Vegetatively-propagated material is bulky and difficult to transport; it is difficult to keep alive, and must be collected at the right stage of maturity or it will not grow. Because these propagates are produced underground no plant is visible at maturity, and it is easy to see what special problem this causes in the collection of wild material. It is less difficult to collect cultivated material because of the better defined population structure within a cultivated field.

Collecting Sites

Most main collection sites can be recognized viz.,

1. Farmers fields
2. Kitchen or orchard gardens
3. Market and
4. Wild habitats

Most sampling strategies have infact been developed for use in farmer's field, and this is where the majority of collection of primitive cultivar of the main food crops have been made. In areas where genetic erosion has been intense and the widespread cultivation of modern cultivars is common, such garden can be an important source of genetic diversity. Sometime a collector may arrive at a particular locality after harvest of the crops, and find it necessary to make collection from a farmer's store. This is often important for vegetatively-propagated crops. Further more it may be necessary to make collections from markets although the amount of information one obtains in market can be quite small compared to that obtained directly from farmers. Some varieties may never be sent to market. Consequently it is very important that as much as possible even the remotest regions are explored as it is never certain what wealth of diversity might be found. Indeed as genetic erosion continues, it is in these remote areas where the last remnants of diversity in a crop may still exist. Wild plants often grow in small disjunct populations and in such circumstances the collection of large samples would most likely wipe out the species in certain localities.

The distribution of collection sites must be governed by climate and ecological considerations, and where these are highly varying, then collections should be made more frequently. When planning a collecting mission to an area, it is desirable to have knowledge of the distribution of variation in the crop or crops to be collected when this is not available, and when it is possible to collect in more than one year, coarse-grid sampling is recommended. In subsequent years area where the crop have shown more variations can be re-visited. Obviously, such decisions can only be made on the basis of evaluation of germplasm.

Damain *et al.* (1983) have shown that polyacrylamide gel electrophoresis (PAGE) of cereal storage proteins, a technique whereby protein are separated according to their net electrical charge and molecular weight, can be used to determine the relative variation in cereal land races. On the basis of such information, areas where diversity is greater can be identified, and collecting plans made or modified accordingly, such procedures are lab based, require plant material and are quite rapid.

Field Documentation

This information will not consist solely of evaluation characteristic, but also of details about where the collection was made, on what date, and what identification numbers and letters were given to it by the collector. It is suggested that the identifying details of any germplasm, collection will be of immense importance if this collection is moved around the international community during its utilization, in order to avoid confusion and duplication of this germplasm and its use.

Other information which needs to be noted at the time of collection may serve to identify the material taxonomically, although if not done with some degree of accuracy should be left until evaluation is carried out at a later stage in the proceedings. Further information which can only be recorded at the time of collection will relate to the source of the germplasm, *i.e.*, wild habitat, farmer's field, farm store, market etc. Similarly, it may be desirable to make note about the local habitat conditions such as soil type and rainfall for instance.

The amount of information which can be recorded during collecting activities is very much dependent upon the time available. There is a minimum of information which must be recorded regardless of time.

Various designs of collecting form sheet or books have been used in different collecting missions (Table 7). Some of which may be fairly specific for certain crop species. It may even be possible to design a form which can be used at a later date for ease of direct input of data into a database.

The development of efficient exploration programme for a particular species depends, according to *Marshall and Brown* on two levels of objective decisions:

1. where significant gene pools of the species occur and where are they most threatened by extinction and
2. the area covered by previous exploration missions and their relative effectiveness as measured by the material held in existing collections.

Table 7. Sites where wild rice germplasm is conserved insitu (From IRRI, 1988)

Country	*Conserved area*	*Species*
India	Parambikulam Game	O officinalis
	Reserve, Kerala	O granulata
	Karulai Range Tea	O. nivara
	Plantation and Forest	O. granulata
	Reserve, Kerala	O. officinalis
		(Tetraploid Form)
Thailand	Pukai Botanic Garden	
	Mae Sai Valley Forest	O. granulata
	Reserve and Botanic	
	Garden near Chian Mai	
	Tong Noga chiang	O. ridleyi
	near Haadyai	
Sri Lanka	Yala National Park	O. nivara
	Ruhuna Wildlife	O. eichingeri
		Hygroryza aristata
Indonesia	Yung kulon National Park	O. officinalis

Conservation

The strategy of conservation depend on the nature of the material and on the objective and scope of the activity. The nature of the material is defined by the length of the life cycle, the mode of reproduction, the size of the individuals and its ecological status, whether wild, weedy or domesticated (*Frankel, 1970*). The purpose to which the material is put will determine the degree of integrity which is essential or desirable to maintain. Conservation strategy must take into consideration the time dimension; whether it be for short-medium-or long-term storage and where the storage will be located.

The genetic make-up of the material and the type of sample collected will reflect the breeding system of the material and how it will be regenerated.

There are two basic approaches to germplasm conservation namely, *in situ and ex situ* conservation.

In Situ Conservation

This is an approach which is applied mainly to wild species related to crop plant, to forest and pasture species. It is often recommended that these species should be preserved, maintaining the genetic integrity of their nature state, as communities in stable environments. Most nature conservation programmes are aimed at the level of ecosystems or multispecies communities. For instance in rice, IRRI has identified certain sites where wild species are conserved in situ (Table 8). Genetic conservation goes further in recognizing the need for a wide genetic base, and as with nature conservation, aspect of population biology such as adaptive radiation, maintenance of variation in population and the long-term stability of population number are important features.

There are areas in Israel, for instance, where diversity in wild wheats, barley, oats and *Aegilops* is great, where they grow among rocks and on poor soils and what the rough terrain create a natural sanctuary. These communities are now protected against heavy grazing pressure from sheep and goats and can continue to evolve.

Table 8. Institutes, which accepted responsibility for long-term conservation. (From IBPGR, 1989)

Crop Species Covered				*Scope of Collection Global Region*	*Institute*
CEREAL	Barley		o		PGR, Ottawa, Canada, NGB.
				European	Lund, Sweden, PGRC/E,
				African	Ethiopia NIAR, Japan,
				Asian	ICARDA, Syria
	Maize			New world	NPGS, USA NIAR Japan
				Asian	TISTR, Thailand, VIR,
				Asian	Leningard, USSR,
				European	Portuguese Gene bank,
				Mediterranean	Portugal.
	Millets	Pennisetum			NPGS, USA PGR, Ottawa,
			o		Canada, ICRISAT, Inida.
		Eleusine	o		PGRC/E. Ethipola
					ICRISAT, India
		Minor Indian Millets		Indian	NBPGR, New Delhi India
		Panicum miliaceum	o		ICRISAT, INDIA
		Setaria italica	o		ICRISAT, INDIA
	Oats	Avena	o		PGR, Ottawa, Canada
			o		NGB, Sweden
			o		FAL, FRG

Rice	Oryza	sativa-	o		IRRI, Phillipines
		indica	o		IRRI, Phillipines
		javanica	o		NAIR, Japan
		japonica		African	IITA, Nigeria
			o		NPGS, USA
	Wild Species		o		NPGS, Nigeria
Rye			o		Polish genebank,
			o		Radzikow
			o		NGB, Sweden.
Sorghum			o		NPGS, USA.
			o		ICRISAT, INDIA
Wheat Cultivated species			o		VIR, Leningrad,
			o		USSR, CNR, Italy
			o		NPGS, USA.
				Asian	CAAS, Beijing
			o		ICARDA, Syria
	Wild Species		o		Plant germplasm
	Triticum and		o		Institute Japan,
	Aegilops		o		ICARDA, Syria.
Food	Chickpea		o		ICRISAT, India
Legumes			o		ICARDS, Syria
Cicer Spp			o		ICARDA, Syria
Faba Bean			o		CNR, Italy
			o		ICARDA, Syria
Groundnut			o		ICRISAT, India
				South America	INTA, Argentina
	Phaseolus wild spp		o		JBNB, Belgium
	Cultivated spp		o		CIAT, Colombia
			o		NPGS, Sweden
			o	European	FAL, FRG
	Pigeonpea		o		ICRISAT, India
			o		NBPGR, India
	Soyabean		o		NIAR, Japan
			o		NPGS, USA
		Wildperennial	o		CSIRO, Australia
	Vigna wild spp		o		JBNB, Belgium
		V. mungo	o		NBPGR, India
		V. radiata	o		IPB, Phillipines
		V. umbellata	o		NBPGR, India
		V. unguiculata	o		IITA, Nigeria
			o		NPGS, USA
Root					

Crops	Cassava (seed)	o		CIAT, Colombia
	Solanum Spp.	o		CIP, Peru
	Sweet Potato (seed)	o		NPGS, USA
		Asian		AVRDC, China
		o		NIAR, Japan
Industrial Crops	Beet	o		FAL, FRG
			Mediterranean	GREEK Gene bank, Thessloniki
	Cotton		Mediterranean	GREEK Gene bank Thessalonki.
	Sugarcane (seed)	o		NIAR, Japan
				NPGS, USA
	Jute and kenaf	o		BJRI, Bengladesh
Others	Sesame	o		KARI, Kenya
		o		RDA, Korea
	Sunflower		European Mediterranean	Research Institute Plant Production, Czechoslovakia

Field genebanks (active collections for vegetative material) which have accepted responsibility for conservation.

Crop	Species Covered	Geographical representation Global	Regional	Institute
Roots & tubers	Cassava	o		CIAT, Colombia
			Central American	INTA, Mexico
		o	African	IITA, Nigeria
	Sweet Potato	o	Asian and Pacific	AVRDC, China
Crops Industrial	Cocoa	o		IITA, Nigeria
		o		Univ. of West Indies, Trinidad & Tobago.
		o		CATIE, Costa Rica
	Sugarcane	o		SBI, Coimbatore, India
		o		USDA, Florida, USA.

The situation of the primitive cultivars or landraces is very different from that of wild species. These categories evolved in association with humans in the agricultural environment, or agro ecosystem. Such systems are continually changing and landraces are being rapidly replaced by modern cultivars. This process of genetic erosion has been underway for decades and for some crops such as wheat and barley, many of the ancient landraces have vanished. It has been suggested by some workers that small area should be preserved in which landraces are cultivated according to traditional agricultural methods, so that the dynamic evolutionary processes can continue within the crop. However, the adoption of such a policy would condemn certain section of communities to social and economic stagnation in a fast changing world, and as such is not often politically acceptable. The application of in situ conservation of landraces is not, therefore, a feasible option, and because of the rapid loss of genetic variation through cultivar replacement, *ex situ* conservation in one form or another is the most practical and safe approach for such material.

Ex Situ Conservation

In the broad sense this form of conservation can include the use of botanic gardens and arboreta on the one hand and genebank on the other. Botanic gardens and arboreta represent the oldest forms of conservation in Europe date back to the 15th and 16th centuries. Little material has survived since then however apart from a few trees, so that most are in fact of later origin.

Guldage (1975) has considered four alternative objectives for *ex situ* conservation:

1. To establish and maintain conservation stands characterised as far as possible by the same genotpye frequencies as the original populations (provenance)-static conservation of genotypes).
2. To establish and maintain conservation stands characterised as far as possible by the same gene frequencies as the original population, thus avoiding total loss of any allele static conservation of gene pools.
3. To establish conservation stands in which gene frequencies are allowed to change freely according to natural selective forces evolutionary conservation.
4. To establish conservation stands in which gene frequencies are deliberately changed by man in order to conserve characteristics important for plantation economy in a region and at the same time eliminate undesirable characteristic-selective conservation.

Guldager distinguishes between conserving genotype frequency and gene frequency, the former only being truly achievable in vegetatively-propagated material. The aim is to prevent the loss of genetic information. Static conservation mainly takes the form of the storage of seeds or vegetative material in genebanks. Conservation carried out in this way by reducing the life processes to a low level is both the safest and the cheapest method.

Role of Gene Banks

FAO and IBPGR have defined two types of conservation centres: long term base collections and medium-term active collections. Base collections will be held in genebank, solely for 'posterity' and will not be drawn upon except for viability testing and subsequent regeneration or if seed is urgently required of any accession which cannot be acquired from any other source. In medium-term active or working collections, samples can easily be sent to scientists working in other centres or issued to breeder in general.

IBPGR has been in the forefront of genetic resource work, particularly through the establishment of a 'global network' of genetic resource centres. This network includes national, regional and international institution working to preserve the world dwindling resources. IBPGR has reached agreements with a number of national genebanks, IARC's regional and other institutions for the conservation of plant genetic resources.(*IBPGR, 1989*) (Table 9).

Table 9. Example of General Collecting form (From Hawkers, 1983)

Expedition Organization: Collector's Number

Country:

Team/Collector(s)

Date of Collection:

Species name:

Vernacular/Cultivar name:

Locality:

Latitude: Longitude:

Material: Seeds Inflorescences Roots/tubers Live Plants Herbaruium

Sample: Population Pureline Individual Random Non-random

Status : Cultivated Weed Wild

Source: Field Farm Store Market Shop Garden Wild Vegetation

Original source of samples:

Frequency: Abundant Frequent Occasional Rare Habitat:

Descriptive notes:

Uses:

Cultural Practices: Irrigated Dry

Season:

Soil Observations: Texture, Stoniness, Depth, Drainage, Colour etc.

Soil pH:

Land Form: Aspect:

Slope:

Topography	Swamp	Flood Plain	Level	Undulating
	Hilly	Hilly Dissected	Steeply	
		Mountainous	Dissected	
			Other (specify)	

Plant Community:

Other Crops grown near or in rotation:

Pests/Pathogens:

Name & Address of Farmer:

Taxonomic Identification: Date:

By:

Name of Institution Accession No.

Once a genebank has been established, and provided that it can satisfactorily fulfil the requirement of medium and long-term storage of seeds or other organs, then its value will depend on the strategy that researchers follow with respect to collection, evaluation and provision of material for utilization. Within each of these three activities, documentation must play an important part as the ultimate effective use of germplasm is totally dependent upon the availability and quality of information associated with it.

Storage of Seeds

In term of static conservation, the most convenient way for maintaining plant germplasm would be by storing seeds. The main exceptions to this are plants which are normally vegetatively propaged and do not produce viable seeds (*e.g.*) banana and crops where the seed produced are very short-lived (*e.g.*) cocoa. The cocoa are often referred to as being recalcitrant because they will not stand drying below some relatively high moisture content without very serious loss of viability (*Roberts, 1973*).

Fortunately the majority of crop plants do produce seeds and these show 'orthodox' behaviour, being tolerant of a disease in moisture content coupled with temperature, so allowing storage for relatively long period. The storage potential of orthodox seeds is influenced by inherent as well as external factors. Genetic differences may exist at the level of genus or species, but even cultivars may vary in their storage characteristics.

Leguminosae seeds are renowned for their longevity by virtue of their hard and impermeable seed coats. (*e.g.*) *Harrington (1972)* records *Cassia* seeds as surviving for 158 years. *Trifolium* for 100 years. *Lotus and Lupinus* seeds thought to have survived for over 500 years. But there are exceptions too: the peanut is notoriously short-lived. Short-lived species of other families includes onion, parshnip and lettuce.

Persistence of fruit or floral structures is another factor affecting storability of seeds. Seeds of many cereals or grasses store better if kept with their 'hulls' or glumes intact rather than if threshed, and this may be due to an inhibitory effect upon fungal growth. Threshing may also have an adverse effect on viability because of possible mechanical damage during processing. *Roberts (1972)* suggested that smaller seeds suffer less in this respect than do larger ones. Seeds which are round are also less likely to be damaged before storage, than ones with more irregular shapes.

Strong correlations between weather condition during ripening and harvesting, and loss of viability have been found in several crops such as barley, wheat and oats. While it is important to appreciate the effects of such environment factors, germplasm preservation has to go on regardless of the weather.

Problem with Recalcitrant Seeds

Many species of fruit and some of the large-seeded species have been described as recalcitrant. Since, as the term implies, they do not obey the 'rules' which can be applied to orthodox seeds. Even under moist storage conditions they are relatively short-lived and last no more than a few weeks or a few months, depending on the species. It is quite clear that recalcitrant seeds have a physiologically distinct behaviour from orthodox seeds and their delimitation rests mainly on their inability to withstand desiccation and their short-lived characteristics even when fully imbibed.

From the genetic resources point of view, recalcitrant crops of interest include such economically important species as cocoa, rubber, tea, most tropical fruits and many timber species. The identification of recalcitrant types is, however, not always easy: *e.g.*, citrus species were once considered recalcitrant, but are now known to be orthodox.

International Board for Plant Genetic Resources (IBPGR), 1976; Munfd and Grout, 1979

Problems of maintaining viability of recalcitrant seeds are not just related to storage for conservation. Often such seeds are so short-lived that major problems arise even during actual collections. The little progress which has been made in the storage of truly recalcitrant seeds towards keeping the seeds viable long enough for them to survive the period of collection. In most cases the best recorded methods involve storage in a fully imbibed state, in a moist medium in aerobic condition (a loosely-sealed polythene bag) where the seed has received a preliminary treatment with hot water or a fungicide to inhibit microbial growth under such conditions, however, germination has often to be suppressed, either by lowering the temperature of storage very slightly, or by osmotic inhibition.

Seed Regeneration

Apart from the need to regenerate seed to maintain viability in collections, the other purpose of regeneration is to provide an appropriate amount of seed for long-term conservation as well as to restore the seed in active collection used for evaluation and utilization. The aim should be, as far as possible, to maintain as many of the original gene as possible within the sample.

Problem of Vegetatively-Propagated Crops

Although potato, casava, yams, sweet potato, sugar-cane are sexually fertile, it is often not convenient to propagate them commercially from seed because of high levels of heterozygosity, and breeders and horticulturists commonly require uniform clones. Many vegetatively propagated crops are, however, sexually sterile, or at the very least have reduced fertility which precludes the possibility of seed storage. The vegetative organs which are stored, namely tubers, rhizomes, corms and cuttings etc. are relatively short-lived and often deteriorate rapidly after harvest, unless ideal storage conditions are provided. The cultivation of collection vegetatively year after year is costly and bring with it dangers of diseases infection and many other hazards. New methods of long-term storage of vegetative organs are urgently required. Storage of vegetative organs of tropical versus temperate crops is quite different.

In Vitro Storage of Germplasm

Much interest has been focussed in recent years on the application of tissue culture or in *vitro* technique to plant genetic conservation because of the special problems presented by vegetatively-propagate and recalcitrant species. The basic aim of such methods is to introduce explants from these plants into sterile culture and maintain them in a pathogen-free environment, making them available for further use. In addition, the introduction of such culture to ultra-low temperature then open the possibility of storing germplasm indefinitely, in "suspended animation" (*Henshaw, 1984*). Many methods have been developed for the propagation of different species, including callus, meristem and anther culture, as well as the maintenance of isolated protoplasts, but there are considerable deficiencies, nevertheless, in our understanding of these system which have serious implications for genetic conservation. There are two major aspects when considering tissue culture for plant genetic conservation namely; (i) the techniques to be used which can be either normal and minimal growth in culture or freeze preservation, often referred to as 'cryopreservation and (ii) the problems related to genetic stability.

Cryopreservation

The potential for the long-term storage of plant germplasm using freeze preservation is vast, although the application of cryopreservation technique is still in its infancy and reliable, routinely-usable procedures for organised vegetative tissue are not yet available for any major crops (*Keefe and Hensaw, 1984*).

During the freeze preservation of plant tissue there is a complete cessation of cell division; it is important that damage by the crystal formation within the cells is prevented altogether or at the very least, minimized as the temperature of the culture is reduced to that of liquid nitrogen (– 190 °C). Two methods have been developed to achieve this namely the ultra-rapid freezing and the slow stepwise freezing of culture. In the former, ice crystal form within the cells but are extremely small and therefore cause no disruption cell must be rapidly thawed to prevent recrystalisation: this method has been successfully applied to potato shoot-tips. The slow freezing method depend for its success on extracellular freezing for protection, and this is the method which has been most widely and successfully applied. In both systems, however, the tissue is often treated with a cryoprotectant such as DMSO. (dimethyl sulphoxide) or glycerol which reduce damage in a number of ways. Cryoprotectant have the ability to reduce the size and growth rate of ice crystal, and lower the freezing point of intercellular content and enable cell to be subjected to very low temperatures without disruption of the cell membrane or contents. Recent research by Keefe and Henshaw (1984) has shown that artificial nucleation of the cryoprotectant medium (DMSO) at a temperature close to it freezing point has a beneficial effect, and explant of Solanun tuberosum suffered less freezing injury when this occurred.

Genetic Stability

The ultimate aim of any of the tissue culture system in plant genetic conservation is the long-term preservation of genetic system which are prone to genetic change are therefore, desirable. The maintenance of a consistent "germ- line' through several generations require an ordered process of chromosome replication in the cells of the apical and other meristem regions. Apical meristems are more favoured for storage and the propagation of species through tissue culture. Cell culture are prone to genetic stability is likely to be maintained however where continued organised growth of the meristems in culture follows culture initiation.

Parent plants in which meiotic and other cytological abnormalities are common will often be more susceptible to genetic changes in culture. Any tissue culture system must also guarantee the retrieval of material at the end of the storage period.

When deciding upon a conservation strategy employing tissue culture system, the relative advantage of each must be evaluated (Fig. 4). The genetic stability and potential for both quantitative losses and qualitative changes are all factors which will determine which tissue to utilize and the most appropriate storage conditions. Undoubtedly freeze preservation offer the best hope for long-term storage. Under such condition there is only one source of genetic variation, namely mutations induced by background radiation may accumulate *(Withers, 1980)*.

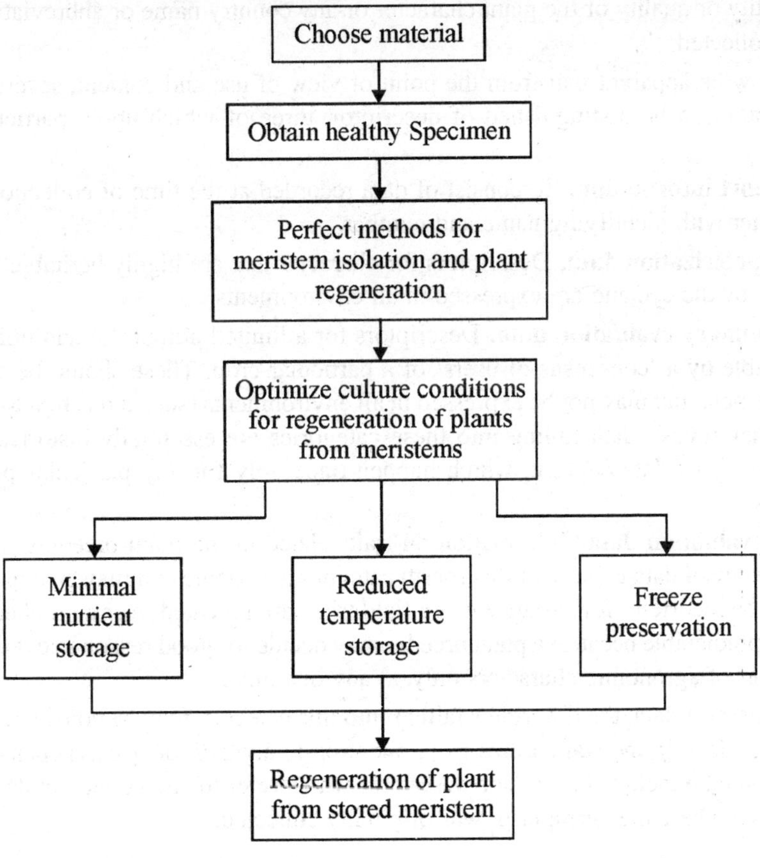

Fig. 4. Flow diagram of the strategy for conservation of any plant (Wilkins et al. 1982).

Management of Genetic Resources Data

The conservation of rapidly disappearing genetic stocks for possible future use and their subsequent utilization in plant breeding are two highly important spheres of activity in any genetic resources strategy. But the success of these is dependent upon the availability of descriptive information which will enable plant breeders to make decision regarding the material to be introduced into breeding programme. Adequate information systems are also essential for the smooth operation of the global network of genebank, while germplasm collections are more useful if reliable descriptions of their content are easily accessible to breeder. One major factor affecting genetic resources information handling and exchange is the sheer volume of existing germplasm collection and their associated description. For instance, International Rice Research Institute (IRRI) hold over 82,268 samples of rice (*IRRI 1989*). The morpho-agronomic data alone run to over 38 traits recorded, giving something in excess of 3 million pieces of information to manage.

1. The form of descriptive information

The information that we will need to handle will serve to describe these plants in terms of form, function and origin. It is these descriptions which are stored and processed in the computer. *'Descriptor'* is now widely accepted as the computer term for the 'character' of a plant as well as for other unit of information such as the country of origin of the plant, or the date of its collection. The descriptor date is then the quantity or quality of the plant character or any country name or abbreviation, or the actual month or year collected.

It may now be apparent that from the point of view of use and content, several categories and types of descriptor can be distinguished of descriptor, three of which apply particularly for use in genebanks.

1. **Passport information.** To consist of data recorded at the time of collection of germplasm, together with identifying name and number.
2. **Characterization data.** Descriptors for character that are highly heritable, that can be seen easily by the eye and are expressed in all environments.
3. **Preliminary evaluation data.** Descriptors for a limited number of additional traits thought desirable by a 'consensus of users' of a particular crop. These should be capable of visual assessment, but may not be expressed in all environments (such a reaction to disease, drought or other stress), data falling into these categories are essentially historical, this relates to frequency of data capture, which happen once only for any particular plant or group of plants.
4. **Full evaluation data.** Information on trait related in the main breeding programmes. The frequency of data capture of this fourth category is variable and may be unpredictable. It will often be recurrent; for instance when field trials are repeated over a number of years. It will be unpredictable because a plant breeder may decide for good reason to evaluate for a limited number of agronomic characters only, at any one time.

For the passport data, the descriptor falling into this category may be considered the easiest and most desirable to identify and standardize from one crop to another, or one genebank to another with this category should be included accession descriptor which refer to and collection descriptor referring to show, when and where the germplasm was originally collected.

2. Genetic Value of Descriptors

The other three categories of descriptor will be used for information regarding specific traits and will be more or less understood. The genetic aspect of any plant character which is of importance to plant breeding. Hence, clearest possible definition of character is needed and to define character as the smallest possible heritable unit.

For many crops and many characters, the results of genetic analysis are just not available. Providing characters are distinct and recognizable they can be represented by clear and distinct descriptor states, expressed for example as a numerical scale of integers. The aim should be to make descriptor more 'genetically accountable', where the descriptor states will be the alleles of a known gene. Based on the wealth of genetic knowledge for the genus *Pisum, Blixt and Williams (1982)* prepared pea descriptors that can be managed at different botanic and genetic levels designated as polygenic, oligogenic and monogenic, one practical difference in descriptor, at these levels lies in their number of descriptor states (*Blixt, 1982*).

3. Models for Data Management

Two basic types of database management system can be defined, namely hierarchical and relational. A hierarchical system tends to be extremely complex because of the 'relationships' or 'connections' which are set up between the data elements. A relation database management system is much simpler and probably more appropriate. Data are represented as they are, and the 'relation' between data elements can be considered in the form of two dimensional table. Each row across the table is called a record (referring to an accession, seed lot plant etc.) and each column is called a field of the record (containing individual descriptors). Information contained in any two or more separate tables can then be related or combined if there are common fields in those tables.

4. Data Preparation

The sources from which data are to be exerted, be the field books, laboratory notebooks, collection forms are evaluation trials sheets, need therefore to be documented. All descriptors of interest need to be listed thoroughly and to be defined, so that further recording of measurement will be standardized, the procedures for data preparation need careful considerations, as future universal use of an existing system for data management will depend upon them. The steps involved are:

1. Formulate descriptor list
2. Define descriptors
3. Revise or check descriptor list
4. Define physical representation of data, *i.e.*, descriptor states,
5. Produce code dictionary
6. Design data table and/or at final data record sheet as permanent record and for data entry into computer.

5. Utilization

The exploitation of genetic diversity for crop improvement should be the ultimate objective of genetic resources exploration and conservation. While it is important to ensure that plant genetic resources are adequately safeguarded for future generations, they cannot just be placed in a genebank and forgotten about. The vital stages of evaluation and incorporation of valuable characters such as disease resistance or tolerance of environmental stress factors into new varieties, are the justification of genetic resources activities.

The immediate value of genetic resources depends to a considerable extent upon the ease with which the breeder can utilize them. For instance, the primitive cultivars of crops are more closely genetically allied to modern varieties belonging to the same biological species than are the wild species. As a consequence the transfer of genetic characteristic is generally much easier from the primitive cultivars. Most plant breeders prefer to cross plants that belong to the same biological species, and which generally produce fertile hybrids and present little or no hindrance to genetic recombination. Zohary et al., (1969) have pointed out the importance of the diploid wheats which have remained unexploited by breeder. As Harlan (1976) has stressed the need for genetic variation and sources of resistance in the future will lead to the exploitation of all available sources of diversity.

Although originally devised as a method of classifying cultivated plants, the gene pool concept of *Harlan and de Wet (1971)* provides a useful scheme for the presentation of data from pre-breeding studies. Species which belong to the same gene pool as the cultivate species can be utilized relatively easily in breeding but the use of species from other gene pools presents some difficulties, or hybrids are sterile or lethal. Manipulation of species in wide crosses through bridging species which are cross-compatible and one way in which genes may be transferred from one species to another. There are several pre-fertilization and post-fertilization isolation mechanisms which have been recognised between plant species. These are described in detail by Stace (1980), and are applicable to cultivated species and their wild relatives.

Future Prospects

While it is clear that IBPGR has played a pivotal role in recent years in the co-ordination of germplasm activities throughout the world, the success of such programmes will be achieved only through international collaboration. Without international collaboration it will not be possible to undertake many important germplasm collecting missions, and certainly such activities generally come down the list of many developing country priorities.

The International Board for Plant Genetic Resources (IBPGR) is an autonomous international scientific organization under the aegis of the Consultative Group on International Agricultural Research (CGIAR), was established in 1974. The basic function of IBPGR is to promote and co-ordinate an international network of genetic resources centre to further the collecting, conservation, documentation, evaluation and use of plant germplasm and thereby contribute to raising the standard of living and welfare of people through the world.

It is a certainty that if plant genetic resources are not, and have been gathered and preserved them their continued existence is unlikely. If genetic erosion is not counteracted by adequate measures of conservation, then the future of plant genetic resources cannot be guaranteed. The effectiveness of the world plant breeding process and consequently the condition of the world human population becomes a matter of much concern. Assuming that measures for adequately conserving plant genes continue on a broad scale and that useful genes or gene combinations can be identified and located, 'genetic engineering' and biotechnology are the new techniques likely to be available for their full exploitation in the future. Plant genetic resources must be used by both plant breeders and genetic engineers for the benefit of mankind.

REFERENCES

Allard, R.W. (1970). Population Structure and sampling methods. In Genetic Resources in plants-their Exploitation and Conservation. Frankel, O.H. and Bennett, E. (eds). IBH Handbook No. 11. Blackwell Scientific Publications, Oxford and Edinburgh.

Anonymous, (1990). Asian Vegetable Research and Development Centre Progress Report AVRDC, Taiwan, China.

Bennet, E. (1970). Tactics of plant exploitation. In Genetic Resources in plants–Their Exploration and conservation. Frankle, O.H. and Bennet, E. (Eds.). IBH Handbook No. 11. Blackwell Scientific Publications, Oxford and Edinburgh.

Blixit, S. (1984). Application of computer to genebanks and breeding programmes. In crop Breeeding: A contemporary Basis, Vose, P.B. and Blixit, S.G. (Eds.), Pergamon Press, Oxford.

Blixit, S. and J.T. Williams, (Eds.). (1982). Documentation of genetic resources: a model. IBPGR secretariat, Rome.

Brian Ford-Iloyd and Michael Jackson (1986). Plant Genetic Resources: an introduction to their conservation and use. Edward Arward Arnold Publishers, London, pp. 146.

Damain, A.B., E. Porceddu, and M.T. Jackson (1983). A rapid method for the evaluation of variation in germplasm collections of cereals using polyacrylamide gel electrophoresis. Euphytica, 32;877-83.

Darwin, Charles, (1859). The Origin of species.

de Candolle, A. (1864). Origin of cultivated plants. Hafner Publishing Co., New York and London.

Elgin H.J., and M.S. Alexanian. (1991). Plant germplasm accession, maintenance and data management at the Vavilov Institute of Plant Industry (VIR), Leningard, UR, In Plant Breeding Abtracts. 1991.61:4;397-401.

Frankel, O.K. (1970). Genetic conservation in perspective. In Genetic Resources in plants: Their Exploration and conservation, Frankel, O.H. and Bennett, E. (Eds) IBH Hndbook No. 11. Blackwell Scientific Publications, Oxford and Edinburgh.

Frankel, O.H. and J.G. Hawkes (1975). Crop genetic resources for today and tomorrow. Cambridge University Press, Cambridge.

Gregor, J.W. (1983). The ecotype concept in relation to the registration of crop plants. Annals of Applied Biology, 20:205-219.

Guldager, P. (1975). Ex itu conservation stands in the tropics. In Methodology of conservation of Forest Genetic Resources, Roche, L. (Ed.) FAO, Rome.

Guzhov, Yu (1989). Genetics and Plant Breeding for Agriculture. Mir Publishers, Moscow.

Harlan, J.R. (1976). Genetic Resources in wild relatives of crops Crop Science, 16:329-333.

Harlan, J.R. and J.M.J. De Wet. (1971). Toward a rational classification of cultivated plants, Taxon, 20:509-517.

Harrington, J.F. (1972). Seed storage and Longevity. In Seed Biology, Kozlowski, T.T. (Ed.) Academic Press, New York and London.

Hawkes, J.G. (1980). Crop Genetic Resources Field Collection Manual, IBPGR/EUCARPIA, 37 pp.

Hawkes, J.G. (1983). The Diversity of crop plants. Harvard University Press, Cambridge, Massachusetts and London.

Henshaw, G.G. (1984). Tissue culture for disease elimination and germplasm storage. In *Crop Breeding A contemporary Basis,* Vose, P.B. and Blixit, S.G. (Eds.), Pergamon Press, Oxford.

International Board for plant Genetic Resources. (1976), Report of *IBPGR*

Working group on Engineering Design and Cost Aspects of Long. Term seed storage facilities. IBPGR, ROME.

International Board for Plant Genetic Resources (1983) *Annual Report.* IBPGR, Rome.

International Board for plant Genetic Resources (1984).

Seed management technique for genebanks in proceedings of a workshop held at the Royal Botanic Garden, Kew, 6-9, July, 1982. IBPGR.

International Board for Plant Genetic Resources (1989). *Annual Report,* IBPGR, Rome.

International Rice Research Institute, (1988).

Annual Report IRRI. Phillipines

International Rice Research Institute (1989).

Program Report IRRI, Philippines.

International centre for the improvement of Maize and Wheat (CIMMYT) (1989), *Annual Report,* CIMMYT, Mexico.

International Crop Research Institute for Semi-arid tropics (1989).

Annual Report, ICRISAT, Hyderabad, India.

International Institute of Tropical Agriculture (1990), *Annual Report,* IITA, Nigeria.

Jackson, M.T., J.G. Hawkes, P.R. Rowe (1980). An ethano-botanical field study of primitive potato varieties in Peru, *Euphytica,* 29:107-113.

Keefe, P.D. and G.G. Henshaw (1984). A note on the multiple role of artificial nucleation of the suspending medium during two-step cryopreservation procedures. *Cryo-letter, 5,* 71-78.

King, M.W. and E.H. Roberts (1980). Maintenance of recalcitrant seeds in storage. In *Recalcitrant crop seeds,* chin, H.F. and E.H. Roberts (Eds.). Tropical Press, Kualalumpur.

Mumford, P.M. and B. Grout (1979). Desiccation and low temperature (– 196 °C) tolerance of *citrus limon seed. Seed science and Technology*, 7: 407-410.

National Bureau of plant genetic Resources *(1989/90) Annual Report*, NBPGR, New Delhi.

Roberts, E.H. (1972). Storage environment and the control of viability. In Seed Viability, Roberts, E.H. (ed.). Chapman and Hall, London.

Stace, C.A. (1980). Plant Taxonomy and Biosystematics. Edward Arnold, London.

Vavilov, N.I. (1920). The origin, variation, immunity and breeding of cultivated plants. Vol. 13, Number 1/6. The Chronica Botanica Co., Waltham Mass, U.S.A.

Wilkins, C.P., T.Bengochea, and J.H. Dodids. (1982). The use of *in Vitro* methods for plant genetic conservation. *Outlook on Agriculture.* 11:67-72.

Withers, L.A. (1980). Tissue culture storage for genetic conservation. Technical Report IBPGR, Rome.

Withers, L.A. (1986). Cryopreservation and Genebank. In, Plant cell culture technology-Botanical monographs vol. 23. Yeoman, M.M. (Ed). Blackwell Scientific Publications, Oxford and Edinburgh.

Zohary, D., J.R. Harlan and A. Vardi (1969). The wild diploid progenitor of wheat and their breeding value *Euphytica,* 18:58-65.

Appendix-I
Important Germplasm Collections Throughout the World

Crops	*Species*	*Collections*	*Institutes/Organisations*	*Reference*
CEREALS	Barley	25, 131	VIR, USSR	Elgin and Alexanian, 1991
		2,337	U.S. Small Grains Collections, Beltsville	Ford-lioqd and Jackson 1986
	Maize	19,309	VIR USSR	Elgin and Alexanian, 1991.
	Pearl Millet	21,231	ICRISAT, India	ICRISAT, 1989
	Rye	3,001	VIR, USSR	Elgin and Alexanian, 1991
	Oat	12,950	VIR, USSR	Elgin and Alexanian 1991
	Millets & relatives	14,967	VIR, USSR	Elgin and Alexanian, 1991

	Minor	6,918	ICRISAT, India	ICRISAT, 1989
	RICE	82,268	IRRI, Philipines	IRRI, 1989
		12,500	IITA, Nigeria	IITA, 1989
		6,377	VIR, USSR	Elgin and Alexanian, 1991
	Sorghum	31,817	ICRISAT, India	ICRISAT, 1989
		12,560	VIR, USSR	Elgin & Alexanian, 1991
	Wheat	61,783	CIMMYT, Mexico	CIMMYT, 1988
		69,387	VIR, USSR	Elgin & Alexanian, 1991
FOOD LEGUMES	Chickpea	15,940	ICRIST, India	ICRISAT, 1989
	Pigeonpea	11,172	ICRISAT, India	ICRISAT, 1989
	Soyabean	1,400	IITA, Nigeria	IITA, 1990
		12,505	AVRDC, China	ADRDC, 1990
		6,618	VIR, USSR	Elgin & Alexanian, 1991
	Pea	7,437	VIR, USSR	Elgin & Alexanian, 1991
	Cowpea	15,200	IITA, Nigeria	IITA, 1990
	Mungbean	5,274	AVRDC, China	AVRDC, 1990
	Black gram	408	AVRDC, China	AVRDC, 1990
	Phasseolus	9,987	VIR, USSR	Elign & Alexanian, 1991
	Wild Vignas	1,500	IITA, Nigeria	IITA, 1990
	Other Legumes	15,960	VIR, USSR	Elgin & Alexanian, 1991
	Groundnut	12,712	ICRISAT, India	ICRISAT, 1989
	Bambara Groundnut	2,000	IITA, Nigeria	IITA, 1990
ROOT CROPS	Cassava	2,000	IITA, Nigeria	IITA, 1990
	Sweet Potato	1,000	IITA, Nigeria	IITA, 1990
		1,437	AVRDC, China	AVRDC, 1990
	Potato & tuber	10,472	VIR, USSR	Elgin & Alexanian, 1991
INDUSTRIAL CROPS	Beet	2,974	VIR, USSR	Elgin & Alexanian, 1991

	Cotton	6,068	VIR, USSR	Elgin & Alexanian, 1991
	Flax	5,256	VIR, USSR	Elgin & Alexanian, 1991
	Hemp & bast Crops	1,670	VIR, USSR	Elgin & Alexanian, 1991
Sugarcane			SBI, INDIA	
OTHERS	SUN FLOWER	2,131	VIR, USSR	Elgin & Alexanian, 1991
	Hort crops	80,368	VIR, USSR	Elgin & Alexanian, 1991

CHAPTER 7

Utilization of Wild Species in Crop Improvement

Introduction

Plant improvement has progressed significantly during the past 10,000-15,000 years. Early cultivators not only exploited their environment but also domesticated most modern crop plants. The harvesting-sowing-harvesting cycle with associated selection pressures transformed low yielding plants into present-day productive cultivators. During the centuries of cultivation, a wild-weed-domesticte species complex have evolved for most groups of cultivated species. Plant breeding techniques have resulted in increased yields and solved many problems associated with quality, diseases, insects and harvest. The plant breeding techniques have resulted in increased yields and solved many problems. The plant breeder has historically, utilized the variability in land races which are no longer grown and the associated wild species are becoming extinct. In addition, the variability and germplasm resources available for many cultivated varieties are extremely limited.

As additional genetic resources are required to fill voids in breeding populations, unique and imaginative procedures are 'required to exploit fully the potential of our crop plants. Utilization of wild species is one method designed to introduce additional germplasm into cultivated species. The first recorded interspecific hybrid was made in 1717 between carnation and sweet william by Thomas Fairchild (Allard 1960) obtaining hybrids between cultivated and wild species often requires a great effort. First generation hybrids are often partially sterile, and many programmes are abandoned after a few cycles of selection because of continued sterility, low yields, or poor quality characteristics of hybrid derivatives, chromosomal, genetic, cytoplasmic, or mechanical isolation barriers can present severe handicaps for utilization. The effort required to transfer even a single gene from a wild to a cultivated species is often very great, and quantitative traits are even more troublesome to transfer.

The integrity of the crop's phenotype must also be maintained to assure high-yielding varieties within a production regime. Although great effort may be required, wild species have contributed germplasm to several crops with great economic rewards.

Bio Systematics

The first concern of the breeder who wishes to utilize wild species is the collection of germplasm and the relationships of the taxon with which he or she is working. This is followed closely by chromosome numbers, crossability, fertility and chromosome homology of species.

Collection and Preservation of Germplasm

While most plant breeders select desired genotypes within the cultivated species, making interspecific hybrids depends on species collections for progress. Vast areas of land are planted with a small number of high-yielding varieties to feed the burgeoning population. Not only are land races being replaced by new varieties, but many wild species are becoming extinct. The loss of some species may be of little importance for crop improvement, but potentials are unknown until collections are screened and tested. For example, only one accession of *Oryza nivara* was found resistant to grassy stunt virus after thousands of cultivated and hundreds of wild accessions were screened (Anonymous, 1974). Primary and secondary gene centres of Vavilov are often the best sources of plant pest resistances.

Species Concepts

Before a group can be fully exploited, the available plant materials and their relationships must be understood. Taxonomy is of central importance in this goal. Numerical, cytological, biochemical, any hybridization tools have been used to delineate groups of organisms. Conceptual problems arise when two or more morphological species are not cross compatible. Because of the inherent problems associated with classical taxonomic classification, biological species concepts have been developed. However, Harlan and de Wet (1971) devised gene pool concept. The primary gene pool includes all forms that hybridize freely, produce viable hybrids, and have chromosomes that pair and exchange genes freely. Most hybrids between subspecies of wild-weed-complexes belong to this group. The secondary gene pool includes species that can be used as germplasm resources ; but hybrids are difficult to make because of polyploid level differences, chromosome alterations, of genetic barriers to hybridization. Some degrees of sterility is associated with the first generation hybrids in the secondary gene pool. Members of tertiary gene pool are difficult to utilize and plants often belong to different genera, crosses are usually made only of high polyploid levels in complex hybrids, and sterility is always associated with off-spring.

Barriers to Hybridization

Reproductive barriers to hybridization can be divided into two broad groups—premating and postmating. Premating barriers due to geographic isolation are usually unimportant for utilizing wild species. However, apomixis of pollen-pistil incompatibilities may restrict the use of some species while a sexual reproduction can often be circumvented by making reciprocal crosses, pollen-pistil incompatibilities usually present greater problems. Screening for compatible parents, applying immunosuppressants, mechanical manipulations of styles, and bud pollination offer avenues to circumvent these problems.

Postmating barriers to hybridization usually present the greatest handicap for obtaining interspecific hybrids such as ploidy differences, chromosome alterations, chromosome loss or elimination, cytoplasmic incompatibilities, seed domancy or hybrid break down (*i.e.,* lethals or lose of vigour in later generations).

Cytological barriers due to chromosome variation between two species caused by inversions, translocations, duplications, deletions, aneuploidy, or polyploidy may be observed under the microscope. Manipulation of cytological barriers to hybridization offers one of the greatest challenges to utilizing wild species for improvement of crop species.

Chromosome elimination or loss in interspecific hybrids can be classified as a secondary barrier to gene transfer, but an important one for consideration in some species. Selective chromosome elimination has also been reported in *Nicotiana.* Hybridizing species with different euploid or aneuploid chromosome series can result in a similar effect. The end product of a crossing programme to incorporate wild species germplasm should be cytologically stable plants at the chromosome number of the cultivar. After species at different ploidy levels are hybridized and gene exchange occurs, then repeated back crossing to the crop species will usually restore stability in hybrid progenies and eliminate most or all wild species chromosomes.

Cytoplasmic incompatibilities can preclude successful reciprocal crosses, lead to sterility and degeneration of F_1 plants. Many authors reported interspecific hybrids being successful, only when one of the parents is used as the female. An explanation for this unidirectional hybridization include nonflowering of hybrids. Seed dormancy or hybrid breakdown, when hybrids are made between members of section *Rhizomatosae* and section *Arachis* of peanuts, most first generation plants are vigorous, but fail to produce flowers.

Although seed dormancy is not encountered as frequently as other barriers to hybridization, it can restrict rapid introgression of germplasm from wild cultivated species. In the genus *Arachis* a few hybrid combinations remain dormant for 5 to 10 years regardless of the techniques applied to initiate germination.

Genotype Buffering and Chromosome Homologies

The difficulties encountered during interspecific hybridization depend on the cultigen's acceptance of foreign germplasm, the chromosome homology between the species, and many other factors.

Polyploids are generally more likely to accept genes from other species than diploids because ploypoloids have greater genetic buffering and because one dose of an additive gene is tolerated more easily than two genes. Out-crossing species are similarly more likely to be successful parents in a crossing programme with wild species because they are heterozygous for many genes (Harlan, 1975). Manipulation of ploidy levels and reproductive systems have played a key role in introducing foreign germplasm into commercially acceptable types.

Methods available for determining chromosome homologies include the observation of chiasma frequency in diploid hybrids, multivalent formation in allopolyploids (Gerstel and Mann, 1965), nuclear gene segregations, and statistical procedures. Bivalent formation is usually a good indication of chromosome homologies in diploids, whereas multivalent formation in polyploids is less definite.

Gap Between Hybridization and Utilization

A key observation on utilizing wild species germplasm for crop improvement is the predominance of sterility in F_1 hybrids. Even when partially fertile interspecific hybrids are produced, linkages with undesirable genes many times limit the usefulness of these hybrids. The most common fertility restoration method is colchicine treatment of sterile F_1 plants, but naturally occuring unreduced gametes have also played a role in fertility restoration for a few groups. Many examples could be cited when the initial hybrids were made, fertility was restored, and the hybrids were abandoned after a few generations, due to addition of sterility barriers. There are several reasons for this-for example, the effort required to propagate species hybrids is often great, and associated problems of unfavourable linkages related to poor quality or yield often restrict the usefulness of wild species germplasm, considering the expenses of resources, the list of commercial varieties with wild species germplasm in their pedigree is meagre for most crops.

The clearest plant breeding objective is to find a single dominant gene in a wild species, hybridize plants, obtain fertile offspring and backcross to the cultivated species.

Approaches for Utilizing Wild Germplasm Resources

The wild relatives of crop plants differ greatly in their potential usefulness as germplasm resource. Many collections are often distantly related to the crop, and many severe problems must be overcome before introgression of germplasm is successful. In order to overcome, sterilly barriers, unique and imaginative methods have been employed to manipulate chromosomes. Many of the success have come only after extensive cytogenetic, genetic and reproductive investigations. Approaches for utilizing wild germplasm are as follows:

Direct Hybridization

Same Chromosome Number

The easiest route for wild species germlasm incorporation into the cultigen's genome is by direct hybridization of two species at the same ploidy level. When both species have homologous chromosomes, success is often accomplished with little difficulty. Although two compatible taxa may be considered members of the same species, the relationships of many crops and related crops are not clearly defined. Many weed-crop complexes exist, and the advantages of transferring germplasm from closely related species rather than distantly related taxa enhance the chances of success.

Different Chromosome Numbers

Differences in the Euploid Chromosome Complement

Great difficulty is encountered when interspecific hybrids are attempted between two taxa with different ploidy levels. Many crop plants are polyploids, such as cotton, potato, peanut, wheat, tobacco, oat, sugarcane, and most forages. Many of the wild relatives of these plants have lower chromosome levels. Even when the chromosomes of diploid and tetraploid species are homologues, some degree of sterility is encountered in tripled hybrids because of irregular chromosome segregation during meiosis. Four basic methods have been employed to overcome sterility arising from ploidy differences.

Direct Hybridizaiton

Direct hybridisation of different ploidy levels is the most common method of hybridizing wild diploids with their polyploid crop counterparts. Numerous examples are found in species of tobacco, potato, peanut, cotton, wheat, and others (Sanchez-Monge and Garcia-olmedo, 1977). A tetraploid *x* diploid species hybrid usually results in a partially sterile 'triploid' F_1 plant. Fertility is most commonly restored by colchicine, heating vegetative tissues, thus producing hexaploid offspring. The triploids may also be produced naturally, unreduced gametes thus creating hexaploids if both sets of naturally, unreduced gametes thus creating hexaploids if both sets of unreduced gametes combine. Sometimes pollinating the F_1 with $2x$ of $4x$ pollen will restore full fertility. Backcrossing the hexaploid with either the crop of wild species is a logical method to re-establish chromosomally stable progenies at the same ploidy level of the crop. The most successful pathway depends on the amount of chromosome pairing between the species and on whether lines with wild species chromosomes substituted for those of the crop species can be tolerated. Plants can then be backcrossed to commercially acceptable varieties. In theory this is a simple procedure, but in practice, overcoming the sterility barriers and breaking adverse linkage groups can make introgression of the desirable traits difficult (Gerstel and Mann, 1966).

Doubling Chromosome Number of the Species of Lower Ploidy Level

Synthesizing autotetraploids or amphidiploids from the wild species before hybridizing with cultivated species offers a second pathway for introgressing wild species germplasm to the cultilgen. When unreduced gametes are frequently produced such as in the genus *Solanum,* direct diploid *x* tetraploid pollinations will often produce the desired tetraploid progenies. In most other species colchicine is used to raise the chromosome number of the diploid taxa or species hybrids to the ploidy level of the crop. Producing autotetraploids or amphi diploids can have the immediate advantage of avoiding sterile triploid hybrids. In potato the diploid species are subjected to colchicine for chromosome doubling before successfully hybridizing with *Solanum tuberosum* in order to transfer frost resistance to the cultivar.

Doubling Chromosome Number of Species of Higher Ploidy Level

Doubling the chromosome number of a polyploid cultivated species before crossing with a diploid species presents another method of utilizing germplasm from species of different ploidy levels. Although first generation hybrids are semi-sterile, the method has the advantage of avoiding colchincine treating vegetative plant parts of sterile F_1 hybrids. Interspecific hybrids thus produced can either be self pollinated or back crossed to the cultivated species. In tobacco, crossing *8xN. tabacum x 2x N.otophora* resulting in pentaploid F_1 hybrid. This pentaploid hybrid is backcrossed with *4xN.tabacum.* Synthetics produced by this method had nearly a 13% yield increase above the *N.tabacum* parent.

Reducing Chromosome Number of Species of the Higher Ploidy Level

Reducing the ploidy level of the cultivated species to that of the wild species is usually more difficult than the reverse process. With the use of anther culture the procedure may have more success in the future. Selecting homozygotes at the diploid level usually is easier than of the polyploid level because simpler genetic ratios exist, dosage effects are absent or lessened and homozygosity can more easily be attained for desired genes.

Haploid plant introduction by reduced parthenogenesis affords a unique and successful method of creating ployhaploids in the genus *Solanum.*

Different Basic Genomes

Hybridizing species with different basic genomes creates added difficulties for utilizing wild species germplasm. Several examples in the genera *Nicotiana, Saccharum, Zea, Sorghum, and Triticum* are found. Extremely rare recombination events (*i.e.*, Zea ($2n = 20$) × *Tripsacum* ($2n = 36$ or 72) or *Saccharum* ($2n = 112$) × *Sorghum* ($2n = 20$) may facilitate gene exchange. Special techniques such as induced tranlocations have been successful chromosome addition or substitution line may also be used, when the chromosomes of two species are only partially homologues.

Bridge Crosses : (Fig. 6)

Alternate pathways for germplasm introgression have been suggested when direct hybridization of the same or different ploidy level is difficult or impossible. Bridge crosses have been used when gene transfer by simpler methods have failed. The method has been shown to work only under the special condition where species 'A' hybridizes with species 'B' but not with species 'C' and species 'B' and 'C' form viable hybrids, and when desired traits are dominant or partially dominant. Bridge crosses have most often been utilized in species of tobacco and wheat. *Nicotiana repanda* wild *et* Lehmann will hybridize with *N.sylvestris* Speg and Comes, but not with *N. tabacum.* Since *N.sylvestris* and *N.tabacum* will hybridize, the potential exists for *N.sylvestris* to serve as a bridge.

Nicotiana tabacum — Cultivated species
N. repanda — Wild Species (Source of mosaic resistance)
N. sylvestris — Bridge Species

Nicotiana tabacum **X** *N. repanda*

↓

F_1 sterile (due to Incompatibility)

N. repanda **X** *N. sylvestris* | *N. sylvestris* **X** *N. tabacum*

↓ | ↓

F_1 Fertile (Compatible) | F_1 Fertile (Compatible)

N. repanda **X** *N. sylvestris*

↓

F_1 Fertile (Screened for mosaic resistance) **X** *N. tabacum*

↓

F_1 Fertile (Selected for mosaic resistance and backcrossed to *N. tabacum*)

↓

Backcross Progenies Selected for mosaic resistance with *N. tabacum*

Fig. 6. Schematic representation showing transfer of mosaic resistance from wild species to cultivated species through bridge species in tobacco. (Burk, 1967).

Attempt has been made to use *N. sylvestris* as an intermediate between *N. repanda* and *N.tabacum* to transfer mosaic resistance to the cultivar, but difficulties were encountered because resistance was a recessive trait.

However, use of bridging species is complicated and the method makes selection for desirable genes difficult. Unless good screening procedure is available, the desired genes will probably be lost. The probability of successfully transferring quantitatively inherited traits has been found to be many fold less than for monogenic dominant characters. When needs are great and other routes for germplasm incorporation into the cultigen are lacking, utilizing bridge crosses deserves attention.

Chromosome Manipulations

Chromosomes Additions and Substitutions

Chromosome addition lines have been usually found to be of little commercial value. Extra univalents or bivalents from either a cultivated or wild species generally have adverse effects on fertility. Polyploids usually tolerate extra chromosomes more successfully than diploids, but the delicately balanced genome complexes which create the agronomically productive genotype are often disrupted by even a single extra chromosome. Several addition lines have been produced in wheat with improved qualities for disease resistance, earliness, winter hardiness, and contents, however, none of these addition lines have had acceptance as commercial varieties. A notable exception is in the genus *Saccharum* where the addition of entire or partial genomes from wild species to *S officinarum* L. ($2n$ = 80) is a common breeding procedure. Modern sugarcane varieties are complex allopolyploids with chromosome numbers ranging from $2n$ = 100 to 125. A heterozygous chromosome addition line has been used to transfer resistance to *Meloidogyne javanica* (root knot nematode) from *Nicotiana longiflora* to *N. tabacum.*

Chromosome substitutions many times have a lesser effect on the plant's phenotype than chromosome addition lines. The value fo interspecific chromosome substitutions depend on the nature of the desired character, adverse linkages of the substituted chromosomes and the divergence of the wild species chromosome from the cultivated one. Utilization of substitution lines has been mostly been restricted to species where extensive cytogenetic knowledge has been accumulated. Substitution lines have been used to transfer rust resistances from *Agropyron* species to *Triticum aestivum.*

Chromosomal Translocations

Extreme measures may be required to incorporate desired genes into cultivated species when species have nonhomologous chromosomes and when chromosome substitutions are undesirable. Utilization of chromosomal translocations provides a mechanism by which this transfer can occur. First successful intespecific gene transfer utilizing a chromosomal translocation was reported in wheat. An exchange was induced from *Aegilops umbellulata* to *T. aestivum* with the aid and radiation, and the transferred segment carried genes for resistance to *Puccinia triticana.* Homozygous lines which were distinguished form the normal wheat only by their resistance to rust are by slightly later maturity were than selected.

Physiological Manipulations

Embryo Culture : (Fig. 7)

By this technique the products of embryo culture fertilization are rescued when they might otherwise degenerate. This has considerable application as a method of obtaining novel gene combinations from interspecific or intergeneric hybridization. (Hadley and Openshaw, 1980). Recent examples in which embryo culture has been used to rescue previously unknown interspecific hybrids are as follows.

Genera	Reference
Aegilops	Raupp, Browder and Gill, 1993
Brassica	Matsusawa, 1983
Glycine	Brove *et al.*, 1982
Gossypium	Liang *et al.*, 1978, Stewart 1981
Helianthus	Chandler an Beard, 1983
Nicotiana	Reed an Collins, 1978
Oryza	de Guzman, 1983
Phaseolus	Alvarez, Ascher and Davis, 1981
Solanum	Sharma *et al.*, 1980
Triticum	Williams, Vetty and Williams 1982
	Oetler, 1983

Tomato *Lycopersicon esculentum* X. L. peruvianum

↓

F_1 Hybrid (Fruit development normal but embryos are underdeveloped)

Conventional Hybridization embryo culture

Embryos are dissected and cultured in artificial medium

↓

Hybrid seedlings raised and nurtured to the stage of flowering

↓

Hybrids are selected for resistant to molds, nematodes etc.

Fig. 7. Schematic representation of embryo rescue technique in tomato. (Reinett, J. and Bajan, Y.P.S., 1990)

The embryo is excised from the plant at the time after pollination when it has reached its maximum size, thus avoiding its inevitable degeneration and transferred to culture. The exact nutritional \requirement depends upon the stage of development. Interest in interspecific hybridization usually related to the need to transfer particular useful characters from wild species into cultivated form. The important examples are the transfer of resistance of black shank.

Lycopercion esculentum—cultivated tomato (susceptible to moulds, nematodes and virus)

L. peruvianum—Wild tomato resistant to these agents.

(Phytophthora parasitica) from *Nicotina nesophilia* and *N. repanda* to tobacco (Reed and Collins, 1978) and the transfer of the perennial habit from *Trifolium* sarosiense to red clover *T. pratense (Phillips, Collins and Taylor, 1982).*

Grafting

Grafting one plant into another does not in itself produce an interspecific hybrid. However, the technique has been applied to exchange the probability for successful hybridization. Grafting has promoted survival of interspecific *Beta* hybrids.

Protoplast Fusion

Somatic cell hybridization also called somatic cell fusion or protoplast fusion, refers to the fusion of plant protoplasts form somatic cells of different species and the subsequent regeneration of hybrid plants form the fused protoplasts. The procedure is proposed for use in plant breeding to form a hybrid by fusing somatic cells where seeds cannot be obtained by sexual hybridization following wide crosess.

Both intergeneric and interspecific fusion has been reported by various workers in oat *x* maize, *Brassica chinensis x Brassica oleracea* (Kameya and Takahashi, 1972) *Vicia* sp, *Glycine* (Miller et al, 1971). From these studies it can be seen that fusion of protoplasts is a physical, phenomenon, and would occur between protoplasts of any two plants. However, it still remains to be seen if there would be an incompatibility reaction among the protoplasts taken from distantly related or unrelated plants.

Examples of Species Used in Wild Species Hybridizaiton Programmes

Man uses 3,000 or more plant species for food and cultivates some 150 of these to the extent that they have entered into commerce. Among these wheat, rice, corn, potato, cotton are some of the important crops. The list of crops in which great effort has been taken to utilize wild species germplasm includes cotton, tobacco, sugarcane, potato, wheat and corn.

Cotton

The genus *Gossypium* has four cultivated species G hirsutum L., ($2n$ = 52); *G. barbadens* L., ($2n$ = 52); *G.herbaceum* ($2n$ = 26) and *G.arboreum* L., plus 30 short-fibered or lintless wild diploid species (Phillips, 1976). The genus has seven genomic groups- six groups of diploids (A, B, C, D, E and F genomes) and one tetraploid group (AD genome). The wild species are scattered in subtropical areas around the world. *G. hirsutum* (upland cotton) is the most important cultivated cotton species and furnishes most of the world's cotton fiber. All tetraploids have one geneome of 13 chromosomes similar to cultivated Old world diploids and another set of 13 chromosomes similar to New world species. The incorporation of germplasm from wild species with either these A or the D genome of the cultivated tetraploids would thus appear to be relatively easy. However, many cotton breeders do not recommend interspecific hybridization programmes for improvement of cultivated cotton. Successful incorporation of wild species germplasm is difficult because of the finely tuned genetic system controlling lint quality strength and texture, hybrid breakdown among interspecific hybrids, high negative correlations between fiber strength and high yield and reduced chiasma frequencies in many of the interspecific hybrids.

One method of utilizing wild cotton species is to hybridize taxa at the diploid level, treat F_1 plants with colchicine and then cross the amphidiploid with the cultivated tetraploid species. After chromosome doubling the amphidiploid *(G.thurberi x G.arboreum) x G. hirsutum* and back crossed to *G.hirsutum.* Lines were developed with superior lint quality but otherwise poor quality would be prevailing.

Introducing cytoplasmic male sterility to cultivars from wild species is another potential use of wild *Gossypium* germplasm. The nucleus of *G.hirsutum* has been incorporated into a number of cytoplasms of wild species, and male fertility restorer genes have been identified in several cases.

Sugarcane

The cultivated species of the genus *Saccharum* are *S.officinarum* L. (noble canes) and *S.edule* Hassn. Plus four wild species *S.officinarum* ($2n$ = 80) are considered as ancient allopolyploids and the world's commercial sugarcane utilization of wild sugarcane species has been an integral part of sugarcane improvement, for incorporating disease resistances into cultivars. Wakker made the first reported interspecific cross between *S.officinarum* and *S.spontaneum* for incorporating disease resistance into noble canes.

All modern sugarcane varieties are derived form interspecific hybrids and contain three to five species in their pedigrees. Chromosome number increases are common when species of *Saccharum* are hybridized, and modern varieties have very high aneuploid chromosome numbers ranging from $2n$ = 1000 to 125. Since vegetative propagation is practiced for commercial production, fertility is not required in hybrid plants. The increase in chromosome number is important for adding desired characters from wild species because there is little gene exchange between *S.officinarum* and other species of the genus. *S. spontaneum x S.officinarum crosses* have led to plants with as many as 160 chromosomes.

Sugarcane has probably been hybridized with species of more genera than any other taxon. Eight genera in which intergeneric hybrids have been made with *Saccharum* including *Eccoilopus, Erianthus, Miscanthidium, Miscanthus, Narenga, Ripidium, Sclerostachya and Sorghum.*

Wheat (Triticum sp)

The genus *Triticum* includes a polyploid series of species and hybrids ranging from diploids ($2n$ = 14) to hexaploids ($2n$ = $6x$ = 42). The cultivated hexaploid species *T.aestivum* contains the genomes of three different species and is a classical example of utilizing wild species for crop improvement through interspecific hybridization and genome building. Out of 22 species in the genus *Triticum,* 2 are most commercially cultivated. They are *T aestivum* and *T.durum.*

Resistance to several diseases has been incorporated into cultivated wheats from closely related species of *Triticum, Aegilops* and *Agropyron.* Mc Fadden was the first to transfer disease resistance to a commercial wheat variety from a related species. According to Sears, three methods have been developed for transferring genetic materials to wheat from wild relatives. These include.

1. Induction of homologous pairing and crossing-over between wild and cultivated species.
2. Use of ionizing radiation to translocate alien chromosomes segments , and
3. Exploitation of the tendency of univalent which reunite and produce a chromosome with an arm from each of the two univalents.

The induction of homologous pairing is, by for, the easiest method of transferring germplasm but induced translocations have also produced favourable results. Substitution lines has also been successfully utilized for transferring germplasm into the cultivated wheat genomes. Several chromosomes of *Agropyron* species carrying genes for disease resistance have been substituted for those of the wheat genomes. In addition to serving as a source of genes conditioning disease resistance, wild species have been used as a source for improved winter hardiness, lodging resistance conditioned by short stature.

Maize (Zea)

The genus *Zea* contains four members :

Zea mays ssp. *mays* (cultivated maize)

Z.mays ssp. *mexicana* (annual teosinte)

Z.mays ssp. *perennis* (tetraploid perennial teosinte)

Z.mays ssp. *diploperenis* (diploid perennial teosinte)

Maize and annual teosinte hybridize naturally where they grow together in Mexico of Central America and teosinte-maize introgression occurs in a few areas. It has been reported that the chromosomes of maize and teosinte pair normally, and many maize-teosinte hybrids have been produced.

Intergeneric crosses with *Tripsacum* were made to improve maize, *Tripsacum* has a base chromosome number of 9 as opposed to 10 for maize. Crosses can be made in either direction, but hybrids are often more easily produced when *Tripsacum* serves as the female parent. When the reciprocal cross is attempted, the maize styles (ear silks) must be shortened to allow the *Tripsacum* pollen to reach the ovules.

Diploid maize *x* 2 *x Tripsacum* produces viable F_1 hybrids, but in the second or third backcross generation, using maize as a pllen parent, 20-chromosome maize plants have been recovered without apparent intergenomic gene exchange.

Pearlmillet (Pennisetum sp). *Pennisetum* is a genus with over 140 species (Brunken, 1977) divided into five sections : *Gymnothrix, Eupennisetum, Penincillaria, Heterostackya and Brevivalvula. Pearlmillet* belong to the section *Penicillaria.* A system of primary, secondary, and tertiary gene pools has been suggested for dividing the wild germplasm (Harlan and de Wet 1971).

In *Pennisetum,* the primary gene pool includes the wild, grassy, and weedy *P. americanum* subspecies ($2n = 14$) *monodii* and *stenostachyum* that readily cross with perarlmillet and produce fertile hybrids. *The subspecies monodii* is a rich source of germplasm that can be used to improve pearlmillet. Data indicate that it is an excellent source of genetic diversity for new cytoplasms, stable cytoplasmic sterility, pest resistance (disease and insect) fertility restoration, hybrid vigour etc.

The secondary genepool has only one known species *P. purpureum* or napier grass ($2n = 28$). It readily crosses with pearlimillet and produces sterile triploid hybrids, but fertility can be induced by doubling the chromosome number of the triploid to produce a hexaploid. Napier grass is a rhizomatous perennial with desirable characteristics, *e.g.*, resistance to most pests, vigorous growth, and outstanding forage yield potential. Inter-specific hybrids with pearl millet have immediate forage potential but can also be used as bridges to transfer genes from napier grass to pearlmillet (Fig. 8).

Specific uses of Wild Species for Crop Improvement

Disease and Insect Resistances

By far the most common reason for attempt of utilizing wild species is to transfer disease resistance. Most success have been with single genes where adequate selection pressures could be applied. Many authors made the following conclusions about using wild species for crop improvement :

1. the most valuable genes conditioning disease resistance are found in species distantly related to the cultivated species with which interspecific hybrids are difficult to make :
2. Gene conditioning resistance that would justify interspecific transfers often behave as monogenic dominant genes for immunity of very high resistance.
3. transfer requires 10 or more years of continuous work and
4. if possible a species that produces a F_1 of allopolyploid which exhibits full resistance to the pathogen should be used. Although horizontal resistance is desirable, transferring gene complexes only compounds the problem of producing commercially acceptable, disease free plant types.

Programmes aimed at breeding for insect resistance have often met with less success than programmes aimed at disease resistance. But now with the appropriate screening procedures for pest resistance some insect resistant lines could be identified and utilized in the breeding programme.

Pigeon Pea

Atylosia caraboides, a wild relative of *Cajanus cajan* possesses several useful genes which can be utilized for pigeon pea improvement. It is found to be resistant to cyst nematode, *Phytophthora* blight, sterility mosaic, *Fusarium* wilt, pod borer and pod fly and wasp. It is also less susceptible to lepidopteran borer than the cultivated one (Saxena et al., 1990).

Chickpea

In *Cicer* spp, it was found that wild species are the only sources of resistance to cyst nematode and seed beetle. *Cicer bijugum* is found to be resistant to Ascophyta blight, leaf miner, unlike cultivated species where resistance to only one stress was obtained, several wild species accessions were resistant to two or more stresses. Thus, lines with multiple resistance have been identified (Anonymous, 1989).

Groundnut

Arachis chacoense, a wild relaive of Arachis hypogaea is found to be resistant to early leaf spot (ELS) (*Cercospora arachidicola*). Embryo rescue technique was adopted to transfer the resistant genes from the wild to the cultivated species (Anonymous, 1989).

Pearl millet X Napier grass hybrids :

Pearlmillet (2*n* = 14, AA genome) X Napier grass (2*n* = 4*x* = 28, A'A BB genomes)

↓

F$_1$ Triploid
(2*n* = 3*x* = 21, AA'B)
(male and female sterile)

↓ Colchicine

Hexaploid
(2*n* = 6*x* = 42. AAA'A'BB)
male and female fertile

X Pearlmillet (2*n* = 14, AA)	X Pearlmillet (2*n* = 4*x* = 28, AAAA)	X Pearlmillet (Cytoplasmic male sterile) (2*n* = 14, AA)
↓	↓	↓
BC$_1$ × pearlmillet (2*n* = 21 to 35) (2*n* = 14, AA) A and B genomes	BC$_1$ (2*n* = 5*x* = 35, AAAA'B) (Range 2*n* = 28 to 46 male and female sterile)	BC$_1$ 2*n* = 4*x* = 28, AAA'B Male and female sterile 2*n* = 14, AA Male sterile, female fertile 2*n* = 14, AA' Male sterile, fertile female 2*n* = 14, AB Male and female sterile 2*n* = 14 A'B Male and female sterile 3*n* = 21 AA'B Male and female sterile
↓		
BC$_2$ × pearlmillet (2*n* = 18 to 33) (2*n* = 14, AA) (A and B genomes)		
↓		
BC$_3$ 2*n* = 14, AA male and female fertile 2*n* = 15, A and B genomes High Sterility 2*n* = 16 or 17, A genomes High sterility		

Fig. 8. Pearlmillet × Napier grass Hybrids and backcross progenies with pearlmillet (Hanna, 1986).

Tobacco :

Nicotiana gossei, a wild relaive, found to be resistant to aphid (*Myzus persicae*) and leaf eating caterpillar (*Spodoptera litera*). Another spp *N. benthamiana* also found to resistant to aphid and leaf eating caterpillar. *N. repanda* is found to be resistant to root knot nematode. *N.benthamiana* is used as bridge species to transfer the resistance from *N. repanda* to cultivated species. *N.longiflora* is another bridging species for root knot nematode resistance. Using tissue culture technique resistance is transfered from wild to cultivated species (Anonymous, 1990/91).

Rice

Oryza longistaminata, O.nivara, O.perennis and O.latifolia were tested for disease resistance at IRRI. *O.nivara* and *O.perennis* were found to be resistant. The patterns of disease reaction in these wild ries resemble those observed among rice cultivars, suggesting that genes conditioning a differential reaction to races of the pathogen are present in these wild rices (IRRI, 1987). In another study, *O.minuta, O.longiglumis* and *O.ridleyii* were found to be resistant to blast and bacterial blight.

Several methods were used to introduce DNA of wild rices into *O.sativa.* Since *O.minuta* is a tetraploid wild relative of rice it can be crossed with *O.sativa,* although the fertility is low. Since *O.longiglumis and O.ridleyii* are not sexually compatible with cultivated rice, pollen-tube pathway transformation technique is employed to introduce total genomic DNA from two species in the disease susceptible varieties (Anonymous, 1988).

To discover possible donors of durable and multiple resistance, 185 wild rice accessions against 19 insect pests were screened at IRRI. The species found to be resistant or moderate to resistant to BPH (Brown plant hopper) biotypes 1, 2, and 3 ; GLH (Green leaf hopper) ; WBPH (White back plant hopper) and RWM (rice whorlmaggot) are *O. alta, O.australiensis O.eichingeri, O.grandiglumis, O. latifolia, O.minuta and O.punctata.* The research on developing resistant varieties utilising these wild species are underway (Anonymous, 1988).

Yield

Increased yield can be separated into two parts—that due to increased production of vegetative plant parts and that due to increased seed yields. When leaf or stem plant parts are of commercial value, such as in forages or tobacco, increasing vegetative structures has great economic importance. Hybrids between wild and cultivated species of wheat, maize, sorghum, pearl millet, rice, tobacco and several forage plants have resulted in increased production of leaves and stems. However, increasing yields while maintaining the required quality associated with commercial varieties in cultivation can be difficult. For example, in tobacco, leaf production could be increased by using progenitor species, but the advantage gained was negated by poor leaf quality. Likewise in cotton, fibre length may be lengthened as a result of interspecific hybridization, but there appear to be severe barriers to increasing both yield and cotton fiber lengh at the same time. The probability of increasing seed yields from utilization of wild species would seem more unlikely than observed gains in vegetative plant parts because of sterility associated with many interspecific hybrids. Although breeding programmes for yield have historically been secondary to breeding for diseases or specific morphological traits when attempting to utilize wild species, the number of success are increasing.

In sweet potato (*Ipomoea batatas)* interspecific hybridization with its wild relative *Ipomoea trifida* followed by backcrossing gave higher yield increase due to better root bulking capacity over their parents (Anonymous 1990).

Possibily the best example of utilizing wild species germplasm to obtain yield increases for commercial production is in the genus *Saccharum.* All modern varieties are progenies of *S.officinarum* ($2n = 80$) × wild species of the genus. With the resulting resistance to sugarcane pests, the high polyploid ($2n = 100$ to 125) derived varieties greatly out yielded noble canes.

Quality

Improvement of quality is of great importance for many crops. Wild species germplasm has had limited use in this area because of the genetic complexity of most quality characters. However, the potential exists for making alterations in chemical composition and neurological traits by using wild species germplsam. Wild grasses usually have a higher protein percentage than related cultivars and protein quality or quantity has been altered in several species of the poaceae. The advantage sought from wild germplsam is often lost by the time large, seeded types are selected from crosses with wild species. In addition, oil quality has been improved by utilization of wild oil palm species. Starch content has also been improved in potato tubers by utilizing wild species germplasm.

Although few forage species are completely domesticated, the quality of plant products has been improved by utilization of wild species in this group. Improvement includes both the manipulation of wild species for economic use and the utilization of species hybrid for crop improvement. Hybrids between widely separated species may be better adapted in forages than in other groups because 50% or more of the grass species are polyploids.

Earliness and adaptation

As land for agriculture becomes more limited and crops are moved into less suitable environments, variability for new adaptation must be exploited in order to maintain high yields. Although breeding programmes utilizing only the cultivated species have done much for moving crops into new environments, the variability found in wild species offers a valuable resource for genes conferring wider adoptions. Cold tolerance has been successfully transferred from wild species of peppermint, tomato, grape, strawberry, wheat, rye, onion, potato to their cultivated relatives. Wild *Brassica* and *Glycine* species have been used to introgress genes for earliness into cultivated varieties. Cicer bijugum a wild relative of *Cicer arietinum* is found to be cold tolerant than its cultivated species.

Modes of Reproduction

Sterility is the most common alteration in reproduction which results from interspecific hybridization. Cytoplasmic male sterility is one economically important derivative of this result, especially for hybrid seed production. List of crops in which cytoplasmic male sterility has been discovered in crosses between wild and cultivated species include wheat, cotton, barley, tobacco, sunflower etc.,

Miscellaneous Uses

Wild relatives of crop plants have been utilized for a number of different traits that do not conveniently fit into other categories of wild species have been used to transfer genes conditioning hard-seededness to cultivated bramble fruits, to introgress dark green colour and excellent leaf texture into lettuce and to transfer a gene for carotenoid synthesis to cultivated tomato. Plant stature and other morphological traits can also be improved by utilising wild species germplasm. For example, interspecific hybrids of oil palms resulted in shorter plants. One byproduct of species hybridizaiton has sometimes been naturally occuring mutations which might serve as new sources of breeding materials.

Summary

Germplasm resources of wild species provide no miracle cures for combating plant pests of improving quality, yield, or specialized traits for crop improvement. Valuable germplasm resources are available in wild species, however, and their exploitation will continue to improve the quantity and quality of cultivated species. The greatest benefits from wild relatives of crops have been in the families of gramineae, *i.e.*, (wheat and sugarcane) and in solanaceae (*i.e.*, tobacco and potato). This has been in part due to the many polyploid species in these groups of ancient hybrid origin and extensive research effort devoted to aspects of crop improvement involving wild species.

As new techniques are developed (protoplast culture, embryo culture) and greater efforts are devoted in improving crop species with the aid of wild species, greater numbers of successful transfers will be made by the wild species.

Germplasm resources in wild species have been most utilized when desirable genes were unavailable in the crop species. Although wild species have commonly been used as a source for disease resistance in some crops, a great void of knowledge exists for most groups. Many plant collections need to be screened for disease and insect resistances, genes for quality and yield and other desirable characteristics.

The use of wild species germplsm holds a dynamic place in crop improvement and will become increasingly important as new variability to meet the needs of conventional breeding programmes.

REFERENCES

Allard, R.W. (1960). Principles of Plant Breeding, wiley, New York.

Alvarez, M.N., P.D. Ascher and D.W. Davies. (1981). Interspecific hybridizaion in *Euphaseolus* through embryo culture. *Hort Science,* 16:541-543.

Anonymous. (1990). Asian Vegetable Research Development Centre (AVRDC). Progress Report, Taiwan.

Anonymous. (1989). International Crop Research Institute for Semi-arid tropics (ICRISAT), Annual Report, Hyderabad, India.

Anonymous. (1990/91). Central Tobacco Research Institute. (CTRI), Annual Report. Rajmundry, India.

Anonymous. (1974). International Rice Research Institute, Annual Report, Los Banos, Philipines.

Anonymous. (1988). International Rice Research Institute, Annual Report, Los Banos, Philipines.

Anonymous. (1987). International Rice Research Institute, Annual Report. Los Banos, Philipines.

Brove, P.J., J.P. Grace, and D.R. Marshall (1982). Interspecific hybridization of Soy beans and perennial *Glycine* species indigenous to Australia via embryo culture. *Euphytica, 31*:715-724.

Brunken, J.N. (1977). A systematic study of *Pennisetum.* Sec. *Penni setum* (Gramineae). *American Journal of Botany. 64* (2) : 161-176.

Burk (1967). Bridge crosses in Nicotiana sp. Tob. Sci. 16:43-45.

Chandler, J.M. and B.H. Beard. (1983). Embryo culture of *Helianthus* hybrids *Crop Sci., 23*: 1004-1007.

De Guzman, E.V. (1983). Recent progress in rice embryo culture at IRRI.I. *Cell and tissue culture Techniques for cereal Crop Improvement* pp. 215-228. Beijing, Science Press.

Dunwell. J.M. (1986). Pollen, Ovule and embryo culture as tools in plant breeding. *In plant Tissue culture and its Agricultural applications* (Withers. L.A. and P.G. Alderson. eds.) pp. 375-404, London, Butterworths.

Erich Hoyt (1988). Conserving wild relatives of crops. IBPRIUCN-WWF.

Gerstel, D.U., and T.J. Mann. (1965). In 'Chromosome Manipulations and Plant Genetics' (R.Riley and K.R. Lewis, eds.) pp. 113-120 Heredity (Suppl.) 20.

Hadley, H.H. and S.J. Openshaw. (1980). Interspecific and Intergeneric hybridizaion In. Hybridization of crop plants. (W.R. Fehr and H.H. Hadley, Eds.). pp. 133-159. Madison, Am.Soc. Agron. Crop Sci. Soc.Am.

Hanna. W.W. (1986). Utilisation of wild relatives of Pearlmillet. In proceedings of International Pearlmillet workshop. ICRI SAT, Hyderabad. India.

Harlan, J.R. (1975). 'Crops and Man'. American Society of Agronomy, Madison, Wisconsin.

Harlan, J.R. and J.M.J. de wet, (1971). Toward a rational classification of cultivated plants. Taxonomy 20:509-517.

Kameya, T., and N. Takahashi. (1972). The effects of inorganic salts on fusion of protoplasts form roots and leaves of *Brassica species, Japan, J. Genet. 47*:215-217.

Liang, C.L., C.W. Sun. T.L. Liv., and J.C. Chiarg. (1978). Studies interspecific hybridization in cotton *Scientia sin., 21:* 545-555.

Mann, S.S. (1977). "In Interspecific Hybridization in Plant Breeding" (E.Sanchezmonge and F.Garcia-Olmedo. eds.), pp. 201-214. Proc. 8th EUCARPIA Congress, Madrid.

Matsuswa, Y. (1983). Studies on interspecific hybridiation in Brassica II. Jap. J. Breeding, 33:321-331.

Millet, R.A., L.L. Gamborg., W.A. Keller and K.N. Kao. (1971). Fusion and division of nuclei in multinucleated Soyabean protoplasts. *Can J. Genet. Cytol. 13:* 347-352.

Oettler. (1983). Crossabiltiy and embryo development in Wheat-rye hybrids. *Euphytica, 32:*593-352.

Phillips, G.C., G.B. Collins and N.L. Taylor. (1982). Interspecific hybridization of red clover *(Trifolium pratense L.)* with I Satosiense Hazsl. Using in Vitro embryo rescue. *Theoret. Appl. Genet., 62: 17-24.*

Phillips, L.L. (1976). In Evolution of Crop plants (N.W. Simmonds, ed), pp. 196-200. Longman, London.

Ranpp, W.J., L.E. Browder, and B.S. Gill. (1983). Leaf rust resistance in *Aegilops squarrosa, its transfer and expression in common wheat (Triticum aestivum L.)* (Abstr.). Phytopathologia, 73:818.

Reed, S.M. and G.B. Collins. (1978). Interspecific hybrids in *Nicotiana* through in *vitro* culture of fertilized ovules. *J.Hered., 69:* 311-315.

Reinett, J. and Y.P.S. Bajaj (1990). Protoplast Isolation, Culture and Somatic hybridization. In plant cell, Tissue and organ culture (Reinett, J. and Y.P.S. Bajaj eds.), Narosa Publishing Housing, New Delhi, pp. 467-496.

Sanchez-Monge, E.E. and F. Garcia-olmedo. (eds.). (1977). Interspecific hybridizaiton in Plant Breeding, Proc. 8th Congress of EUCARPIA, Madrid.

Saxena, K.B., L.Sing., M.V. Reddy., U.Sing, S.S. Latccf., S.B. Sharma and P. Ramanandan. (1990). Intra species variation in *Atylosia scarabaeoides* (L) Benth., a wild relative of pigeonpea. *(Cajanus cajan* (L) Millsp.) *Euphytica, 49:* 185-191.

Sharma, H.C., J.B. Chowdhury, U.Ahuja, and B.S. Dhankhar (1980). Interspecific hybridization in genus *Solanum Z.Pflanzenziichu., 85:* 248-253.

Sears, E.R. (1948). The cytology and Genetics of the wheats and their relatives. *Advances in Genetics.* 2 : 239-270.

Singh, B.D. (1983). Plant Breeding, Kalyani Publishers, Ludhiana. pp. 457-495.

Stalker, H.T. (1980). Utilisation of wild species for crop improvement In. *Avances in Agronomy.* Vol. 33. pp 111-147. Academic Press, New York.

Stewart, J.M. (1981). In vitro fertilization and embryo rescue. *Env. Exp. Bot., 21:* 301-315.

Williams, E.G., I.M. Vetty., and W.M. Williams. (1982). Use of embryo culture in interspecific hybridization. In *Plant improvement and somatic cell genetics* (I.K. Vasil, W.R. Scow croft and K.J. Frey., Eds.), pp. 119-128. New York, Academic Press.

CHAPTER 8

Interspecific Crosses

The genetic improvement of some cultivated plants has involved incorporation of useful genes from other species or genera. New types of plants have been developed and are being contemplated that represent the merger of the chromosome compliments of two existing cultivated species. The unique opportunities and challenges presented by interspecific hybridization can be important for plant breeders interested in cultivar development.

Interspecific hybridization refers to crosses between different species, and it includes crosses between different species of the same or different genera.

History of Interspecific Hybridization

The first authentic record of the hybridization work for crop improvement is the production of a hybrid between Carnation (*Dianthus caryophyllus*) and Sweet William *Dianthus barbatus* by Thomas Fairchild in 1717. Subsequently a large number of interspecific hybrids were produced. Most of the interspecific hybrids were of no agricultural value, and were of academic interest only. But many interspecific hybrids, particularly in case of ornamentals served as commercial varieties. Though many intergeneric hybrids, like *Triticale* and *Raphano Brassica* were developed, at present only improvement of a few crops are done based on interspecific hybridization. For example, in sugarcane transfer of a few desirable genes, particularly those for disease resistance was done. In many crop species, the genes for disease resistance transferred from the wild species are extremely useful.

Objectives of Interspecific Hybridization

Improvement of Cultivars of a Species

The first choice of parental material for most plant breeders is cultivars and experimental lines with desirable quantitative and qualitative characteristics. If genes for a desired characteristic cannot be found in elite parents a breeder will attempt to find them in a genotype having a minimum number of undesirable traits. Genotypes within the cultivated species generally are evaluated first. If the characteristic cannot be found within the species, the breeder will consider related species within the primary gene pool then within the secondary genepool and finally within the tertiary genepool.

Formation of New Species

The formation of new cultivated species with desirable characteristics from two or more existing species can be a goal of interspecific hybridization.

Techniques for Gene Transfer

The development of improved cultivars of a cultivated species by the transfer of genes from another species involves two factors; obtaining viable seeds or vegetative propagules in the F_1 and later generations and eliminating undesirable characteristics of the donor species. One or both of these factors may be a consideration in gene transfer between species (Hadley and Openshaw, 1980).

Securing viable seeds and plants

The primary gene pool of a cultivated species includes related taxa with which it can be hybridized to obtain viable hybrid seeds and progeny without the use of special techniques. Because non-cultivated types commonly have unusual plant or floral characteristics making crosses between species in the primary gene pool may be more difficult than crossing genotypes within the cultivated species. Non-cultivated types also may have chromosomal aberrations that reduce seed set of plants in the F_1 or later generations. In *Soybean, Glycine soja* is a member of the primary gene pool of the cultivated species *Glycine max. G. soja x G. max* crosses readily.

Barriers to Interspecific Hybridization

The first concern in breeding programmes involving interspecific hybridization is whether or not a viable F_1 hybrid plant can be obtained. In general, however, close relationship as indicated by taxonomic status is the most useful criterion of crossability. The correlation is far from complete.

1. **Failure of Zygote Formation.** For example, the common cabbage, *Brassica oleraceae* will cross with the Radish, *Raphanus sativus*, but not with the more closely related turnip, *Brassica rapa* or with *Brassica napus*. This may be due to cytoplasmic incompatibilities that are expressed either in failure of fertilization or in death of the zygote at any stage between early cleavage divisions and maturity. In *Nicotiana* and in hybrids between *Zea mays* and *Tripsacum species*, failure of fertilization sometimes result from the fact that the style of the maternal species is much longer than the style of the pollen parent. In addition, the thick pollen tubes of polyploid species sometimes have difficulty in growing in the slender styles of diploids or lower polyploids. This type of barrier to fertilization can sometimes be overcome by making the cross in the reciprocal direction.
2. **Lethal Genes.** Barriers to interspecific hybridization that act by preventing the growth in F_1 hybrids are common in plants. Hybrid inviability can be caused by single genes, by general incompatibility of the genotypes of the parents, or by disharmonies between the developing embryo and endospem. A few examples of these types of barriers follow :

   In 1930, Holling Shead reported that hybrids between crops *Crepis tectorum* and *C. capillaris* died in the cotyledonary stage or reached maturity, depending on whether the *C. tectorum* parent carried the lethal gene '*l*' or its nonlethal allele 'L'. The gene '*l*' also caused the early death of *Crepis tectorum* hybrids with *C.bursifolia*. The gene '*l*' had no discernible effect on C.tectorum itself. In the cross between Einkorn wheat and *Aegilops umbellulata*, Sears (1944) distinguished three alleles, Le for early lethality, 'L*l*' for late lethality and '*l*' for viability.
3. **Genotypic Disharmony.** Lethality is common in hybrids between species of *Gossypicum.* In an extensive study of the wild American species, *G-gossypioides*, Brown and Menzel (1952) found that all but one of the interspecific hybrid that could be obtained with this species, died in some stage ranging from early embryogeny until after meiosis had taken place. Hybrids of another wild American species, *G.davidsoni*, with G. barbadense, G. hirsutum, G. stocksi and *G. thurberi* die in the cotyledonary stage.

Stephens (1950) and Gerstel (1954), however have been able to demonstrate that hybrid inviability is controlled by simple genetic mechanisms in some of the other interspecific hybrids in cotton.

4. **Endosperm Abortion.** Differences between reciprocal crosses are often observed when the species that are hybridized do not have the same number of chromosomes. In the cross of hexaploid *x* tetraploid wheats, the combination ($4x$) × ($6x$) often produces badly shrivelled seeds, which fail to germinate. The seeds from the combination ($6x$) × ($4x$) are reduced in size but nearly normal in germination. Thompson (1930)attributed the difference to the chromosome balance of the endosperm. The endosperm in the combination $6x$ by $4x$ has 21 + 21 + 14 chromosomes or 88% of the chromosome complements of the hexaploid endosperm. In the reciprocal cross, the endosperm has 14 + 14 + 21chromosomes, or only 78% of the hexaploid complements. The abnormal endosperm may be due in part to quantitative relationships between the chromosome numbers of the endosperm and embryo, but it seems likely that unbalance of gene dosage may be responsible for it. If the growth of the embryo is inhibited by the endosperm, the hybrid can sometimes be obtained by dissecting out the young embryo and growing it in nutrient culture. Success was obtained in this way by Brink, Cooper, and Aushermann (1944) in the hybrid between Secale cereale and Hordeum jubatum.
5. **Failure of Hybrid Seedling Development.** Certain interspecific hybrids in Melilotus fails because the F_1 hybrid seeds are albinistic derivatives. Smith (1945) overcame this barrier by grafting the albino seedlings to one of the parental species, where upon they grew to maturity and produce viable seeds.

   In a number of genera there are barriers of hybrid weakness that result from the interaction of certain hybrid genotypes and the cytoplasm of one of the parental species. This subject has been reviewed in detail by Caspari (1948). Good reviews of the general subject of hybrid inviability and its genetic basis are those of Stebbins (1950, 1958).

Cytological and Genetic Basis of Hybrid Sterility

Although the causes of hybrid sterility is numerous, the genetic unbalance is mainly responsible for hybrid inviability. Sterility that becomes apparent only at the time of the formation of gametes may be divided into two types: chromosomal and genic. The disorganized disjunction in structurally heterozygous hybrids usually causes less than a full complement of chromosomes to be partitioned to each gamete in the meiotic process and, as a result many or all gametes are non-functional. The demonstration that amphidiploids derived from certain hybrids become fertile once each chromosome has a fully homologous mate with which to pair shows that the sterility of the F_1 is sometimes due entirely to chromosomal sterility and not to genic unbalance.

Structural differentiation of chromosomes has been observed in all degrees in all interspecific hybrids. In some hybrids, the differentiation has proceeded to the point where pairing fails completely. At the other extreme, the F_1 hybrid may be highly sterile, even though bivalents are formed regularly at meiosis and disjunction is normal. Sterility associated with completely normal pairing of chromosomes is called genic sterility. In most interspecific hybrids chromosome pairing has been analysed at the first meiotic metaphase, when the chromosomes are strongly contracted and are associated at only a few points along their length. Even quite large structural differences may not be detected at this stage. Stebbins has presented evidence for the widespread existence of chromosomal sterility due to heterozygosity for structural differences too small to affect pairing maternally and has proposed a name for this phenomenon cryptic structural hybridity.

Perhaps the best evidence for genic sterility has been presented by Oka (1955). In cultivated rice, hybrids between different varieties, and especially hybrids between the varieties of the subspecies indica and japonica showing varying degrees of sterility, Oka studied segregation ratios in a number of hybrids and concluded that a series of genes affecting development of the gametophyte are responsible for the sterility. These genes appeared to act in a complementary fashion so that the presence of two recessive allels to one megaspore or microspore inhibits full development. Mashima and Uchiyamada (1955) however, have reported that induced tetraploids have a smaller number of quadrivalents and a higher frequency of bivalents than autotetraploids derived from a variety. So it is difficult to distinguish between genic and chromosomal sterility.

Cytoplasmic Basis of Sterility

In some interspecific hybrids, sterility is produced by the cytoplasm. In such cases, the reciprocal crosses, produce fertile hybrids, that fertility of the hybrid depends on which of the two species was used as female.

Segregation in Hybrids. Segregation as a result, would produce a very wide range of phenotypes in F_2, and the later generations. Many of the segregants would have entirely new characteristics distinct from those of the two parental species. As a result, a majority of segregants from interspecific hybrids are much weaker and less adapted than the parental species. The recombinants recovered represent only a small portion of the total recombinants that are possible. Thus, recombination in interspecific hybrids produces only a limited range of variability from the total range possible. Therefore, segregation for different characteristics does not fit the Mendelian ratios in F_2s from wide crosses. This may be due to irregular meiosis, impaired recombination and reduced survival of many recombinants and zygotes. Generally, crosses more difficult to make than *Glycine max* with *G.max* mattings are in *G.soja* due to the small flower size and prostrate growth habit. Genotypes of *G.soja* may have chromosomal translocation not found in cultivars of *G.max* which causes partial sterility in the secondary and tertiary gene pools may require special techniques to be successful. Failure to obtain viable seeds or plants in the F_1 or later generations can occur at various phases of the transfer programme. Various techniques have been used to overcome these barriers.

Lack of fertilization. Male and female gamete fail to unite to form a zygote. It may be due to lack of pollen germination, insufficient pollen tube growth, or inability of male gamets that reach the embryosac to unite with the egg cell. A number of techniques are used to make it successful.

Parent selection. The probability of mating genotypes that are compatible is a function of the number of different genotypes of each species that are used as parents.

Reciprocal crosses. When we cross two species that differ in chromosome number it is usually recommended to choose the female as the one with greatest chromosome number. In some cases use of the species with the lower chromosome number as the female has been more successful. So, reciprocal crosses are recommended for interspecific mating that have not been attempted.

Modification of ploidy levels. When two species of a cross differ in ploidy level, modifying the ploidy of one species to match that of the other may improve the success of obtaining a zygote.

Pollen mixtures. Cross incompatibility is associated with proteins of the pistil that interact unfavourably with proteins of the pollen to prevent normal pollen tube germination and growth. This unfavourable reaction has been avoided in certain interspecific matings by mixing the pollen from a compatible species with pollen from an incompatible parent.

Chemical treatment of the pistil or pollen

The compounds in the stigma or pollen that prevent pollen tube growth in an incompatible mating have been removed by application of an organic solvent to the stigma or the pollen. Application of growth regulators has been used to promote more rapid pollen tube growth or increase the length of time the pistill remains viable.

Protoplast fusion. It is used to produce hybrids that cannot be obtained through sexual fertilisation.

Gene Transfer Techniques

Genes affecting a variety of characters have been transferred from wild relatives; these include disease resistance, wide adaptation, improved quality, mode of reproduction, yield etc.

1. **Disease Resistance.** Disease resistance is by far the most common characteristic transfered to the cultivated species from their wild relatives. Some important examples are, rust resistance in wheat, late blight resistance in potatoes; resistance to bacterial wilt, bacterial canker, curly top virus, mosaic virus etc., in tomato; resistance to many insects and diseases in sugarcane and so on. A glaring example is the potato famine in Europe during 1840's caused by late blight; about one million Irish died as a result of the famine. Solanum demisum provided genes for late blight resistance, which served as a miracle saver of the potato crop, although only temporarily.
2. **Wider Adaptation.** Wild species have served as useful sources of genes for earliness and wider adaptation. Cold tolerance has been transferred from wild relatives to wheat, onion, potato, tomato, grape, rye and peppermint. Earliness has been transferred to cultivated species of *Brassica* and soybean from their wild relatives. Wild species of wheat and peas have contributed genes for drought and heat tolerance. Other similar gene transfers are: salt tolerance to tomato, tolerance to calcareous soils in grape, and lack of photo sensitivity to *Pennisetum.*
3. **Quality.** Genes for increased protein content have been transferred to rice, soybean, oats and rye. Oil quality in oil palm has been improved by genes from its wild relatives. Other examples of genes for improved quality are: Carotenoid content (B genes from *G.hirsutum*) in tomato; improved leaf quality in tobacco (from *N.debneyi*) and improved fibre strength in *G.hirsutum* (from the lintless G.thurberi; the allotetraploid from the cross *G.thurberi* with *G.arboreum* was crossed with *G.hirsutum* since the two species are incompatible).
4. **Mode of Reproduction.** In some cases, genes affecting the mode of reproduction have been transferred from wild relatives to the cultivated species. Self-incompatibility alleles from *B.campestris* have been transferred to the self compatible *B.napus* for the production of hybrid seed. Genes for apomixis have been transferred to maize (*from Tripsacum*) and sugarbeet (from wild Beta sp.), cleistogamy and self fertility traits of wild *Secale sp.,* have been transferred to *Secale cereale.*
5. **Yield.** Encouraging results were obtained from interspecific hybridization in *Cicer.* The F_2 from the cross *C.arietinum x C. reticulatum* showed very large trangressive segregation for yield and several yield traits. Selections for increased 100 seed weight, pod number and seed yield per plant were effective; F_6 progenies showing upto 40% improvement to 100 seed weight and 30-40% increase in seed yield per plant over the *C.arietinum* parent have been isolated.

Interspecific Hybridization Work in Different Crops

1. **Rice.** Co.31,IJ.1 and ADT.27 are the varieties released through interspecific hybridization. ADT-27 was obtained by crossing Norin-8 (indica) X GEB-24 (Japonica) varieties, ADT-27 yields 5t/ha of medium fine rice of good cooking quality. It is early maturing and photo period insensitive.
2. **Cotton.** It is done under two headings: firstly the transference of unit genes or chromosome segments from the wild to the cultivated species, and secondly quantitative changes due to the integration of an indeterminate amount of foreign chromatin. In the first group, a desirable character is found to exist in the wild diploid but not in the cultivated tetraploid. The transfer entails the progressive elimination of foreign chromatin from the hybrid by repeated back crossing to the cultivated cotton, the new character being retaind in each generation.

   Successful transferences of this kind have been the integration of 'smooth leaf' from the D genome of *Gossypium armourianum* into *G.hirsutum.* The transfer of the blight resistance gene B6 to Sudan Sakel, *G.barbadense,* from the A genome in *G. arboreum;* the transfer of nectariless from *G.tomentosum* and lint strength from *G.thurberi;* a restorer gene for cytoplasmic male sterility transferred from *G.hartnessii* and hairness from *G.anomalum* to *G.barbadense.*

   On the Ivory Coast, French Workers have produced the HAR varieties from synthetic tetraploids of *G.raimondii* and *G.arboreum* crossed with Acala varieties (*G.hirsutum),* and similarly the ATH varieties from *G.thurberi* and *G. arboreum.* Varalakshmi is the interspecific hybrid obtained by crossing *G.hirsutum x G.barbadense.* CBS-156, Savithri, Jayalakshmi are the other varieties.
3. **Durum wheat.** To diversify the genetic base on Indian *drums,* an interspecific hybridization programme is in progress to introduce genes from five *Triticum* species, namely *T.dicoccum T.carthlicum, T.turgidum, T.timopheevi,* and *T.aestivum,* into *T.durum* for various yield and yield contributing traits, besides disease resistance. For back crossing the F_1, hybrids between durums and donor species, for each back cross generally a different drum variety was used as recurrent parent to further diversify the genetic base and to increase the chances of desirable recombinants.

   Most of the dwarf *durm* cultivars are susceptible to lodging while the new ideotypes evolved through interspecific crosses are non lodging having very strong straw and slightly reduced plant height. Derivatives from crosses involving, *timopheevi, dicoccum,* and *carthlicum* have high grain weight combined with high kernal protein IQP.5019, a derivative from *durum x* dicoccum, cross is early in maturity having tolerance to high temperature during grain filling period in addition to its high grain weight and high kernel protein. The derivatives of *timopheevi* are giving exceptionally high grain weight, uniform and vitreous non-mottled grain (IWP 5052-1 and IWP 5052-2) in addition to high grain protein; carthlicum derivatives are also having non-mottled and vitreous grain INP 5011; a turgidum derivative has shown very wide adaptability in addition to its high grain yield. A newly evolved *timopheevi* derivative (B-138) has given higher grain yield establishing its statistical superiority over the best *durum* and *aestivum* checks and is being put under multilocation yield evaluation.
4. **Pearl millet.** The major achievement is the production of pearl millet (*Pennisetum typhoides)* x Napier (*P.purpureum)* hybrids. The first controlled *P.purpureum x p.glaucum* hybrids were made in 1941. Burton (1944) and Patil (1963) described two vegetatively propagated *P.purpureum x p.typhoides* hybrids named Pusa Napier-1 and 2 that gave higher forage yields.

The cytophasmic male sterile Tift-23A or other available cytoplasmic male sterilies could be seeded between rows of the perennial napier grass to produce F_1 hybrid seed. Burton (1944) working in the U.S. was the first to make interspecific hybrids which were later produced in India. (Krishnaswamy and Raman, 1949, 1953 a,b).

5. **Cucumis.** (M.Chatterjee and T.A. More) The objectives of the study were, to attempt interspecific hybridizaiton between C.melo (PM and M4) and four wild species *C.figoree, C.zyheri, C.meeusii,* and *C.myricocarpces* to find out barriers involved in the failure of interspecific hybridizaiton between *C.melo* and these wild species and to overcome them by employing various techniques. No fruit set was obtained in any of these crosses and no pollen germination was noticed even 72 hours after pollination indicating a prefertilizaiton barrier was operating in these crosses. Based on these results some techniques like budpollination, use of mentor pollen, brush pollination, application of growth regulator (GA) and use of organic solvent (n-hexane) were employed to overcome these barriers. Umber of pollen technique was found to be useful to obtain 8% fruits, Benzyladenine-4% fruits and n.hexane to the stigma 2% mature fruits etc. So, it was concluded that there was possibility of obtaining interspecific hybrids in *Cuccumis.*
6. **Sesamum.** (Amal K. Biswas and Ashis K.Mitra) *Sesamum indicum* L., suffers considerable loss due to attack of the insect *Antigastra catanaulis.* So steps were taken to incorporate the resistance into the recommended variety B-67 of sesame through interspecific hybridization with *S.mulayanum* and *S.laciniatum.* Fruit setting was obtained when B-67 a cultivar of *Sesamum indicum* was used as female in the crosses with *S.mulayanum* ($2n$ = 26) and *S.laciniatum* ($2n$ = 32). The F_1 hybrids of B-67 x *S.laciniatum* died before flowering, while those of B-67 x *S.mulayanum* were characterized by pinky violet corolla colour (intermediate between parents) having mostly four univalent at M_1 (n = 13) and resembling the wild parent in insect resistance, black reticulated seed coat, and violet marking along longitudinal anther wall. Like B-67, seed shape was normally ovate. It is indicated that *S.indicum* and *S.mulayanum* are closely related and insect resistance of the latter is transferable.
7. **Sugarcane.** The main achievement in sugarcane is the nobilisation of canes, *i.e.*, the transfer of disease resistance, hardiness and other characters from the wild species *Saccharum spontaneum* to *S.officinarum.* In sugarcane mostly all the varieties were released through interspecific hybridization. Some prominent varieties of sugarcane are CO-1148, CO-1158, COS 510, CO 975, COS 109, CO 541 etc.
8. **Potato.** Potato variety Kufri Kuber was developed from a complex cross (*Solanum curtilobum)* × *(Solanum tuberosum)* × *Solanum andigena.* Generally, interspecific crosses are made to transfer specific characters, such as, disease resistance from the wild species to the cultivated potato, *S. tuberosum.* For example *Solanum demissum* has been extensively used as a source of late blight resistance. Kufri Alankar, Kufri Sindhuri, Kufri kundan, Kufri chandramukhi etc., are the other varieties.
9. **Tomatoes.** The genus *Lycopersicon* is naturally divided into 2 sub-genera on the basis of colour of fruit, position of the stigma and some other characters. The red fruited sub genus *Eulycopersicon* includes two species, *L esculentum* and *L.pimpinellifolium.* The green fruited sub-genus *Eriopersicon* includes the species *L. hirsutum, L.peruvianum* and *L.chilense* etc.

The most extensively studied hybrid is that between the two species in *Eulycopersicon.* Resistance to *Cladosporium, Fusarium and Stemphylium* have been achieved. Pusa red plum is the variety released by interspecific cross.

10. **Tobacco.** Interspecific hybridization followed by chromosome doubling of the sterile hybrids produces allopolyploids that may serve as new crop species. More commonly, these synthetic allopolyploids serve as bridging species in crosses between two species with different levels of ploidy. For example, the hybrid between *Nicotiana tabacum* ($2n = 48$) and *N. glutinosa* ($2n = 24$) is sterile and cannot be used in gene transfers. But the amphidiploid *N.digluta* ($2n = 72$) form the cross *N.tabacum x N.glutinosa* is reasonably fertile and produces partially fertile bybrids with *N.tabacum.* Thus, the synthetic alloplyploid *N.digluta* serves as a bridging species between *N.glutinosa* and *N.tabacum* and makes it possible to transfer genes and chromosomes from *N. glutinosa* to *N.tabacum.*
11. **Soyabean.** The cultivated varieties of *G.max* in India have the problem of low productivity due to various reasons, and the inherent poor yielding ability of the present varieties is one of them. Interspecific hybridization provides greater opportunities of enhancing soybean productivity as they possess genes for resistance to different biotic stresses as well as for some yield components that are found lacking in the cultivated species. Among the related species, *G.formosona* exhibits desirable features of plant height, primary branches, pods/ plant, protein content, etc. When two species, genetically diverse for a number of important yield traits, are crossed, F_2 population is expected to release enormous variability transcending even the limits of the parents for different yield traits. In such a situation, it becomes all the more important yield, through a multiple trait selection scheme by utilizing information obtained on all the yield influencing traits.

    The selection indices by discriminant function analysis were constructed based on the data of a population of 1600 F_2 plants developed by crossing a wild species, *Glycine formosona,* with five cultivars of *G.max.* Majority of selection indices were found to be more efficient than straight selection based on yield alone. The selection index consisting of seed yield, pods/plant, seed weight, and oil content gave the highest relative efficiency, followed by the index consisting of seed yield, pods/plant and seed weight. Inclusion of pods/plant in selection index appears to increase its relative efficiency. (Mannur *et. al.*, 1991)
12. **Brassica.** While all the important oleiferous *Brassicas* are susceptible to leaf blight *(Alternaria brassicae),* white mustard (*Brassica hirta* Moench. syn. *Sinapsis alba,* $2n = 2x = 24$,SS) is highly resistant and holds the greatest promise of contributing leaf blight resistance to the adapted cultivars of the cultivated species. Although in vitro hybrids between *Brassica juncea* and *Brassica hirta* have been obtained, the commercial production of such hybrids by conventional breeding methods is practically impossible due to interspecific incompatibility. Therefore, transfer of leafblight resistance from *B.hirta* to the cultivated species has been a cherished goal for many years.

Over 25 manually emasculated flowers of *B.juncea* cv. Varuna ($2n = 4x = 36$, AABB) were pollinated by *B.hirta.* Neither hormonal treatment nor embryo culture was necessary. Only one out of the three seeds obtained at maturity gave rise to viable platn. Pollen fertility was determined by counting more than 500 pollen grains stained with acetocarmine. For meitotic studies, flower buds were fixed in 1:3 acetic acid-ethanol mixture, stored in 70% ethanol overnight, and squashed in 1% acetocarmine.

Morphological features of the trigenomic (ABS) polyhaploid essentially resembled the female parent varuna more closely than *B.hirta.* Hybrid vigour was apparent in the hybrid plant, manifested as greater height, profuse branching and larger siliquae than either parent. Pollen stainability exceeded 90% and the allotriploid F_1 hybrid produced a large number of F_2/BC_1 seed on selfing or back crossing to varuna (Bijral et at., 1991).

Cotton is a secondarily balanced polyploidy

The basic chromosome number 13 of the species *Gossypium* suggests the evolutionary process as being different from euploidy. Aneuploidy might have played an important role in this process. Secondary pairing which is an indication for the homology of chromosomes and secondary polyploidy has been observed in this species. The evidences point out the possibility of the species being a secondary ployploid. Skovested (1937) observed secondary association in wild American species indicating 6 as the basic number. Webber (1934) observed 1 or 2 quadrivalents in certain hybrids between species with $n = 26$ and $n = 13$ and confirmed, the hypothesis that the basic number 13 arose through modified tetraploidy. Sastry and Swaminathan (1960) observed specific secondary association in a plant belonging to *G.barbadense* with ($n = 26$) and attributed the reason for the association as probably homology between a few A and D genome chromosomes. The presence of secondary pairing indicates that the diploid cotton might not be a true diploid but possibly a secondary polyploid. It was postulated that two malvaceous plants, one with $n = 6$ and another with $n + 1 = 7$ crossed to form a trisome and in the course of evolution, due to non reduction of gametes the chromosome complement was doubeld to form 2n=26. Accordingly,

(n = 6) A B C D E F × (n plus 1) AA B C D E F = 7

(2n plus 1) = 13 AAA BB CC DD EE FF

(non-reduction of gametes)

(4n plus 2) = 26 AAAAAA BBBB CCCC DDDD EEEE FFFF

The name hexasomic tetraploid may be attributed to the existing diploid cottons, and the amphidiploid cottons are complex octoploids. By the time the natural crossing of Old world diploid and New world diploid cottons took place, the chromosomes were differentiated. Let us assume the chromosome complement of the old world cotton (AAAAAA BBBB CCCC DDDD EEEE FFFF) as A genome and the chromosome complement of New world cotton (A’A’A’A’A’A’ B’B’B’B’C’C’C’C’D’D’D’D’E’E’E’E’ F’F’F’F’) as D genome.

The natural hybrid, AAA BB CC DD EE FF
A’A’A’B’B’C’C’D’D’E’E’F’F’

was sterile.

The entire chromosome complement was doubled.

AAAAAA BBBB CCCC DDDD EEEE FFFF
A’A’A’A’A’A’B’B’B’B’C’C’C’C’D’D’D’D’E’E’E’E’F’F’F’F’

The resulting hybrid was fertile and the new world allotetraploid cottons are in fact 12 somic allo octoploids. Thus, in the diploid cotton due to the presence of more homologous partners, secondary association would have occurred. During the course of evolution, diploidization took place by micro inversions and mutations under selection pressure and hence there was no multivalent formation and meiotic sterility.

(Vijendra Das, and Mensinkai 1968*a*).

One meiocyte with ten bivalents was observed by Gerstel (1955) in the material (**Gossypium arboreum** x **Gossypium herbaceum**) x **Gossypium arboreum** instead of 13 bivalents in the North Carolina State College, U.S.A. Satisfactory reason as to the reduction in the chromosome complement could not be offered by him. In a similar study Vijendra Das and Mensinkai (1968 *b*) observed, secondary pairing in which three bivalents participated. It is possible that the material examined by Gerstel exhibited secondary pairing or association. In the pre-meiotic mitosis the "three homologous bivalents" or the six homologous chromosomes as shown above might have formed a micro-nucleus due to secondary association and been lost. The other daughter cell with 20 somatic chromosomes might have become differentiated in the another lobe and formed an archesporial cell. This cell could cut off daughter cells by periclinial divisions, the outer ones forming the primary parietal layers and the inner ones giving rise to the primary sporogeneous cells and further meiosis. Another possibility would be that the secondary association resulting in a micronucleus might have taken place in the archesporial cell itself and the primary sporogeneous cell might have been formed with 20 chromosoes, and it might have undergone meiotic division with 20 chromosomes. Thus Gerstel might have observed 10 bivalents at meiosis. The above study also confirms the hypothesis that "Cotton is a secondarily balanced polyploidy". This kind of abnormality might result in a new form in the evolutionary process of **Gossypium**.

REFERENCES

Amal K. Biswas and Ashis K. Mitra, (1990). Interspecific hybridization in three species of sesamum. Indian *J.Genet., 50(4):*307-309.

Anderson, E.J. and Edgar (1939). Recombination in species crosses. Genetics 24:668-678.

Bijral, J.S., B.B. Gupta., K.Singh and T.R. Sharma (1991). Interspecific Hybridization between Brassica juncea (L.) Czem and Brassica hirta Moench. Indian J. Genet. 51(4):476-478.

Brink, R.A., D.C. Cooper, and L.E. Ausherman. (1944). A hybrid between *Hordeum jubatum and Secale cereale* reared from an artificially cultivated embryo. *Jour. Hered.*, 35:67-75.

Brown, M.S. and M.Y. Menzel (1952). Additional evidence on the crossing behaviour of *Gossypium gossypioides. Bot. club 79:* 285-292.

Burton, G.W. (1944). J. Hered. 35:227-232.

Chatterjee, M. and T.A. More., Interspecific Hybridization in Cucumis. Indian *J.Genet.,* 50(4):285-287.

Gerstel, D.U. (1954). A new lethal combination in interspecific cotton hybrids. Genetics, *39:*628-639.

Gerstel, D.U. (1955). A pollen mother cell of Asiatic Cotton with only ten pairs of chromosomes. Cytologia, 20 : 197–198.

Hollingshed, L. (1930). A lethal factor in Crepis effective only in an interspecific hybrid. Genetics, *15:*114-140.

Krishnaswamy, N. and V.S. Raman. (1949). Curr. Sci., 18:15-16.

Krishnaswmay, N. and V.S. Raman. (1953a). Proc. Sci. Wrokers. Conf. 1st, *Agric. Coll. Res. Inst., CBE* pp:43-71.

Krishnaswamy, N. and V.S. Raman. (1953b). Cytologia, *18:*305-321.

Mannar, D.M., P.M. Salimath., S.S. Patil and R. Parameshwarappa (1991). Genetic studies in interspecific crosses of soybean Glycine max (L.) Merill. X Glycine formosona. III selection indices. Indian J.Genet. 51(4):471-475.

Mashima, I. and H. Uchiyamada. (1955). Studies on the breeding of fertile tetrapoid plants of rice.

Oka, H. (1955). Population genetics of rice and barley. *Ann. Rept. Nat. Inst.* Genetics (Japan), 1954(5):40-50.

Patil, B.D. (1963). Indian Farming 12(11):20-23.

Pandey, H.N. (1991). Role of interspecific hybridization in improving Durum wheat. *Indian J. Genet.,* 2:65-67.

Sastry, G.R.K. and Swaminathan M.S. (1960). Chromosome association in haploid Gossypium barbadense. Curr. Sci. 29:398-400.

Sears, E.R. (1944). Inviability of intergenetic hybrids, involving *Triticum monococcum* and *T. aegilopoides.* Genetics, *29:*113-127.

Skovested, A. (1937). Chromosome numbers in the Malvaceae. Indian J. Genet. 31:263-296.

Smith, W.K. (1948). Tansfer from *Melilotus dentata to M. alba* of the genes for reduction in coumarin content. Genetics *33:*124-125.

Stebbins, G.L. Jr. (1950). Variation and Evolution in plants Columbia University Press, New York.

Stebbins, G.L. Jr. (1958). The inviability, weakness and sterility of interspecific hybrids. Adv. in Gen., *9:*147-215.

Thompson, W.P. (1930). Causes of differences in success of reciprocal interspecific crosses. *Amer. Nat.* 64:407-421.

Vijendra Das, L.D. and S.W. Mensinkai (1968a). Pollen mother cell analysis of Intraspecies hybrid of *Gossypium arboreum* Linn., indicating secondarily balances polyploidy. Cytologia, 33(2):188-194.

Vijendra Das, L.D. and S.W. Mensinkai (1968b). A modified interpretion to the occurrence of ten pairs of chromosomes in the PMC of Asiatic cotton as observed by Gerstel. Cytologia, 32 : 320.

Webber, J.M. (1934). Interspecific hybridization i.. *Gossypium* and the meiotic behaviour of F_1 plants. J. Agric. Res. 51: 1047-1070.

CHAPTER 9

Genome Analysis and Evolution of Polyploid Crops

WHEAT

Introduction

Triticum is the classical example of a allopolyploid series. There are 3 groups of species, with chromosome numbers of $n = 7$, 14 and 21 with few exceptions. The species within each group have homologous chromosomes ; and each of the polyploid group apparently has the chromosomes of additional pairs of non-homologous with any of these already present. Hybrids of *Triticum* species are readily made with certain members of the other genera of the sub-tribe Triticianae-namely, *Aegilops, Agropyron*, Secale and Haynaldia, and close relationships are indicated with certain species of *Aegilops* and *possibly Agropyron.*

Relationships in the Triticinae

The chromosome complement of each polyploid species of groups (or genomes) of 7 pairs of chromosomes each. Each of these genomes is considered an entity homologous or homologous with the chromosomes of a particular diploid species. Relationships are indicated by giving each genome a symbol as Kihara and Lilienfeld (1932) and Kihara (1940) especially have done.

Where good pairing at meiosis in the F_1 hybrids of species belonging to same species or different genera is observed, there is sufficient reason for postulating relationship. Where pairing is incomplete, other criteria of relationship particularly morphology and hybrid fertility are considered carefully.

Genome Homologies in Triticum

The correspondence of genome A of the AB tetraploids with A of the diploids is by no means perfect but there seems little doubt that the former is derived from letter; that is the tetraploid group arose as an amphidiploid hybrid of *T.monococcum* or *T.aegilopoides* and some other species with $n = 7$. The other species has not been identified and indeed positive identification may never be possible, because of the many changes that must have occurred in the B genome since its incorporation into the allotetraploid. The letter D is used to represent the third genome of the hexaploid wheats.

After much speculation and research by many investigators (reviewed by Mc Fadden and Sears 1944, 1946) as to the origin of D genome of the hexaploid wheats, this chromosome set has finally been identified as homologous with the genome of *Aegilops squarrosa.* The allopolyploid of T.dicoccoides x Ae.squarrosa has been produced and it closely resembles *T.spelta* leading to the hypothesis that T.spelta arose as an allopolyploid of either *T.dicoccoides* or *T.dicoccum* x Ae.squarrosa. *T.vulgare* is assumed to be a segregate of cross of *T.spelta* with the now extinct Lake Dweller wheat which is believed to have been a tetraploid.

Table 10. The species of Triticum and their Genomic Formulae, (Sears 1948).

S.No.	*Species*	*Chromosome number(n)*	*Formula*
1.	T.aegiolopoides (Link) Bat.	7	A
2.	T.monococcum.L.	7	A
3.	T. dicoccoides Korm	14	AB
4.	T.dicoccum Schrank	14	AB
5.	T.turgidum L.	14	AB
6.	T.perricum(Boiss) Aitch and Hemsl.	14	AB
7.	T.polonicum L.	14	AB
8.	T.durum Dest.	14	AB
9.	T.timopheevi Zhuk	14	AG
10.	T.armeniacum Makushi	14	AG
11.	T.spelta L.	21	ABD
12.	T.vulgareVill	21	ABD
13.	T.macha Dekab and Menab	21	ABD
14.	T.vavilovi (Tuman.) Jakubz	21	ABD
15.	T.compactum Host	21	ABD
16.	T.sphaerococcum Perc	21	ABD

Genome Homologies in Aegilops

The genome *Aegilops* is of interest here chiefly because it has provided the D and possibly the B.genome of common wheat. Two tetraploid species have been synthesised artificially. These are *Ae.cylinderica* from *Ae.caudata x Ae.squarrosa* and *Ae.triuncialis from Ae.caudata x Ae.umbellulata.* The former amphidiploid approaches *Ae.cylinderica* fairly closely in chromosome homology but *Ae.caudata* × *Ae.umbellulata* differs considerably from *Ae.triuncialis* in chromosome constitution.

The species of Aegilops and their Genomic Formula

Section	*Species*	*Chromosome Number*	*Formula*
Anathera	Ae. mutica Boiss	7	Mt
Platystachys	Ae.bicornis (Forsk) Jaub et Spach	7	Sb
			—
	Ae. sharonensis Eig	7	
	Ae.longissima Schw et Muschl	7	S^1
	Ae.speltoides Tausch	7	S
Pachystachys	Ae.squarrosa.L	7	D
	Ae.ventricosa Tausch	14	DM^U
	Ae.crassa Boiss	14	DJ

	Ae.crassa Boiss	21	DJ
	Ae.juvenalis (Thell) Eig	21	-
Monoleptathera	Ae. cylinderica Host	14	CD
Macrathera	Ae.caudata L.	7	C
	Ae.comosa Sibth. et sm	7	M
	Ae.uniaristata Vis	7	M^u
Plionatera	Ae.umbellulata Zhuk	7	C^u
	Ae.triuncialis. L.	14	C^uC
	Ae.variabislis Eig	14	C^uS^v
	Ae.columnasis Zhuk	14	C^uM^c
	Ae.biuncialis Vis	14	C^uM^b
	Ae.ovata L.	14	C^uM^o
	Ae.triaristata Willd	14	C^uM^t

Genome Homologies in Agropyron

There are about 100 species in this genus according to Tzitzin (1940). The degree of polyploidy attained is higher here, extending to $n = 35$. Auto polyploidy is apparently rather frequent.

Of all the wheat x *Agropyron* hybrids produced only *T.vulgare* or *A.elongatam* consistently sets selfed seeds. Tzitzin (1936) obtained hybrids of *T.vulgare x A.intermedium* which set some seed on selfing, and Khijinyak (1937) got selfed seed from several tetraploid wheats x *A.intermedium.* Peto with his cytological observations suggests that *A.elongatum* is AXXXY (or B XXXY), with only one genome homeologous with any in wheat.

Love and Suneson (1945) found an average of about 7 pairs in hybrids of *A.trichophorum* with tetraploid and hexaploid wheats. This suggests that in this species one genome may be homeologous to the A or B of the wheats.

However, there is no direct evidence that any species of *Agropyron* has been involved in the synthesis of wheat species; but Mc Fadden and Sears suggest that the free- threshing characteristic of the hexaploid wheat might have been derived from *A. triticium* following incorporation of this species into an allotetraploid hybrid.

Genome Homologies in Haynaldia: H. villosa ($n = 7$) and *H.hordeacea* ($n = 14$) are crossed with *Secale cereale* ($n = 14$). This genus is obviously closely related to *Triticum, Agropyron and Aegilops.*

Amphiploids in the Triticinae

Combination	*n*	*Amphiploid produced by*
T.monococcum x Ae. uniaristata	7 + 7	Sears 1939b
Ae.speltoides x T.monococcum	7 + 7	Sears 1941
Ae.bicornis x T.monococcum	7 + 7	Sears, 1941
T.aegilopoides x Ae.squarrosa	7 + 7	Sears, 1941
T.aegilopoides x Ae.umbellulata	7 + 7	Sears, 1941
T.aegilopoides x Ae.uniaristata	7 + 7	Sears, 1941
T.aegilopoides x H. villosa	7 + 7	Sears, 1941
Ae.speltoides x Ae. umbellulata	7 + 7	Kihara, 1937b
Ae.speltoides x Ae. sharonensis	7 + 7	Sears, 1941

Ae.speltoides x Ae.uniaristata	7 + 7	Sears, 1941
Ae.speltoides x Ae. uniaristata	7 + 7	Sears, 1941
Ae.caudata x Ae.umbellulata	7 + 7	Sears, 1941
Ae.caudata x Ae. speltoides	7 + 7	Sears, 1941
Ae.caudata x Ae.squarrosa	7 + 7	Sears, 1941
Ae.comosa x Ae.uniaristata	7 + 7	Sears, 1941
Ae.sharonensis x Ae. caudata	7 + 7	Sears, 1941
Ae.sharonensis x Ae. umbellulata	7 + 7	Sears, 1941
Ae.sharonensis x Ae.uniaristata	7 + 7	Sears, 1941
Ae.umbellulata x Ae. uniaristata	7 + 7	Sears, 1941
Ae.umbellulata x Ae.uniaristata	7 + 7	Sears, 1941
Ae.umbellulata x Ae.villosa	7 + 7	Sears, 1941
Ae.bicornis x Ae.squarrosa	7 + 7	Sears, 1941
Ae.caudata x T.dicoccum	7 + 14	Oehler, 1934
T.dicoccoides x Ae.caudata	14 + 7	McFadden & Sears, 1936
Ae.longissima x T.polonicum	7 + 14	Sando, 1935
Ae.longissima x T.durum	7 + 14	Sorokina, 1937
T.turgidum x Ae.speltoides	14 + 7	Britten and Thompson 1941
T.dicoccoides x Ae.speltoides	14 + 7	McFadden and Sears, 1946
T.dicoccoides x Ae.squarrosa	14 + 7	McFadden and Sears, 1946
T.dicoccoides x Ae.umbellulata	14 + 7	McFadden and Sears, 1946
T.dicoccoides x Ae.comosa	14 + 7	McFadden and Sears, 1946
T.dicoccoides x Ae.sharonensis	14 + 7	McFadden and Sears, 1946
T.persicum x Ae.sharonensis	14 + 7	McFadden and Sears, 1947
T.dicoccum x Ae.sharonensis	14 + 7	McFadden and Sears, 1947
T.dicoccoides x Ae.sharonensis	14 + 7	McFadden and Sears, 1947
T.turgidum x H.villosa	14 + 7	Tschermak, 1930
T.dicoccum x H.villosa	14 + 7	Kostoff and Arutinova, 1937
T.dicoccoides x H.villosa	14 + 7	McFadden & Sears, 1946
T.durum x Secale montanum	14 + 7	Derghavin, 1938
T.timopheevi x Ae.caudata	14 + 7	McFadden & Sears, 1946
T.timopheevi x Ae.speltoides	14 + 7	McFadden and Sears, 1946
T.timopheevi x Ae.bicoruis	14 + 7	McFadden and Sears, 1946
T.timopheevi x Ae.uniaristata	14 + 7	McFadden and Sears, 1946
T.timopheevi x Ae.squarrosa	14 + 7	Zhebrak, 1939
T.monococcum x T.persicum	7 + 14	Kasparyan, 1940
T.timopheevi x T.monococcum	14 + 7	Kostaff, 1936b
T.vulgare x S.cereale	21 + 7	Rimpau, 1935
Ae.ovata x T.dicoccoides	14 + 14	Tschermak & Bleier 1926
Ae.ovata x T.durum	14 + 14	Tschermak & Bleier 1926

Ae.ovata x T.dicoccum	14 + 14	Tschermak, 1930
Ae.triuncialis x T.dicoccum	14 + 14	Oehler, 1934
Ae.triuncialis x T.durum	14 + 14	Oehler, 1936
Ae.ventricosa x T.durum	14 + 14	Sears, 1944c
T.durum x T.timopheevi	14 + 14	Zhebrak, 1939
T.persicum x T.timopheevi	14 + 14	Zhebrak, 1939
T.persicum x T.timopheevi	14 + 14	Zhebrak, 1941a
T.polonicum x T.timopeevi	14 + 14	Zhebrak, 1944b
T.timopheevi x T.vulgare	14 + 21	Zhebrak, 1944a
T.durum x T.vulgara	14 + 21	Khijnyak, 1937
T.durum x A.intermedium	14 + 21	Khijnyak, 1937
T.dicoccum x A.intermedium	14 + 21	Khijnyak, 1937
T.polonicum x A.intermedium	14 + 21	Khijnyak, 1937
T.vulgare x A.intermedium	21 + 21	Peto, 1938

The above table gives a list of the known amphiploids in *Triticinae.* No attempt is made to summarize the data on fertility, chromosome pairing etc., in the amphiploids. Pairing is usually poor and fertility is low where the amphiploid chromosome number is 28 or more.

Comparative Gene Analysis

Comparative gene analysis is an analytical study carried out in parallel with a given crop species and relatives in order to elucidate the origin of genes in the former.

1. A single dominant gene Hg in chromosome 1A was found to control glume hairiness of common wheat. The same dominant gene, Hg in chromosome 1A controls the glume hairiness in common wheat, emmer wheat and probably einkron wheat. Overall frequency of the Hg carrier was 43% in emmer wheat and 11% in common wheat. In einkorn wheat the carrier was found only in wild species. From these facts, the wild einkorn is assumed to be a parent of emmer wheat.
2. Waxiness of common wheat and its ancestors is controlled by genes at 4 loci w1 in chromosome 2B (XIII), W2 in D genome, I1-W in 2B(XIII) and I2-W in 2D. The progenitor of common wheat must be of the genotype W1 W2-Wi2-W, consequently its emmer and its squarrosa parents must have had the genotype W1i1-W and W2i2-W respectively. Since the occurrence of *Ae.squarrosa* with the genotype, W2i2-W is restricted to Northern Iran, the birth place of common wheat is assumed to be in this region.
3. The growth habit of common wheat and its ancestors is controlled by genes belonging to three loci sg1, sg2 and sg3 in chromosome 5D,5A and 2B(XIII) respectively. The homologous alleles were found in ancestors of common wheat, *i.e.,* sg2, sg C2, Sg2 Sg3 and sg3 in emmer wheat and Sgc, and sg1 in *Ae.squarrosa.* All waxy strains of the latter had sg1 suggesting that common wheat received its most powerful winter habit gene sg1 from its *squarrosa* parent.
4. Necrosis is controlled by two complementary genes, Nel in chromosome 5B and Ne2 in 2B (XIII). Frequencies of Ne1 and Ne2 genes were 20 and 25% respectively in common wheat.

In emmer wheat, Ne1 was found 64% of the varieties but none of them carried Ne2. It was assumed that the Ne1 gene of common wheat was derived from emmer wheat while Ne2 originated at the 6x level. Asian *T.spelta* was assumed to be in the 6x progenitor of common wheat while the origin of European *T.spelta* was ascribed to an introgression into European *T.aestivum* gene from *T.dicoccum*

5. Chlorosis is controlled by 2 complementary genes, ch1 in chromosome 2 A(II) and Ch2 in 3D. Almost all the varieties of common wheat except those to *T.macha* and all strains of *Ae.squarrosa* parent. The Ch1 gene was found only in *T.macha, T.dicoccoides* and *T.dicoccum.* In this case, *T.macha* seems to have a different origin probably from a hybridization between Asian type *T.aestivum* and Ch1 carrying spelt emmer.

II. Groundnut

Introduction

The cultivated *Arachis hypogaea* is an allotetraploid $2n = 40$ with the basic chromosome number of the genus *Arachis* as 10 and most wild species being diploid with $2n = 20$. In the evolutionary history of peanut species the rate of partioning to fruit was related to survival in different environments. The taxonomic, cytogenetic and biosystematic related to *Arachis* species shall be discussed here.

Taxonomy

Arachis is a large genus with 50 or more species belonging to the family Leguminosae. It is placed in the tribe *Aeschynomeneae,* subtribe Stylosanthinae.

Species of Arachis and their sectional designations

Species	*Section*	*Series*
1. A.batizocoi Krap et.Greg	Arachis	Annuae
2. A.duranensis Krap et Greg.nomnud		
3. A.spegazzinii Krap et Greg.nom nud	"	"
4. A.stenosperma Krap et Greg.	"	"
5. A.ipaensis Krap.et Greg.nom.nud	"	"
6. A.helodes Martius ex Krap.et Rig	"	Perennes
7. A.villosa Benth Var Villosa	"	"
8. A.diogoi Hoehne	"	"
9. A.cardanasii Krap.et Greg.nom.nud.	"	"
10. A.chacoense Krap.et Greg.nom.nud.	"	"
11. A.hypogaea L.	"	Amphiploides
12. A.monticola Krap et. Rig.	"	" "
13. A.guaranitica Chod.et Hass	Erectoides	Trifoliobatae
14. A.tuberosa Benth	"	"
15. A.benthamii Handro	"	Tetrafoliolatae
16. A.martii Handro	"	"
17. A.paraquarieasis Chod et Hassl.	"	"
18. A.oteroi Krap et Greg.nom.nud	"	"

19.	A.rigonii Krap et Greg.	"	Procumbensae
20.	A.lignosa (Chod. et Hassl) Krap. et Greg.nom.nud	"	"
21.	A.repens Handro	Caulorhizae	"
22.	A.pintoi Krap. et Greg. nom.nud.	"	
23.	A.burkartii Handro	Rhizomatosae	Prorhizomatosae
24.	A.glabrata Benth	"	Eurhizomatosae
25.	A.hagenbeckii Harms	"	"
26.	A.marginata Gard	Extranervosae	"
27.	A.lutescens Krap.et Rig.	"	"
28.	A.villosulicarpa Hoehne	"	"
29.	A.macedoi Krap. et. Greg. nom. nud	"	"
30.	A.prostrata Benth	"	"
31.	A.pusilla Benth	Triseminalae	
32.	A.angustifolia (Chod et Hassi) Killip	"	"

The centre of origin of the genus *Arachis* is probably in the 'Planaltine ellipse' remarked by above 550m in the Brazilian shield. However, *A.hypogaea* most likely originated in the North-western Argentine-southern Bolivia region of South America. *Arachis monticola* the only known tetraploid species which is cross compatible with *A.hypogaea* is found in this area. Hammons (1982) also concluded that the southern Bolivia region is an important centre of variability of *A.hypogaea* sub species *hypogaea.* Members of this subspecies are believed to be the most closely related to wild *Arachis* species because they have runner habit, similar branching pattern to other Arachis species and lack compound spikes. However, based on karyological symmetry, Stalkar and Dalmacio proposed that members of *A.hypogaea* subsp. *fastigiata* var. vulgaris comprise the most ancient cultivated group. Further more, members of var.*vulgaris* have karyotypes very similar to *A.monticola.* Gregory and Gregory (1976) described six secondary centres of variations in South America. Africa is an important tertiary centre of diversity of the cultivated species.

Although Husted (1936) proposed that *A.hypogaea* was of direct amphidiploid origin, Krapovickas and Rigoni (1957), Smartt and Gregory (1967), suggested that the cultivated groundnut evolved form a pre-existing wild allotetraploid. Further more, an annual and perennial interspecific hybrid within section *Arachis* was considered important in the evolution of *A.hypogaea* (Gregory and Gregory, 1976). Several diploid species of section *Arachis* have been suggested as possible progenitors including *A.villosa, A.duraensis, A.batizocoi, A.duranensis* and *A.cardenasii* (Gregory and Gregory, 1976), 1976), *A.cardenasii* and *A.batizocol* (Smart et al. 1978) and *A.cardenasii* (Singh and Moss). Smart and Stalker (1982) concluded the most likely ancestors are *A.cardenasii* and *A.batizocoi* among the species currently in collections. Presently the diploid progenitor species of *A.hypogaea* have not been verified.

Cytogenetics of Arachis Species

The first comprehensive reports of *Arachis* cytogenetics was by Husted (1933, 1936) for *A.hypogaea.* He confirmed the somatic chromosome complement of six cultivars and *A.nambyguarae* (synonymous with *A.hypogaea)* as $2n = 4x = 40$. In most metaphase I pollen mother cells, chromosome pairing was 20 II but a few multivalents were also observed. Hybrids among the varieties of *A.hypogaea* showed the same patterns as parental materials with mostly bivalents and a few multivalents. Stalker (1980) observed cultivated lines of spanish (subsp. *fastigiata* var *vulgaris*), valencia (subsp, *fastigiata* var *fastigiata*) and virginia (subsp.*hypogaea* var *hypogaea*) are found similar patterns of chromosome pairing furthermore, Gregory *et at.* (1980) reported reduced fertility in F_1 hybrids between alternately and sequentially branching genotypes which correspond to the sub species *hypogaea* and *fastigiata* of *A.hypogaea* respectively.

The chromosomes of *A.hypogaea* are small and difficult to karyotype. Husted, (1933) was able to distinguish a chromosome pair with a secondary construction and one pair was distinctly smaller than the others.

Several types of secondary constriction chromosomes have been reported in *A.hypogaea* by Baba (1955). Groundnut cultivars can be distinguished based on differences among chromosomes.

The first chromosome report for a wild species was $2n = 40$ for *A.glabrata* (Gregory, 1946). Mendes reported $2n = 20$ for *A.diogoi, A marginata, A.prostrata* and *A.villosulicarpa.* At least two chromosome series $2n = 2x = 20$ and $2n = 4x = 40$ are present in the genus. Polyploidy arose twice once in section *Arachis* and a second time in section *Rhizomatosae* (Smartt and Stalker). However, most species in the genus are diploids.

Meiotic studies of *Arachis* species have revealed that chromosome pairing in diploids is normally 10 II. However, quadrivalents have been reported at a low frequency in the two diploid species *A.villosa* and *A.spegazzinii.* Meiosis in *A.monticola* is generally normal with 20II and occasional multivalents. Meiosis in tetraploid rhizomatous taxa is less regular than in *A.monticola* or *A.hypogaea* which may be due to vegetative reproduction.

Raman (1959) and Smartt (1964) reported a distinctively small chromosome in most diploid species of section *Arachis* except *A.batizocoi.* This was reported by many other investigators and it was concluded that the genome of *A.batizocoi* was significantly different from the other species of the section.

Interspecific Hybridization

To realize the potentials for germplasm introgression to *A.hypogaea,* biosystematic relationships among a broad spectrum of species must be known. Gregory and Gregory, (1976) reported that the cultivated groundnut will only hybridize with members of this section namely *Arachis.* Twelve diploid species have been successfully hybridized with *A.hypogaea.*

Smartt (1964) reported an average of 0.95 III for an *A.hypogaea* x *A.villosa,* var corentina hybrid, 3.40 III for *A.hypogaea x A.helodes* and 2.15 III for *A.hypogaea x A.durensis.* Because a common *A.hypogaea* parent was used for all 3 hybrid combinations, the chromosomes of *A.villosa* can be inferred to be more similar to the cultivated parent than the other two species.

To restore fertility, in sterile interspecific hybrids with *A.hypogaea* plants have been obtained at the hexaploid ($2n = 60$) level by obtaining naturally occurring polyploids or by colchicine treating vegetative tissues. Meiosis was very irregular in hexaploids. Hexaploid *x* tetraploid crosses resulted in pentaploid progenies in which the meiosis was more regular than expected.

In addition to hybridizing *A.hypogaea* with diploid members of section *Arachis* the cultivated groundnut hybridizes with *A. monticola.*

Genomic Evolution and Germplasm Pools

Interspecific hybridization in the genus *Arachis* is difficult and most interspecific hybrid combinations are sterile or only semifertile. Although cytological analyses have not been completed in the genus, a series of genomes for diploid species was proposed by Smartt and Stalker based on extensive hybridization programmes of Gregory and Gregory (1976) as follows:

A = Section *Arachis,* perennials and most annuals.
B = Section *Arachis (A.batizocoi)*
Am = Section *Ambinervosae*
C = Section *Caulorhizae*
E = Section Erectoides (Subgenomes E1, E2, E3 corresponding to series?)
EX = Section *Extranervosae.*
T = Section *Triseminalae*
R_1 = Section *Rhizomatosae* series *Prorhizomatosae*

Another genome for section *Arachis* was proposed by Stalker and Cross (1983) for taxa *Arachis* Sp.GKSSc 30091 and *Arachis* sp GKSSc 30099. The Karyological data indicate that the species is of recent origin. The D genomic designation is proposed for the genome. As cytological analysis is performed in interspecific hybrids and species relationships are better understood, the numbers of genomic groups will probably increase. For example, cytological analysis of (*A.rigonii,* GKP 10034 (E3) x *A.paraquarieasis,* GKP 9646 (E2) indicated that the chromosomes in different species of section Erectoides series are non-homologous and that series designations probably represent distinctive genomes.

Gregory and Gregory (1979) concluded that diploid ancestors of section *Rhizomatosae* probably hybridized with a diploid member of section *Erectoides* and after chromosome doubling, gave rise to the tetraploid rhizomatous species.

Section *Arachis* also has two polyploid members, *A.hypogaea* and *A.monticola*, but the origin of polyploidy in this group was independent from *Rhizomatosae. Arachis hypogaea* is an allotetraploid probably with A and B genomes. Eventhough diploid members of section *Arachis* will hybridize with *A.hypogaea* isolation due to polyploidy restricts genome introgression. The polyploids of section *Arachis* are apparently completely isolated from other members of the genus.

Based on the cross-compatibility relationships of *A.hypogaea* and other species, a germplasm pool system can be hypothesised. *Arachis hypogaea* and the closely related wild species *A.monticola* are in the primary gene pool and the other diploid members of section *Arachis* are in the secondary gene pool, and other members of the genus are in a tertiary gene pool.

Polyploidy in Evolution

Cyto-taxonomical studies and correlation of polyploidy and geographical distribution of the species have revealed that the polyploids are more frequent in the Northern latitudes and higher altitues than in the tropical zones. It has also shown that pereniality is more common in the polyploids. Darlington gives the following in regard to rhizome in relation to polyploidy.

Species	2n	4n	6n	8n
	no rhizome	little rhizome	much rhizome	extensive rhizome
	↓	↓	↓	↓
	Chrysanthemum coronarium	*C.indicum*	*C.morifolium*	*C.deciseanum*

Interspecific and intergeneric species hybrids exhibit generally various degrees of sterility. Unless the hybrid is perennial or it can be vegetatively propagated, its perpetuation would be difficult and it may disappear. Autoploidy increases sterility due to multivalent chromosome association especially in long chromosomes, whereas autoploids with short chromosomes form bivalents due to pairing conditions, Immediately after the origin of autopolyploidy they are generally sterile but in later generations fertility increases due to elimination of chromosomal irregularities, *e.g., Cicer arietinum.*

In amphidiploids a small percentage of sterility may be met with in initial stages but this is overcome in later generations, *e.g.*, amphidiploids. *B.campestris x B.nigra.* Newly evolved amphidiploids are not very stable under natural conditions and they are subjected to certain changes before they form stable species in nature.

The secondarily balanced polyploids may ultimately come to have a new basic number eg., pomoideae-Darlington showed that the apple is fundamentally a complex polyploid. It is supposed to have been derived from an original basic number of 7 and now shows $n = 17$. In evolution of species the chromosomal changes have proceeded both in plus and minus direction, *i.e.*, by addition and loss of chromosomes.

In an autotetraploid for, *e.g.*, the two extra sets of genes are available for mutation without seriously affecting the existing characteristics. If the mutation is deleterious it is at once reflected in the progenies of diploids whereas in the polyploids the deleterious genes are sheltered by the presence of other normal allelic genes. If the mutation is in favourable direction it adds to the selection value. In the course of time autotetraploids behaves like a diploid by a series of mutational changes in the extra genes. By differential changes between the extra chromosomal pairs, in course of time behaves functionally diploid. Polyploids increase variation and make it possible for a number of sub-species.

Origin of Some Crop Plants

1. In banana *Musa paradisiaca*, autotetraploids are inferior to triploids in that they have weaker leaves and increased fertility. But, they offer the only available chance of adding disease resistance to commercially successful varieties. In banana, autotetraploids are produced by chance fertilization of an unreduced triploid egg (AA) by a haploid pollen from the disease resistant diploid parent. A large number of such tetraploids have been produced, but they have not yet gained any commercial success.
2. The tetraploids maize produced by colchicine treatment in USA has 43 per cent more carotenoid pigment and vitamin A activity than the diploid.
3. Certain distant crosses are not successful at the diploid level, but are relatively successful at the autotetraploid level. For example, 4 *x Brassica oleracease x B.chinensis* is successful, but when *B.oleraceae* is diploid it is unsuccessful.

4. The autotetraploid varieties of forage crops have been considerably successful. The most successful examples are, tetraploid red clover (*Trifolium pratense*) are ryegrass (*Lolium perenne*), other examples are tetraploids of alsike clover (*Trifolium hybridum, variety tetra*) and berseem (*Trifolium alexandrium*), variety pusa giant. Autotetraploid red clover and ryegrass are more vigorous, digestible and palatable, and have greater resistance to nematodes as compared to the diploids.
5. The experimentally produced distant hybrids were highly sterile, occasional branches were fertile and had twice as many chromosomes as present in the F_1 hybrid, *e.g.*, F_1 from the cross *Primula verticillata x P.floribunda* produced a fertile branch with twice the chromosome number. The progeny obtained form this fertile branch were allotetraploid and were named as *P.kewensis.*
6. Allotetraploids generally combine the morphological and physiological characteristics of the parent species. But it is difficult to predict the precise combination of characters that would appear in the allopolyploid species. For example, the aim in producing *Raphano Brassica* was to synthesis a crop species that would combine the root of radish (*R.sativus*) and the leaves of cabbage (*B.oleraceae*). *Raphano Brassica* did combine the characteristic root and shoot systems of the two parental species, but in the opposite direction, that is, it had leaves like radish and roots like cabbage. (Experiment of *Russian Botanist Karpechenko*), *Brassica juncea* ($2n$ = 36) was synthesised by Frandsen during 1943 through crosses between *B.nigra* ($2n$ = 16) x B.campestris ($2n$ = 20). The synthesised amphidiploid closely resembled *B.juncea* and crossed readily. The offspring were fertile and meiosis was normal with regular bivalents.
7. Another feature of allopolyploids is that many of them are apomictic. Apomiotic species are common in grasses, *e.g.*, poa. Apomicts are known in many other groups of plants, *e.g.*, *Taraxacum, Parthenium, Rubus, Fritillaria, Tulipa, Solanum.* Apomixis arises in response to near complete sterility of interspecific hybrids. Such sterile plants are able to reproduce asexually through apomixis.
8. Polyploid series of some of the crop plants

Scientific name	*Common name*	*Gametic chromosome number (n)*
Brassica nigra	Black mustard	8
B.oleraceae	Cauli Flower or Cabbage etc.	9
B.campestris	Sarson	10
B.carinata	—	17
B.juncea	Indian Mustrad	18
B.napus	Rape	19
Gossypium arboreum	Asiatic cotton (desi)	13(A2)
G.herbaceum	Asiatic cotton	13(A1)
G.thurberi	Wild American cotton	13(D1)
G.barbadense	Sea island cotton	26(A2D2)
G.hirsutum	American upland cotton	26(A1D1)
Nicotiana sylvestris	Wild tobacco	12
N.tomentosa	Wild tobacco	12

N.tabacum	Cultivated tobacco	24
N.rustica	Cultivated tobacco	24
N.bigelovii	Wild tobacco	24
N.debneyi	Wild tobacco	24
Saccharum officinarum	Nobel canes	40
S.barberi	Indian canes	41, 45, 46, 58, 62
S.sinense	Indian canes	58, 59
S.spontaneum	(Kans) wild canes	20-64
S.robustum	Wild cane	30-74
Sorghum versicolor	Wild sorghum	5
S.bicolor	Jowar	10
S.halepense	Johnson grass	20
Triticum boeticum	Wild eincorn	7(A)
T.monococcum	Eincorn wheat	7(A)
T.dicoccoides	Wild emmer	14(AB)
T.dicoccum	Emmer wheat	14(AB)
T.turgidum	Solid stem wheat	14(AB)
T.carthlicum	Persian wheat	14(AB)

9. **Triticum aestivum.** Evolutionary history of wheat has been the most extensively investigated. Identity of the diploid species contributing the 3 genomes (A, B and D genomes) of *T.aestivum* has been investigated by many workers more notably by Sears, Kihara and others. It is generally accepted that the A genome present in diploid wheats is the same to those present in tetraploid and hexaploid wheats. Further, the B genome of tetraploid emmer wheats is similar to that found in hexaploid wheats. This is evident from chromosome pairing in crosses among diploid, tetraploid and hexaploid wheats. Hybrids between diploid and tetraploid wheats show about 7II and 7I, while those between tetraploid and hexaploid wheats show about 14II and 7I.

The possible evolutionary history of hexaploid wheat is as follows:

T.monococoum	×	*unknown species* *(Aegilops speltoides)*
$2n$ = 14(AA)		$2n$ = 14(BB)
F_1 $2n$ =		14AB (sterile)
		Chromosome doubling
Allotetraploid	×	*Aegilops squarrosa*
$2n$ = 28 (AA BB)		$2n$ = 14(DD)
F_1 $2n$ =		21(ABD)
		Chromosome doubling

Hexaploid wheat
$2n$ = 42(AA BB DD)

10. **Nicotiana tabacum.** *N.tabacum* (n = 24) is most likely an amphidiploid from the cross *N.sylvestris x N.tomentosa;* both the species are diploids with n = 12. The interspecific hybrids *N.tabacum x N.sylvesteris* and *N.tabacum x N.tomentosa* produce 12II and 12I at metaphase I. This indicates homology between chromosomes of *tabacum* and those of *sylvestris* and *tomentosa.* The amphidiploid from the cross *N.sylvestris x N.tomentosa* is similar to *N.tabacum* in many characteristics, which further supports the above conclusion. The species *N.tabacum* has undergone considerable differentiation during its evolutionary history, mostly due to accumulation of gene mutations and to some extent due to loss of some duplicated segments of the two genomes.
11. **Gossypium hirsutum.** The 9 Old world and the 8 New world species of *Gossypium* have n = 13, but the chromosomes of the New world species are smaller than those of the Old world species. The other species, *G.hirsutum, G.barbadense* and *G.tomentosum* (wild Hawaii cotton), have n = 26; in these species, 13 chromosomes are relatively larger than the remaining 13. A possible origin of *G.hirsutum* is from the cross between Asiatic cotton *G.arboreum x G.thurberi* (American wild cotton), followed by chromosome doubling of the interspecific hybrid. According to *q* more recent scheme *G.hirsutum* has originated from the cross *G.herbaceum* var. *africanum x G.raimondii,* followed by chromosome doubling of the F_1.
12. **Brassica species.** *B.juncea* (n = 18) is an amphidiploid from *B.nigra* (n = 8) *x B.campestris* (n = 10), *B.napus* (n = 19) is an amphidiploid from the cross *B.oleracea* (n = 9) *x B.campestris* (n = 10); *B.carinata* (n = 17) is an amphidiploid from the cross *B.nigra* (n = 8) *x B.oleracea* (n = 9). The synthetic allopolyploids produced, resemble the natural amphidiploids, cross easily with them, and the hybrids between synthetic and natural amphidiploids are reasonably fertile.
13. **Difficulties in tracing the origins.** The determination of parental diploid species of the present day allopolyploids presents many difficulties.
    (*i*) The parental diploid species may have become extinct, that is, they may not be present in the nature any more.
    (*ii*) The parental diploid species may be expected to have differentiated considerably in comparison to when they entered the interspecific hybridisation to produce the allopolyploid in question.
    (*iii*) The allopolyploids themselves may have differentiated to a great extent from the state when they were first produced. Thus, the genomes in allopolyploids may have changed to various degrees from those of the parental species.
14. **Bridging species.** Amphidiploids serve as a bridge in transfer of characters from one species to a related species, generally from a wild species to a cultivated species. Use of an amphidiploid as a bridging species becomes necessary when the hybrid between the cultivated species (recipient species) and the wild species (donor species) is sterile.

Examples

Transfer of resistance of tobacco mosaic virus

Nicotiana tabacum	×	*N.glutinosa*
$4n$ = 48 (AA BB)		$2n$ = 24(CC)

$3n$ = 36F_1ABC 1 to 2 IIs

N.tabacum	×	*N.digluta* (AABBCC) 1 to 2 IVs
$4n = 48$ (AA BB)		($6n = 72$)
$5n = 60$		AA BB C 12II + 11II + 1III + 11I

Self

Select a trisomic carrying one chromosome of C
genome + 48 chromosome of tobacco

Self the trisomic (23 pairs + 1 III) = 46 + 3= 49

Select a 4n=48 plant. The C chromosome has substituted one of B.

Self (recombination between B & C)
Select a plant showing resistance to mosaic

15. **The addition of chromosome of Hayanldia villosa to common wheat**

Stage I parent:	*Triticum dicoccoides*	×	Haynatdia villosa
	$2n = 28$ AA BB		$2n = 14$ VV
	F_1 hybrid		ABV $2n = 21$
	Amphidiploid		AA BB VV $4n = 42$
Stage II parent:	Wheat	×	amphidiploid
	AA BB DD		AA BB VV
	(21 II)		
			AA BB DV ($4n = 42$) 14 II +7 ID + 7 IV
Stage III 1st BC to wheat:	Wheat	×	F_1 hybrid
	AA BB DD		AA BB DV 14 II + 14I
Selected plant from BC.	AA BB DDV		21 II + 7I
	(21 II)		
			(7I villosa chromosome)
2nd BC to whet	Wheat	×	Selected BC plant
	AA BB DD		AA BB DDV
	(21 II)		(21 II + 7I)

AA BB DD + 0 to 7 'V' chromosome
$4n = 42$ to 49

Addition of trisomic	43 Chromosome plant selected
from 2nd B.C.	21 II + 1I (V)
	Selfed
	44 chromosome plant
	21 II + 1 II

Final addl. line: Wheat chromosome *Haynaldia* chromosome pair

16. **Gene substitution. Transfer of leaf rust resistance from Aegilops umbellulata to wheat**

In crosses where the parent species differs in chromosome number and homology, the crosses are more difficult. In crosses between widely differing species, the breeder usually wishes to transfer only a single gene for a superior character like disease resistance from a wild to the cultivated species. This can be accomplished only by an exchange of a very small segment of a single chromosome from the wild species, which bear the desired gene, with a corresponding segment of a homologous chromosome from the cultivated species. In this exchange it is important that deleterious and undesirable genes should not be brought in with the desired gene. A successful exchange might thus be limited to a segment of the chromosome bearing a single gene.

REFERENCES

Baba, C.N., (1955). Cytogenetical investigations on groundnuts. The somatic chromosomes Ind. J. Agric. Sci. 25:41-46.

Britten, E.J. and W.P. Thompson, (1941). Science 93:479.

Derzhavin, A., (1938). Social Reconstr. Agric. 213-219.

Gregory, M.P. and W.C. Gregory. (1979). Exotic germplasm of Arachis L. Interspecific hybrids. J.Hered. 70:185-193.

Gregory, W.C. (1946). Peanut breeding programme underway. Research and farming. Annual Report, North Carolina Agricultural Experiment Station 69:42-44.

Gregory, W.C. and M.P. Gregory, (1976). Groundnut Arachis hypogea (Leguminosae-Papilionatae). Pages 151-154 in Evolution of crop plants (Simmonds. N.W. ed.) London, U.K. Longman.

Gregory, W.C. A.Krapovickas and M.P. Gregory. (1980). Structure, variation, evolution and classification in Arachis. Pages 151-154 in Evolution of crop plants (Simmonds, N.W. ed.) London, U.K. Longman.

Gregory, W.C., A.Krapovickas and M.P. Gregory. (1980). Structure, variation, evolution and classification in Arachis. Pages 469-481 in Advances in legume science. Proceedings of the International Legume Conference, 24-29 July 1978. Kew, Surrey, U.K. (Summerfield, R.J. and A.H. Bunting, eds.) vol. 2 Kew, Surrey U.K. Royal Botanic Gardens.

Hammons, R.O., (1982). Origin and early history of the peanut. Pages 1-20 in Peanut science and technology (Paltee, H.E. and C.T. Young, eds.) Yoakum, Texas, U.S.A. American Peanut Research and Education Society.

Husted, L. (1933). Cytological studies of the pea nut Arachis. I Chromosome number and morphology, Cytologia, 5:109-117.

Husted, L. 1936. Cytological studies of the peanut Arachis. II. Chromosome number morphology, and behaviour and their application to the origin of cultivated forms. Cytologia 7:396-423.

Kasparayan, A.S. (1940). C.R. Acad. Sci. U.R.S.S. 26:166-169.

Khijnyak (Hizniak), V.A. (1937). C.R. Acad. Sci. U.R.S.S. 17:489-490.

Khihara, H., (1937). Bot. Mag. Tokyo, 51:584-589.

Khihara, H. (1940). Zucbter, 12:49-62.

Kihara, H. and F. Lilienfeld, (1932). Cytologia, Tokyo, 3:384-466.

Kostoff, D. (1936). Z.Pflanzenz. 21:41-45.

Kostoff, D. and N.Arutiunova (1937). Curr. Sci. 5:414-415.

Krapovickas A. and V.A. Rigoni (1957). Neuvas especies de Aaachis Vinckredas al problema del origin delmani. (In. Eb). Darwiniama 11:431-455.

Love, R.M. and C.A. Suneson. (1945). Amer J.Bot. 32:451-456.

Mc. Fadden, E.S. and E.R. Sears (1944a). Rec. Genet. Soc. Amer. 13:26-27.

Mc.Fadden, E.S. and E.R. Sears. (1944b). J.Hred. 37:81-89, 107-116.

Mc. Fadden, E.S. and E.R. Sears. (1947). J. Amer. Soc. Agron. 39:1011-1026.

Oehler, E. (1934). Zuchter, 6:263-270.

Oehler, E. (1936). Zuchter, 8:29-33.

Peto, F.H. (1936). Canad J. Res. C. 14: 203-214.

Raman, V.S. (1959). Studies in the genus Arachis VII. A natural interspecific hybrid. Indian Oil seeds Journal 3:226-228.

Sando. W.J. (1935) J. Hered. 26:229-232.

Schiemann, E. (1932). Handb. Vererbungsw 3(1), pp. 377.

Sears, E.R. (1939). J. Hered. 30:38-4.

Sears, E.R. (1941). Res. Bull. Mo. Agric. Exp. Sta. 336: p. 46.

Sorokina, O.N. (1937). Bull. appl. Bot. Genet, Plant Breed. Ser. 2(7): 161-173.

Tschermak, E.V.on. (1930). Ber. dtsch. bot. Ges. 48:400-407.

Tschermak, E.Von. & H. Blaier. (1926). Ber. detsch. bot. Ges: 44: 110-131.

Tzitzin, N.V. (1940) Breed. Seed Growing, 10:4-7.

Zhebrak, A.R. (1939). C.R. Acad. Sci. U.R.S.S. 25:56-59.

Zhebrak, A.R. 1941. a. C.R. Acad. Sci. U.R.S.S. 31:485-487.

Zhebrak, A.R. 1941. b. C.R. Acad. Sci. U.R.S.S. 31:617-619.

Zhebrak, A.R. 1944. a. C.R. Acad. Sci. U.R.S.S. 43:120-121.

Zhebrak, A.R. 1943. Social Agric. 8:51-57.

Zhebrak, A.R. 1944. b. Nature, London 153:549-551.

CHAPTER 10

Cytoplasmic-Genic Male Sterility Systems in Hybrid Seed Production

Introduction

Hybridization either for the production of hybrid varieties or for the introduction of genetic variation as a basis for selection involves crossing female plants with male plants. In order to prevent self pollination, it is convenient that the female parent is devoid of functional pollen. This is difficult in plants because most are normally bisexual. Emasculation by hand with tweezers is extremely laborious and time consuming and seed set is often low as a result of mechanical damage to ovaries. Alternative methods of emasculation, like removal of anthers by vacuum suction or killing the pollen grains with hot water treatments in most cases do not appear reliable. Chemical induction of male sterility may be useful but the chemicals are not cheap and they are potential environmental pollutants and are carcinogenic. There are 3 types of male sterility namely,

1. Genetic or genic male sterility in which pollen sterility is governed by one or occasionally more genes.
2. Cytoplasmic male sterility in which pollen sterility is brought about by extrachromosomal hereditary particles.
3. Cytoplasmic/Genetic male sterility in which pollen sterility results from the concerted action of both extrachromosomal factors and genes.

Among these 3 types, cytoplasmic/genetic male sterility is more useful for hybrid seed production. The male sterility was first reported by Jones and Davis (1944) who found C type of male sterility in onions.

Cytoplasmic-Genetic Sterility (Fig. 8)

This type of sterility depends upon the interaction of a particular gene with a particular cytoplasm. Neither the genetic factor nor the cytoplasmic factor alone is able to cause male sterility; both are required to cause sterility.

Where the gene (ms) stands for sterility and (MS) for fertility, the gene (MS) is dominant to (ms). A plant will be male sterile only if it carries sterile cytoplasm (s) and when both genes are also msms. Plants with sterile cytoplasm will be fertile if they have MSMS or MSms genotype. Plants with fertile S cytoplasm will produce viable pollen whatever may be the genotypes *viz.* MSMS or MSms or msms.

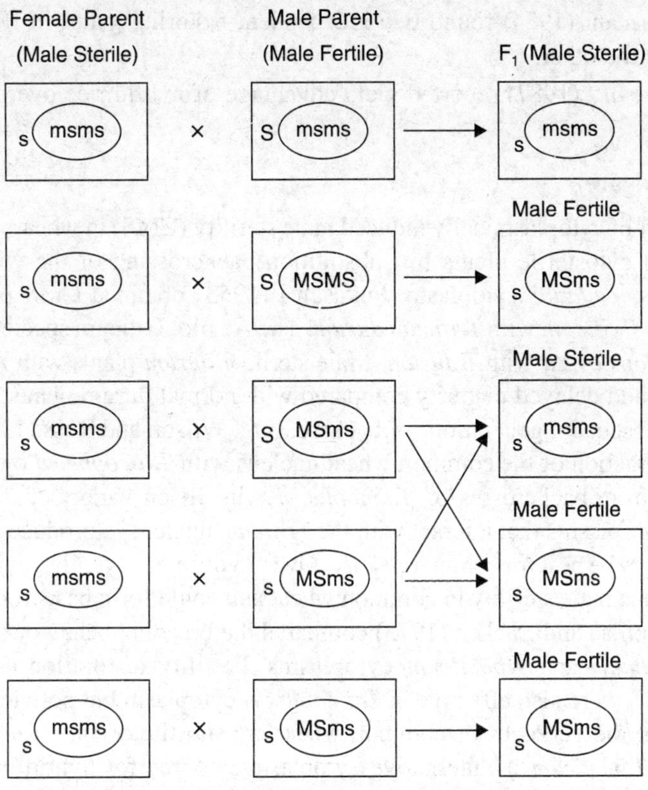

Fig. 8. Cytoplasmic—Genetic male sterility.

Cytoplasmic-Genetic male sterility in Rice: (Oryza sativa)

The role of cytoplasm in causing male sterility in rice was first reported in 1954 by Weeraratne, Sampath and Mohanty. Shinyo and Omura (1966) developed the first cytoplasmic male-sterile line in cultivated rice by substituting nuclear genes of a Japonica Variety, Taichung 65, into the cytoplasm of Indica variety Chinsurah Boro II. Watnable *et al.*, (1968) observed male sterility in the progeny of Indi-Japonica cross (Lead X Fujisaka 5), but those of the reciprocal cross was fertile. However, no male sterile line was developed. Carnahan *et al.*, (1972) developed cytoplasmic male sterile lines from crosses of an Indica variety, Birco (PI279120) with Californian Japonica rice varieties calrose, caloro, and colusa, the F_1 plants were almost completely sterile. Watanable (1971) reported development of cytoplasmic-genetic male-sterile lines by means of Indica-Japonica crosses.

Athwal and Virmani (1972) developed a cytoplasmic male sterile line at the IRRI by substituting nuclear genes of Indica rice variety Pankchari 203 into the cytoplasm of a semidwarf Indica variety, Taichung Native 1. The first cytoplamsic male-sterile line used to develop commercial F_1 rice hybrids was developed in China in 1973 from a sterile plant (wild-aborted) occurring naturally in a wild rice population *(Oryza sativa f. spontanea or O. perennis)* on Hainan Island in 1970. Rutger and Shiny (1980), studied the distribution of male sterile cytoplasms in various geographical forms of *O.perenis.* In Asian and American strains, the frequency of male-sterile cytoplasm were about 64 and 4% respectively. No male sterile cytoplasm was found in the African and oceanean strains.

Young *et al.*, (1983) classified six CMS lines of different cytoplasm sources into four categories. They included pankhari 203A and Wu10A in S1, V20A and Zhenshan 97A in S2, Yar-Ai-Zhao A in S3 and MS577A in S4 categories.

Young and Virmani (1984) found two independent restoring genes in IR54, one of them with stronger fertility than the other.

He Gui-Ting *et al.,* (1987) reported yield advantage of hybrid rice over commercial rice is 15 percent.

Wheat (Triticum vulgare)

The first report of cytoplasmically induced male sterility (CMS) in wheat was reported by Kihara (1951) who obtained cytosterile plants by substitution backcrossing of the whet genome *(Triticum vulgare)* into *Aegilops caudata* cytoplasm. Fukasawa (1953) obtained CMS plants from successive backcrosses of *Aegilo-Triticum* with *Triticum durum.* The Aegilo-Triticum species had been synthesised from a cross of *Aegilops ovata* with *T.durum.* Male sterile *T.durum* plants with *A.ovata* cytoplasm had reduced plant height and delayed maturity compared with normal durum plants and these effects were consistent through subsequent generations of backcrossing. Wilson and Ross (1962) established a male sterility from the interaction of the common wheat nucleus with *T.timopheevi* cytoplasm and identified pollen sterile plants from backcrosses of *T.timopheevi* with Bison variety of *T.aestivum* as recurrent parent. Among the cytoplasms that interact with the *Triticum* nucleus to produce a male sterile reaction, *T. timopheevi* appears to be best for commercial use. Over 15 different cytoplasms have been 'recognised, several of which induce male sterility in common wheat and could form basis for alternative systems of hybrid production. Gomaa and Lucken (1973) compared the breeding behaviour of the restores R5 and BR 4704 in *T.timopheevi* and *T.boeticum* cytoplasms. Fertility restoration (Rf) genes effective in *T.boeticum* cytoplasms were also effective in *T.timopheevi* cytoplasm but not vice-versa. The reduction in vigour noted when the genomes of common wheat are substituted into *T.bocticum* cytoplasm has curtailed the use of *T.boeticum* as alternative cytoplasmic source for hybrid breeding. Mann (1973) observed that number of factors will influence the expression and detection of cytoplasmic effects. They are: (1) Stability of the cytoplasm; (2) Stability of the nuclear genome; (3) Choice of cytoplasm and genome donor; (4) Persistence of nuclear genes from the cytoplasm donor and (5) Genotype-environment interactions. Ghiasi and Lucken (1982) compared reactions of *A.speltoides* and *T.timopheev* cytoplasms to various restorer gene combinations and examined a number of agronomic quality traits and concluded that A. speltoides cytoplasm can be used interchangeably with that of *T.timopheevi* in hybrid wheat breeding.

Sorghum (Sorghum bicolor)

The discovery of cytoplasmic-nuclear male sterility in Sorghum was initiated by Stephens and its development for hybrid seed production by Stephens and colleagues (1954). The male sterility in this system is caused by the interaction of milo cytoplasm and Kafir nuclear genes. The milo cytoplasm which induces male sterility in the female parent is passed into hybrid. Here the use of milo cytoplasm restricts nuclear diversity. The use of female parents with different cytoplasms would provide hybrids with different cytoplasms and a measures of security associated with diversity (Table 11).

Mital *et al.,* (1958) reported the occurrence of a male sterility in line IC 2360. A male sterile line in this line was subpollinated to produce the sterile line. Rao (1962) reported male sterile plants in varieties W.E. 1, Blichigan, Red Jonna, Indore local, G.O. 103, B.D.8, Burma black, Norghum and C.10-2. Hussani and Rao (1964) reported two additional male sterile plants in PJ 22K and a white seeded variety from Warangal. Appathurai (1964) isolated a sterile line from G.1 and designated it G. 2-S, Nagur and Menon (1974) reported six male sterile lines including CK 60. Rao (1972) in the *Sorghum* in seventies symposium reviewed the work on male sterility and provided information of new male sterile lines developed in milo cytoplasm. Tripathi *et al*., (1980) provided additional information regarding sterility as well as other characters on six male sterile lines some of which differed only in cytoplasms.

Besides milo cytoplasms in U.S.A., cytoplasms from diverse source are classified on the basis of fertility restoration into four groups viz. A1, A2, A3 and A4 (Worstel et al., 1984). Indian male steriles were grouped under three categories viz. A1, A2, and A3 (Rao et al., 1984).

Table 11. Some of male sterile lines and incompatible lines studied by different authors:

Authority	Line	Source
Mital et al (1958)	Ic2360	
Rao (1962)	WE1 Bilichigan Red Jonna Indore local	
Hussani and Rao (1964)	PJ 22K White seeded variety from Warangal	
Appathurai (1964)	G 2S	G1 Strain Guntur, India
Rao and Gouda (1966)	M 312	M31-2
Nagur (1971)	ms CK 60 milo cytoplasm msG1 Durra cytoplasm msVZM1 durra cytoplasm msVZM2 durra cytoplasm msM35 1 dochha cytoplasm msM31-2 Cernum cytoplasm	USA, Guntur, India. from Vizanagaram, India from Vizanagaram, India from Kovilpatti, India from Dharwar, India
Nagur and Menon (1974)	msCK60A (S_1) msG-1-G(S2) msM-35-1-G(S_2) msVZM-1-V(S_3) ms VZM-2-V(S_3) msM-31-2-R(S_4)	from Guntur, India from Vizianagaram, India from Vizianagaram, India from Raichur, India
Rao et al., (1971)	Swarna Indigenous maldandi cytoplasm IS84 Indigenous maldandi cytoplasm M35-1indigenous maldandi cytoplasm Kafir B indigenous maldandi cytoplasm	

Tripathi (1979)	CK60A milo cytoplasm	Maharashtra, (India)
	Nagpur A based on milocytoplasm	
	GIA-unidentified cytoplasmic source	
	VZM2A unidentified cytoplasm source	Vizianagaram (AP), India
	M35-1A unidentified cytoplasmic source	Maharashtra, (India)
	CK60A Same sterility Nagpur A	Raichur (Karnataka), India

Radish (Raphanus sativus)

Production of hybrid cultivars using self-incompatible parent lines is not possible in radish because of the absence of an effective self incompatibility mechanism. Ogura (1968) reported a form of male sterility in a Japanese radish type, governed by a cytoplasmic factor and a recessive gene. To multiply male sterile (ms) lines, male fertile (mf) maintainers are needed. No research has been reported on identification of maintainers of the Japanese radish material. Bonnet (1975) found maintainers among some European radish cultivars. He also reported that male sterility is developed by one recessive and number of dominant genes. Humaydan and Williams (1976) also found a type of male sterility in radish, which was determined by one recessive gene.

Rape: (Brassica spp.)

Heterosis for seed yield in F_1 hybrids of rape seed is achieved by effective pollination control system. One such system known as CMS system is used in rape in combination with appropriate maintainer and fertility restorer lines. Several cytoplasmic male sterility inducing cytoplasm are available in rape seed. They are Nap, Mur, Ogu and Pol. Among them 'Nap' system is important since it could potentially used for production of hybrid seed. Sigha and Baba (1973) reported a naturally accuring male sterility inducing cytoplasm in *B.napus.* Bannerort et al., (1974) transferred *B.napus* nucleus to the radish cytoplasm through intergeneric crossing followed by backcrossing. The resulting male sterile *B.napus* plants showed complete male sterility over several generations. Konno (1979) transferred the genome of *B.campestris* CV Yukina to the cytoplasm of *Diplotaris muralis* through intergeneric hybridization followed by backcrossing. Crosses between cytoplasmic male sterile *B.campestris* carrying the sterility inducing *D.muralis cytoplasm* and the four rape seed cultivars 'Cresor', 'Brutor' 'Tandem' and 'Bienvenu' always produced a male fertile interspecific hybrid plants which indicated that these *B.napus* cultivars carried restorer genes for the male sterility inducing *D.muralis* cytoplsm if this cytoplasm could induce male sterility in *B.napus.* Sigha (1980) observed male sterile 'F_1' plants from the cross between *D.muralis and B.napus* cv 'Norin 16' while in cross between cytoplasm male sterile *B.campestris* with *D.muralis* cytoplasm and 'Norin 16' produced male fertile F_1 plants indicating that 'Norin 16' carried restorer genes for the *D.muralis* cytoplasm. Pelletier *et al.*, (1983) reported male sterile chlorophyllous nondeficient plants hav been obtained but fertility restoration in F_1 hybrids has not been achieved. Rousselle *et al.*, (1983) crossed cytoplasmic male sterile B.oleracea with *B.napus* and backcrossed the interspecific hybrid plants with *B.napus.* The cytoplasmic male sterile plants that were obtained exhibited complete male sterility but no maintainers has been found for this cytoplasm.

Mathias (1985) transferred the nucleus of *B.napus* to the male sterility inducing *B.juncea* cytoplasm and obtained cytoplasmic male sterile *B.napus* plants. No restorer has been reported for this male sterility. Fan *et at.*, (1985) observed male sterile plants in the progeny of crosses between *D.muralis* and *B.napus* followed by backcrossing to *B.napus.* However, the presence of a *D.muralis* chromosome in the nucleus of these plants seemed to the cause of male sterility and no cytoplasmic male sterile plants were obtained. Fu (1987) found male sterile plants occurring in the *B.napus* cultivar 'polima'. The expression of male sterility in this cytoplasm is influenced by environmental conditions. The sterility of pol CMS is sensitive to temperatures changes depending upon the nuclear genotype of he CMS lines. There are three groups: high temperature sterility, low temperature sterility and stable sterility. If the cytoplasm of a variety is sterile and no Rf gene occurs in the nucleus, the variety would be partially sterile in low temperature conditions when the nucleus carries the low temperature sensitive genes (LTS) or in high temperature conditions when the nucleus has the high temperature sensitive genes (HTS). If no TS gene exist, compete male sterility would occur. The pol CMS has only one pair of major genes (Rf) governing fertility restoration.

Sunflower (Helianthus annus)

In sunflower Vranceanu (1967) identified monogenic male sterility and this stimulated interest in developing commercial hybrids. Leclercq (1966) detected linkage between male fertility, MS, and the dominant gene T, which determines the red anthocyanin coloration of the plant. This lead to production of first commercial hybrids in Romania and France.

The breeding of hybrid sunflowers has been greatly assisted by the discovery of cytoplasmic male sterility among progenies of the interspecific cross *H.petiolaris x H. annus.* The main difficulty in the exploitation of this male sterility is the low frequency of genes for pollen fertility restoration (Rf) genes) in the cultivated sunflower. Kinman (1970) identified genes for restoration in wild sunflower species and also in certain obsolete sunflower cultivars. Vranceanu and Stoenescu (1978) introduced, Rf genes into the genotypes of best inbread lines with a view to the large scale production of hybrids based on cytoplasmic male sterility, Giriraj *et al.,* (1987) conducted line X tester analysis with seven converted CMS lines F-44, F-48, F-50, F-50, F-73 and F-97 as well as CMS-234 and 3 testers PR-1, RHA-274 and RHA-801.

Sugarbeet (Beta vulgaris)

Male sterile plants are to be found in all diploid, open pollinated varieties and also in *B.maritima.* In some cases this male sterility is purely genetic origin, while in others it is based on combination of cytoplasmic and genetic factors. This latter type of sterility is suitable for the production of hybrid varieties. For practical purposes, it is in accordance with Owne's hypothesis, that this sterility is based on cytoplasmic factor (plants carrying this factor designated (S) whereas with normal cytoplasm are designated (N) and on atleast two recessive genes designated X and Z). A male sterile plant thus has for male (S) XXZZ. When such a plant is crossed with male-fertile plant carrying the recessive genes for sterility and designated O-type, the offspring of the male sterile plant will also be male sterile.

Cytoplasmic male sterility has been induced in fertile cytoplasms by mutagenesis. In sugarbeet, Kinoshita (1976) starting with monogerm and multigerm non 0-type pollinators carrying the N cytoplasm produced a series of male sterile lines by X-ray irradiation. Mikami *et al.,* 1980 made attempt to induce cytoplasmic male sterility in both O-type and non O-type pollinators of N cytoplasm with chemical mutagenic agents. Using EMS (at 12500ppm) it was possible to induce cms in a restoring genotype and using acriglavine (1500 ppm), Sterpto mycin (1500 ppm) and ethium promodids (2000 ppm) cytoplasmic male steriles were induced in O-type lines. Owen (1959) states that O-types are very common in most diploid open-pollinated varieties show that it not easy to find O-types capable of maintaining 100% male sterility. To identify O-types it is necessary to make experimental crosses between a male sterile plant and single plants selected from a population.

Maize (Zea mays)

Maize cytoplasms that exhibit male sterility are classified as T (Texas), C (Charrna) or S(USDA) on the basis of the capacity of specific nuclear (Rf) genes to restore male fertility. Cytoplasmic male sterility was extremely successful as a technique to eliminate detassling in Maize, when a race of *Helminthosporium maydis* specific to T cytoplasm spread over most of the eastern United States in epidemic from causing yield reductions.

Hybrid rice production in China

Because rice is a self-pollinated plant, hybridization requires a male sterile plant. This sterility may be induced chemically, but commercially, cytoplasmic male sterile lines are used.

In rice 19 sources of cytoplasmic male sterility had been identified. Among them 5 are widely used for hybrid seed production. They are (1) Wild rice designated as wild aborted or WA type. (2) *O.sativa f.spontanea* (3) Chinsurah Boro II (BT type). (4) Gambiaca (Gam type) and (5) Shan-Tao-Bai. (Table 12).

In the world China is the only country which produces hybrid rice commercially. Hybrid rice was commercially introduced to Chinese farmers in 1976. The area under hybrid rice was 9 million hectare and accounting for one-third of Chinas rice production during 1976. The yield advantage of hybrid rice was about 15 percent compared to conventional rice. The male sterile lines from China are classified into three basic groups according to genetic properties and relation between restorer and maintainer lines (Lin and Yuan, 1980).

(*a*) **Major Sources of Cytoplasmic Male Sterility**

Group I The WA cytosteriles are typical of this group, but Gam type and some male sterile lines derived from *O.sativa f. spontanea* also being here. The function of the male sterility gene is *sporophytic;* pollen grain's abort at the uninucleate stage. Maintainer lines are found in both Indica and Japonica rice.

Group II This group consists of BT type male sterile lines developed by Shinjyo and Omura (1966). The function of the male sterility gene is *gametophytic;* pollen grains abort between the binucleate and triangulated stages. The restoration spectrum is wider than group I.

Group III The Cytoplasmic genetic male sterility mechanism of this group is derived from *O.sativa f.spontanea* lines. The Hong-lien is typical of this group. Pollen grains abort at the binucleate stage. The restorers and maintainers in this group is in contrast or reverse to that of group I. The maintainers of group I, such as Zhen Shan 97 and Er-Jiu-Ai 4 become restorer's of this group, and the restorers of the group I, such as Tai-Yin 1 are good maintainers of this group. Among these three groups cytoplasmic male sterile lines derived from WA cytosterility system have been to be the most stable and more than 95% of the hybrid rice varieties used now in production in China are derived from this type. According to Yuan (1980), the probability of developing stable male sterile line is higher from relatively wider crosses where the female parent is primitive line and the male parent is an advanced line. The closer the relation between the two parents, the harder it is to obtain a stable male sterile line.

In the IRRI hybrid rice breeding programme 11 cytosterile lines representing 5 cytoplasmic sources is Gabiaco, Birco, *O.sativa f.spontanea,* Taichung Native I and BT types are available. Among them only 6 are stable and used for hybrid seed production. They are Zhen Shan 97A, V20A, Er-Jiu Nan1A (WA type), Yar Ai Zhao A (Gam type), Pankhari 203A (TN type) and Wu 10A (BT type).

(*b*) **Additional Sources of Cytoplasmic Male Sterility**

More than 90% of hybrid rice are derived from WA cytosterile lines. This makes hybrid rice vulnerable to disease or insect epidemics. So work is in progress to diversify usable sources of cytoplasmic male sterility for hybrid rice. Rice species *(i.e., Oryza glabarrima, O.fatua,* Asian forms of *O.perennis, and O.rufipogen)* and varieties (i.e., PTB 16, Tadukan, Lead 35, Akeborao, IARI 10061, IARI 10560 and jeerage Samba) that are known to induce cytoplasmic male sterility, may result in cytosterile lines that possess different cytosterility systems.

(*c*) **Techniques for Cytoplasmic Differentiation**

1. Substitution backcrossing
2. Use of Cytoplasm differentiating genes
3. Study of pollen abortion patterns
4. Interaction of restorer genes (Rf) with the cytoplasm

1. **Substitution Backcrossing.** In substitution backcrossing of the genome of one species into the alien cytoplasm of another, sufficient backcrosses are required to eliminate all nuclear genes derived from the cytoplasm donor species and produce a new nucleo-cytoplasmic combination. If this results in relatively stable male sterility, the cytoplasms of the two species involved in the cross may be considered distinct.

Table 12

Cytoplasmic sources identified to induce male sterility in rice.

Cytoplasm Source	Nuclear Source	Male Sterile lines developed (number)	Reference
PTB 16		—	Weeraratne (1954); Sampath and Mohanty (1954)
Oryza sativa f.spontanea	Fujisaka 5	—	Katsuo and Mizushima (1958)
Oryza fatua	Fujisaka 5	—	Katsuo and Mizushima (1958)
Tadukan	Norin 8	—	Kitamura (1962)
Chinsurah Boro II	Taichung 65 Wu10 Several Japonico Rice varieties in China	1 Several	L.P. Yuan (Personal communication)

Lead Birco (P1 279120)	Fujisaka 5 Calrose, Caloro	— 2 or 3	Watinable *et al.*, (1969) Casnahen *et al.*, (1972)
Oryza glaberrima	Colusa IR 36	1 1	Carnahan *et al.* (1972)
Taiching Native 1	Pankhari 203	—	Athwal and Virmani (1972)
Wild rice with aborted pollen	O.glaberrima Er-Jiu-Nan	1	Yaburo (1977) Yuan (1972); Lin and Yuan (1980).
O.rufipogon (KR 7) Gambiace	Taichung 65 Gang-Yi-Ya-Ai Zhao, Yar-Ai Zhao	Several	Cheng and Huang (1979) Lin and Yuan (1980)
O-Shan-Tao-Bai	Toride 1 and Several other Indica and Japonica rices	Several	Lin and Yuan (1980)
IARI 10061	IARI 11445	—	Parmar *et al.* (1981)
Jeeraga Sambaa	IR24	—	
O.perennis	Taichung 65	—	
K 1080			
O.perennis (W 1092)	Taichung 65	—	Shino and Motomura (1981)

2. **Use of Cytoplasm-differentiating genes.** Cytoplasm-differentiating nuclear genes include male fertility restoring genes, male fertility inhibiting genes and genes affecting plant vigour in various ways. Research during past led several hybrid breeding programmes to apply the terms "hard to restore" and "easy to restore" to genotypes in which male fertility inhibiting genes are present or absent.
3. **Study of Pollen Abortion Patterns.** Xu (1982) found that pollen of WA cytosterile lines abort at uninucleate stage and those in BT cytosteriles abort at the binucleate and trinucealate stages. Chaudhary *et al.,* (1981) suggested that the pollen abortion stage in a CMS line depended on the distance of relation of its cytoplasmic and nuclear donor parents. It appears that cytosterile lines with pollen abortion at the uninucleate state are more stable for complete pollen sterility than lines with pollen abortion at the binucleate or trinucleate stage.
4. **Interaction of Restorer (Rf) Genes with the Cytoplasm.** Shinio (1975) and Kinoshita *et al.,* (1980) reported that WA sterility system is different from the BT system because the restorer gene (S1 for) WA system is sporophytic action and those for BT system is gametophytic action.

(*d*) **Fertility restoration**

(*i*) **Sources of Restorer Genes**

The practical use of cytoplasmic-genetic male sterility in developing hybrid varieties is possible only when the effective lines are identified and/or developed. In rice, effective restorer lines for the WA, Gam and BT cytosterility systems have been identified but effective restorer lines for cytosteriles pankhari 2034 and MS 577A have not been identified.

The first set of effective restorer lines used in China in commercial F_1 hybrids involving the WA cytosterility system was identified in 1973. The frequency of restorer lines was even less among varieties from China, eastern Europe, Japan and Korea and the highest frequency was found from varieties originating in the lower latitudes of south and south-east Asia.

The restoring ability of rice varieties has been related to their origin. The frequency of restorer lines is higher and restoration ability is stronger among varieties closely related to WA cytosterility source; it is higher among Indicas because Indica varieties originated earlier than Japonica varieties and are closely related to wild rice. The frequency of restorer lines among Japonica varieties is negligible. Among Indica, restorer lines have been more frequent among late maturing than among early maturing varieties because late maturing Indicas are primitive, relatively closer to wild rice. (Table 13).

Table 13. Restorer lines used widely in commercial F_1 Rice hybrids in China

Line	Restorer of CMS System	Origin
IR24	WA, Gambiaca	Cuba/IRRI
IR26	WA, Gambiaca	Cuba/IRRI
IR30	WA, Gambiaca	Cuba/IRRI
IR661	WA, Gambiaca	Thailand/IRRI
IR665	WA, Gambiaca	Indonesia
GU54	WA, Gambiaca	China
GU223	WA, Gambiaca	China
GUi630	Chinsurah Boro II	China
Tai Yin 1	(BT)	
Yinni Ai He	Chinsurah Boro II	China
Ko-Zhen 145	(BT)	
Zhao Hui 3533	Chinsurah Boro II	China
C55	(BT)	
C57	Chinsurah Boro II	China
Beijing 300	(BT)	
5350	Chinsurah Boro II	China
5154	(BT)	
Hong Zhao Nou	Chinsurah Boro II	China
Nanguang Zhan	(BT)	

(*ii*) **Inheritance of Restoration**

Shinyo (1969) identified a fertility restoring single dominant gene, Rf in the variety Chinsurah Boro II for the BT cytosterile line, its effect was gemetophytic in the male sterile inducing cytoplasm

(MS-Boro). Kitamura (1962) reported that high fertility in the F_1 hybrids of cytosterile Ta 820 was conditioned by a recessive gene combined with modifiers or polygenes. It indicated that cytosterile Ta 820 possessed functional male sterility, because there was no abortion of male and female gametes but the anthers were non-dehiscent. Shinyo (1975) studied the genetics of fertility restoration in the male sterility inducing cytoplasm of rice variety lead. Two restorers Fukuyama (Japanica variety) and a strain of MS-Boro Rf, Taichung 65 possessed Rfx and the Rf gene and had gametophytic effects in the lead rice cytoplasm. The Rfx gene gave weaker restoration than did the Rf gene in the cytoplasm of chinsurah Boro II, Gao (1981) observed that restoration in the WA cytosterile line is controlled by one or two pairs of dominant genes and the effect of the restoration gene is sporophytic.

(*iii*) **Environmental Effects on Male Fertility Restoration**

The effects of environmental factors on male fertility restoration in hybrid rice would appear to be genotype specific. The restoration ability of restorer lines IR 24, IR 26, IR 661, Gui 154 is stable under various environments. Lin and Yuan (1980) reported that the rate of seed set in hybrid rice particularly under unfavourable climate conditions was only about 80% less than that of best conventional varieties in China. A joint study between IRRI and China aimed at identifying effective restorer lines indicated that about 15 and 24% of these lines were effective restorers in China and IRRI respectively, but only 6% were effective restorers at both sites. The restoration ability of the remaining restorer lines was site specific; the frequency of restorer genotypes was higher in tropical than in a subtropical environment.

(*iv*) **Influence of Female Genetic background of Fertility Restoration**

Widespread occurrence of intervarietal hybrid sterility attributable to gametic development nuclear genes randomly distributed among rice varieties suggests that genetic background of the female parent could influence the pollen and spikelet fertility of F_1 hybrids. There is an evidence that stable restoration and consistent seed set under different environmental conditions result from balance between Rf genes and modifier genes (enhancing fertility).

(*e*) **Factors Affecting Cross Fertilization**

(*i*) **Flowering Behaviour**

Varietal differences in the duration of flower opening from 28 to 93 minutes have been reported in rice. Floret opening in CMS rice plants ranged from 105 to 280 minutes Parmar *et al.,* (1979) noted a longer duration of flower opening in late maturing varieties (50-70 min) compared with early maturing varieties (28-35 min). Delay or failure in pollination was found to prolong flowering and consequently CMS plants have a longer duration of floret opening than fertile plants. A mean temperature of 25-30 °C and a relative humidity of 75-90% was found to provide optimum conditions for anthesis in male-sterile, maintainers and restorer lines with the flowering period lasting 6-7 days. Higher or lower temperatures affected synchronisation; both low (18-20 °C) and high (27-35 °C) temperatures affected a number of agronomic and floral characters that influence out-crossing. High temperatures delayed anthesis and slightly increased the period of floret opening.

(*ii*) **Floral Structure**

Stigma size, style length, stigma exertion, stigma receptivity, anther size, filament size and pollen number are important floral characters that influence outcrossing in rice. Inheritance studies (Virmani and Athwal 1974) for floral traits such as anther length, stigma length and stigma exertion in rice indicated that these traits were governed by polygenes. Huang and Huang (1978) reported that the stigma exertion was dominant, or recessive depending on the cross. They found negative correlation between stigma exertion and pollen fertility. Both additive and non-additive gene effects were important in the inheritance of floral traits. The prevalence of duplicate types epistasis was considered a barrier to fixing these traits at higher level of manifestation through conventional breeding. Therefore, recurrent selection in biparental progenies was proposed to improve these characters they may have to be transferred from wild rice in which undesirable linkage relationship exists. Attempts to transfer the long stigma (5mm) trait of genetic stock designated as 6209-3(IRRI, 1983) into genetic background of selected maintainers lines from which it would be transferred to cytoplasmic male sterile lines in rice are being made at IRRI.

These studies indicate the anther length and extrusion score are the simplest measurements of pollen shedding capacity for R line breeding. After the confirmation of restoration ability based on test cross fertility, the yield evaluation of new R lines will have duel significance. The economics of hybrid production will be improved and high yielding lines will have more fertile florets per unit area and an increased pollen load.

(*iii*) **Effect of Pollinator Distance**

Provided that an adequate pollen load is present and that there is correct synchronisation of flowering between male and female parents, seed set on male sterile plants is influenced by the distances that pollen can travel and remain viable. Rodrigo (1925) observed that in cultivated rice varieties pollen grains are carried by the wind, in the same plane as the flower itself, as far as 1.5-2.0 m from the source depending on wind velocity. Virmani *et al.*, (1980) found that seed set from cross-pollination was higher on plants located 30 cm from the pollen source than on those located 30-110 cm from the source. This differences were less pronounced when the CMS line is located down wind from the pollinator, and it was concluded that satisfactory seed set could be obtained in hybrid seed production by growing five rows of the CMS line (1-1.25 m wide) alternately with one row of the restorer line. With the isolation distances of 10, 20, 30 and 40 m, contamination resulting from foreign pollen was 5.2, 1.0, 0.2 and 0% respectively. So, a minimum isolation distance fo 40m is set for hybrid seed production in China.

(*iv*) **Effect of Plant Height and other Morphological Traits**

Because wind velocity will affect the rate at which pollen descends following dispersal from an extruded anther, researchers suggested that there would be a positive effect on seed set if the male sterile parent were shorter than pollen parent. Chinese researchers have used rice restorer lines that are 10-20m taller than the CMS line. Use of 'recessive tall' gene in rice (Rutger and Carnahan, 1981) should enable the use of even taller R lines with semidwarf CMS lines to develop semidwarf hybrids. Other traits that might hinder Cross-pollination in rice include long and broad flag leaves and incomplete panicle exertion. Consequently, selection of CMS and restorer lines with short, narrow flag leaves and good panicle exertion would be advantageous.

(*v*) **Disadvantages of using chemical Pollen Sterility CPS method for hybrid seed production (Advantages of CMS system over CPS system)**

1. Pollen abortion is incomplete and erratic in CPS system.
2. Treatments are only effective for short periods and at a particular stages in the development of plant.
3. Because of the short duration of the effect, treatments have to be repeated.
4. Harmful side effects are evoked, such as leaf deformation, stunting of the plant or wilting.

5. The high dosage, the many repetitions needed and the high cost make the treatments uneconomic.
6. Chemical dosage levels sufficient to ensure male sterility often induce some female sterility and reduced seed set.
7. Extended rainfall and prolonged winds can prevent field application of the chemical at the optimum time.
8. Genotype-environment-chemical interactions must be evaluated.
9. Male fertility or selfing in hybrid fields can result in seed that does not conform to seed law specifications for a hybrid.

Hybrid Seed Production

Hybrid seed production using the cytoplasmic-genetic male sterility system involves three steps :

1. Multiplication of CMS (A) lines
2. Multiplication of maintainer (B) and Restorer (R) lines
3. Production of hybrid seed (A x R)

A-line:

A line is the male sterile parent lien in a cross being made to produce hybrid seed.

B-line:

The fertile counterpart of the A-line. The B line does not have fertility restoring genes and is used as the male parent to maintain A line.

R-line:

The male fertile line carries genes for fertility restoration.

A. Multiplication of Cytoplasmic Male Sterile and Maintainer Lines:

Four rows of A line are alternated with 2 rows of B line (2 : 1 ratio), and to improve synchronization of flowering, a split seeding of the maintainer is carried out at 4 and 8 days after seeding male sterile line. Both the A and B lines are transplanted on the same day and plants from early and late B line planting are alternated in the rows. Cross pollination is improved by planting the rows perpendicular to the prevailing wind direction, and flag leaves of both the A and B lines are clipped at the boot stage to aid pollen movement. This procedure require 25 man-days per hectare and has been found to increase seed yield by 42.9%. One or two applications of 20 ppm Gibberlic acid are applied to the A and B lines after clipping to improve panicle exertion. Pollination is improved by (i) open pulling (2) Shaking of the pollen parent with a pole. Isolation in time (21 days) or distance (100 m) from other rice crops prevents contamination by foreign pollen. By this method, the average seed yields per production hectare in China are 400-900 kg of A line and 1000-1500 kg of B line.

B. Hybrid Seed Production

The selected R line serves as the pollen donor parent for the male sterile A line, and seed harvested from the A line constitutes the commercial hybrid seed. Seed harvested from the R line is the result of self fertilization and may be used again in hybrid seed production.

The effects of 'Strip Width' and parent ratio on pollen load and distribution are basic considerations in hybrid seed production. Increasing the ratio of female to pollinator decreases pollen production per unit area and also increases the distance over which pollen must travel. A change to narrower drill strip width without altering the ratio alters only the distance pollen must travel and may result in a more homogenous pollen distribution.

Hybrid rice seed production in China involves six to eight rows of A line and one to two rows of R line. These parental lines are seeded and transplanted on different dates, depending on their growth duration, to synchronize their flowering. The R line is seeded on three or four different dates, but seedings of different ages are transplanted the same day in the field to increase the duration of pollen availability at flowering.

Difficulties in using CMS for Hybrid Seed Production

1. Male sterility must be stable. Instability would lead to partial self fertilization or intercrossing of the mother line, giving rise to heterogeneity and lower yield of the commercial seed.
2. Fertility restoration is generally under the control or more than one gene. Consequently more than one Rf gene must be introduced into the breeding material.
3. The expression of Rf genes is affected by the genetic background. This means that after the introduction of a restorer gene into a particular inbred, the Rf gene may fail to be expected or it may show incomplete activity.
4. The susceptibility of restorer genes to environmental influence. The fertility expression of restored wheat plants has been shown to vary with day length, temperature, relative humidity, light intensity and plant density. In garden petunia both low and high temperatures adversely affect pollen development. Heterozygous restored plants are sensitive to changes. Since hybrids are heterozygous for restorer character, it threat to CMS hybrid production.
5. In some crops abnormalities in growth, development or plant morphology are observed in connection with the introduction of (S) cytoplasm or Rf genes. For example, some petunia varieties exhibit flower bud abortion (blindness) and erect instead of bushy plant habit. Susceptibilities of maize with Texas cytoplasm to southern corn leaf blight in US corn belt in 1970.

CMS lines in Crops

1. Rice

Some Male Sterile lines released in China for hybrid seed Production

	Male Sterile line	**Maintainer**
1.	V20A	V20B
2.	Yar-Ai-Zhao-A	Yar Ai-Zhao-B
3.	Pankhari 203A	Pankhari 203-B
4.	MS 577A	MS 577B
5.	MS 519A	MS 519B
6.	Wu10A	Wu10B
7.	Zhen Shan 97A	Zhen Shan 97B

Some Male Sterile lines released in IRRI for hybrid seed production

	Male Sterile line	Maintainers lines
1.	IR 46826A	IR46826B
2.	IR46827A	IR46827B
3.	IR46828A	IR46828B
4.	IR46829A	IR46829B
5.	IR46830A	IR46830B
6.	IR46831A	IR46831B

2. Sorghum

1.	MSCK 60A	from USA
2.	MS G1 A MS M-35-1-G	from Guntur India
3.	MS VZM-1-V	from Vizanagaram, India
4.	MS VZM-2-V	from Vizanagaram, India
5.	MS M-31-2-R	from Raichur, India
6.	M35-1-A-Maldandi	Maharashtra
7.	Nagpur A	Maharashtra
8.	MS 2219A	Tamil Nadu
9.	MS 2077A	Tamil Nadu
10.	MS 8271A	Tamil Nadu
11.	MS 3660A	Tamil Nadu
12.	MS 2947A	Tamil Nadu

3. Rape

1. pol CMS line xiani A

4. Sunflower

1. CMS 234

5. Wheat

1. Minn II-54-30
2. Deklb 3047
3. MS Butte
4. MS-Maris Hobbit
5. MS-Abe

Conclusion

A. Current outlook

Rice Researchers played a major role in increasing understanding of nucleocytoplasmic interactions. Several usable cytoplasmic male sterile systems are available in diverse genetic backgrounds and may be used to stabilise maintainers varieties (B lines) that possess desirable agronomic characteristics, insect an disease resistance and adaptability to local conditions. The frequency of rice maintainers lines is adequate and these and restorer lines have been found to possess multiple disease and insect resistance. Genetic variability in floral characters influencing outcrossing have been obtained in rice and breeders should be able to further improve outcrossing potential of maintainers and restorer lines.

B. Future Strategies

Hybrid rice breeding should provide farmers with an opportunity to improve productivity, particularly in potential high yield areas and where conventional breeding has apparently reached a yield plateau. The approach of the breeder to stabilizing production will be to continue identifying floral features that improve pollen load and female receptivity, to establish variation for these traits and to incorporate favourable genes into new inbred lines. To protect F_1 hybrids against genetic vulnerability to disease and insect epidemics which may be linked to the use of a certain CMS systems, studies on the effects of existing cytoplasms on resistance or susceptability are essential and alternative sources of CMS should be sought out. In rice more than 95% of CMS lines were derived from WA type cytoplasmic which makes it susceptible to disease. Hence, suitable alternative to the WA cytosterility systems need to be discovered. There is an improved understanding of the importance of a balance between genes for pollen fertility restoration in the male parent and fertility enhancing genes that affect the case of restoration in the female. It may be possible to identify CMS female that exhibit partial fertility in a favourable (Shallow Sterile) environment. Such lines in hybrid combination may interact with restorer genes to provide a highly fertile hybrid that is stable over a range of climatic conditions. The molecular basis of cytoplasmic genetic male sterility and fertility restoration in rice is very limited with clear understanding of the CMS mechanism, the search for diverse male sterile cytoplasm(s) and the donor(s) for their restorer genes would become easy. Developments in protoplast fusion and cell culture would also helpful to transfer the available cytosterility system(s) expeditious with different nuclear genotypes.

REFERENCES

Appathurai.R, (1964). Diverse Plasmons in male sterile Sorghum. Madras Agric. J. 51:276-278.

Athwal, D.S., and S.S. Virmani, (1972). In 'Rice Breeding', pp 615-620. Int. Rice. Res. Inst. Los Banos, Philippines.

Bannerot, H., L. Boulidard, Y. Cauderon and J. Tempe. (1974). Cytoplasmic male sterility transfer from Raphanus to Brassica. Proc. Eucarpia Crop. Sect. Cruciferae. 25, 52-54.

Bonnet, A. (1975). Introduction et utilisation dune Sterilite male cytoplasmique dan des varieties process europeones de radis, *Raphanus sativus* L.Ann Amelior Plantes 25:381-397.

Carnahan, H.L., J.R. Erickson, T.S. Tseng., and J.N. Rutger (1972). In 'Rice breeding' pp. 603-607. Int. rice Res. Inst., Los Banos, Philippines.

Chaudhary, R.C., S.S. Virmani and G.S. Khush (1981). Oryza 18, 142-149.

Cheng, Y.K. and C.S. Huang (1979). Chung-Hua Nung Hsueh Hui Pao. (J.Agric. Assoc. China) 106, 11-22.

Fan, Z., and B.R. Steffanson, (1986). Influence of temperature on sterility of two cytoplasmic male sterility in *Brassica napus* L. Eucarpia Crucarpia Curciferae Newslett. 12, 46-47.

Fu, T.D., G.S. Yang, and S.N. Yang (1987). Some investigations on polima Cytoplasmic male sterility in *Brassica napus* L. Eucarpia Couciferal Newslett, 12, 211-217.

Giriraj, K., R. Hiremath and K. Seenappa (1987). Combining ability for Converted male-sterile lines of sunflower. Indian J. Genet., 47, (3) 315-317.

Gao, M.V. (1981). Acta genet Sin. 8. 66-74.

Gomaa, A.S., and K.A. Lucken (1973). Poc. Int. Wheat Genet. Symp. 4th pp. 345-350.

Ghiasi, H., and K.A. Lucken (1982). Crop Sci. 22, 527-537.

Heu, M.H. and Y.A. Chae, (1970). Res. Rep. Rural Dev. (Crop) Korea 13, 9-13.

He Gui-Tang, Zhu xi Gang and J.C. Flinn (1987). Hybrid Seed Production in Jiangsu Province, China, Oryza, 24, 297-312.

Hussani, S.H., and P.V. Rao. (1964). A note on the spontaneous occurance of cytoplasmic male sterility in Indian sorghums. Sorghum Newsletter, 27:27-28.

Huang, C.S. and H.H. Huang (1978). Publ. Agric. Res. Inst. Taipei, pp. 212-255.

Humaydan, H.S. and P.M. Williams, (1976). Inheritance of seven characters in *Raphanus sativus* L.Hort. Science 11:146-147.

Jones, H.A., and G.N. Davis. (1944). Inbreeding and heterosis and their relation to the Development of new varieties of onions US. Dept. Agric. Tech. Bull, 874:1-28.

Kastsuo, K., and U. Mizushima, (1958). Ikushugaku Zasshi (Jap. J. Breed) 8:1-5.

Kihara, H. (1951). Cytologia, 16, 117-193.

Kinman, M.L., (1970). New developments in the USDA and State experiment station sunflower breeding programs. Proceedings of the fourth International sunflower conference, Mephis, Tennessee, 181-183.

Kinoshita, T. (1976). Genetical studies on Cytoplasmic male sterility induced by Gamma-ray irradiation in sugarbeets Japan. J. Breed, 26:256-265.

Kinoshita, T., K. Mori and M. Takahashi, (1980). J. Fac. Agric. Hokkaids Univ. 60, 23-41.

Kitamura, E. (1962). Ikushugki Zassi (Japan J. Breed.) 12, 81-84.

Konno, N. (1979). Studies on a male-sterile srain having the Brassica campestris nucleus and the *Diplotaxis muralis* cytoplasm, Japan J. Breed, 29, 305-311.

Leclercq, P. (1966). Use sterility male utilisable pour la production d'hybrides simples de tournesol. Annales de L'Ameliorationdcs planetes 16:135-144.

Lin, S.G., and L.P. Yuan (1980). In Innovative Approaches to Rice breeding, pp. 35-51. Int. Rice Res. Inst., Los Banos., Philippines.

Mathias, R., (1985). Transfer of Cytoplasmic male sterility from brown mustard *(Brassica juncea L. Coss)* into rape seed *(Brassica napus L.)* Z. Pflalnzenzuchtg. 95, 371-274.

Mikami, T., T. Kinoshita and H. Takahashi, (1980). Induction or cytoplasmic mutations on male sterility by acridine dyes and streptomycin in sugarbeets. Pro. Sugar Beet Res. Assoc., Japan 22:48-54.

Mital, S.P. Y. Swarur and A.B. Joshi, (1958). Cytoplasmic male sterility in Jowar *(Sorghum vulgare pers)* Current science 27:314.

Mukai, Y., and K. Tsunewaki. (1979). Theor, Appl. Genet., 54, 153-160.

Mann, S.S. (1973). Proc. Symp. Genet, Breed. Durum wheat Basi, Italy, pp. 117-138.

Nagur, T. (1971). Studies on fertility restoration and combining ability in relation to Genetic diversity and cytoplasmic constitution in sorghum. Ph.D Thesis Tamilnadu Agricultural University, Coimbatore, India.

Nagur, T., and P. M. Menon, (1974). Characterization of different male sterility inducing cytoplasms in sorghum, Sorghum Newsletter, 17:18.

Ogura, H., (1968). Studies on the new male sterility in Japanese radish, with special reference to the utilization of this sterility towards the practical raising of hybrid seeds. Mem. Fac. Agri. Kagoshima Univi 6:38-78.

Parmar, K.S., E.A. Siddiq, and M.S. Swaminathan (1981). Indian J. Genet, 39, 551-559.

Pelletier, G., C. Primard, F. Vedel, P. Chetrity, R. Remy, P. Rouselle and M. Renard, (1983). Intergeneric cytoplasmic hybridization in cruciferae by protoplast fusion. Mol. Gen. Genet 191:244-250.

Rao. N.G.P. (1972). Sorghum breeding in India recent developments pp: 101-142.

Rao, N.G.P. (1962). Occurrence of Cytoplasmic-Genetic male sterility in some Indian Sorghums. Indian Journal of geneties and Plant Breeding, 22:257-259.

Rao, N.G.P. D.P. Tripathi and B.S. Rana, (1984). Genetic analysis of cytoplasmic systems in sorghum. Indian J. Genet, 44:480-496.

Rao, M. and S.Gouda (1966). Occurrences of male sterile characters in Maldandi M31-2. Sorghum Newsletter 9:43-44.

Rodrigo, P.A. (1925). Philipp. Agric. 14, 155-171.

Rousselle, P. (1983). Premiers results d'um programme d'introduction del' and rosterilite Ogura du radis Chez le Colsa. Agronomie 2, 859-864.

Rutger, J.N., and C. Shinyo (1980). In Innovative Approaches to Rice Breeding, pp. 53-66 Int. Rice Res. Inst., Los Banos, Phillippines.

Sampath, S., and H.K. Mohanty, (1954). Curr. Sci. 23, 182-183.

Stephens, J.C. and R.F. Holland (1954). Cytoplasmic male sterility for hybrid seed production. Agronomy Journal 46:20-23.

Sigha, T. and S. Baba (1973). Cytoplasmic male sterility in oil seed rape, and its utilization to breeding. Japan J. Breed, 23, 187-197.

Shinyo, C. (1969). Jap. J. Genet. 44, 149-156.

Shinyo, C. (1975). Sci. Bull. Coll. Agric. Univ. Ryukus 22, 1-57.

Shinyo, C., Y. Ishimine, and M. Tanaki, (1981). Ikushugatu Zasshi (Jap. J. Breed) 1 (Suppl. 1), 238-239.

Shinyo C., and K. Motomura, (1981). Ikushugaku Zasshi (Jap. J. Breed) 31 (Suppl. 1) 240-241.

Shinyo, C., and T. Omura (1966). Ikushugaku Zasshi (Jap. J. Breed) 31 (Suppl. 1) 240-241.

Tripathi, D.P., S.L. Mehta., B.S. Rana and N.G.P. Rao (1980). Characterization of diverse cytoplasmic genetic male steriles in sorghum (*Sorghum bicolor* (L). Moench). Sorghum Newsletter 23:107-108.1

Virmani, S.S., and D.S. Athwal (1974). Crop. Sci. 14, 350-353.

Virmani, S.S., G.S. Khush and R.C. Yang (1980). Int. Rice Res. Newsl. 5, 4-5.

Vranceanu, A.V. and F.M. Stoenescu (1978). Genes for pollen fertility restoration in sunflower. Euphytic 27:617-627.

Vranceanu, A.V. (1967). Inheritance of sources of Male steriltiy sunflower, Probleme Agricole 19(2):28-39.

Watanable, Y. (1971). Oryza 8, 9-16.

Watnable, Y. S.Sakaguchi, and M.Kudo (1968). Ikushugaku Zasshi (Jap. J. Breed) 18 (Supple. 2), 77-78.

Weeraratne, H. (1954). Trop. Agric. (Colombo) 110, 93-97.

Wilson, J.A., and W.M. Ross (1962). Wheat Inf. Serv. 14, 29-30.

Wrostel, J.V., A.J., Kidd and K.F. Schertz, (1984). Relationships among male sterility inducing cytoplasms of Sorghum. Crop. Sci. 24:186-189.

Xu, S. (1982). Sci. Agric. Sin 2, 9-14.

Yaburo, T. (1977). Eupytica 26, 451-463.

Young, J.B., S.S. Virmani and G.S. Khush, (1983), Cytogenetic interrelationship among cytoplasmic genic male sterile maintainer and restorer lines in rice Phil. J.crop Sci. 8:110-124.

Young, J., and S.S. Virmani (1984). Inheritance of fertility restoration in a rice cross Rice Gen-News, 1:102-103.

Yuan, L.P. (1980). Sci. Bull. 4, 32-34.

CHAPTER 11

Somaclonal Variation in Crop Improvement

Introduction

Somaclonal variation refers to genetic, and therefore heritable, variation that arises following cell culture and plant regeneration (Larkin and Scowcroft, 1981). It is now about a decade since genetic variation arising from the culturing of plant cells began to be seriously and vigorously studied. There is no longer any doubt as to its existence and we are now much clearer as to the variety of genetic effects responsible for the altered phenotypes (Larkin, 1987).

Somaclonal Variation in Culture

There is now sufficient evidence, direct as well as circumstantial, to indicate that most of the variation seen in culture or in the regenerants occurs during the culture phase. This does not precludes pre-existing variation in the original explant as providing some of the variation, but this seems to be small component. Barbier and Dulieu's (1980) extensive analysis of variation among plants regenerated from protoplasts or leaf callus showed that about 3% of regenerated plants carried a mutation at either one of two defined loci affecting chlorophyll synthesis. A genetic analysis of sample of the altered genotypes indicated that the alleles had either reverted to wild type or had been deleted. From this analysis they could not discriminate between pre-existing and culture induced variation.

A similar type of experiment in tobacco used the semidominant aurea mutant, sulphur (Lorz and Scow croft, 1983). Both of the homozygotes and the heterozygotes can be readily distinguished by leaf phenotypes as normal green (su/su), albino (Su/Su) and yellow-green (Su/Su). In leaves of the Su/su heterozygote, single green or albino spot and twin spots are normally observed at a low frequency as consequence of somatic genetic events. Protoplasts were isolated from leaves of the heterozygotes and cultured to produce 2156 morphogenic colonies each of which was derived from a single protoplast. From observations on the regenerated shoots, colonies were classified as either parental where all shoots were Su/su, homogeneous variant where all shoots were either Su/su, or su/su, or heterogeneous variant where colonies give rise to both Su/su, and su/su, or Su/su and Su/Su shoot. The homogeneous variants represented pre-existing variations, whereas the heterogeneous ones reflected genetic alteration occuring after protoplast isolation. The frequency of this later class of colony was three times that of the homogeneous variant class, indicating that the occurrence of the somatic genetic event was significantly enhanced during tissue culture.

Sources of Variation

Serially propagated cultures often undergo changes in morphology, capacity for specialised function and potential for organogenesis. The major source of this variation-at least during the early stages of cultures—is likely to be cell selection. Because, in general, different cell types proliferate at different rates, cultures, established from complex tissue explants will tend to change in cellular constitution. On prolonged culture, tissues probabally reach a histogenetic equilibrium; cellular interactions, difference and in proliferation rates, and the inter conversion of cell types balance one another, and the relative abundance of the cells remains constant. In principle, even at equilibrium cultures can change. Local fluctuations in the abundance of different cell types within a tissue can generate sectors that retain their special character when subcultured separately.

Of particular interest is changes that results from heritable variation, *i.e.*, cellular alterations that are transmitted to daughter cells. Heritable variation can be detected by using tissues derived from single-cell clones that do not exhibit the new trait and then cloning a second time after the trait is detected to verify that it is heritable. Unfortunately, this analysis has not been made in most cases, and heritable changes in the development potential are commonly inferred by persistence of a new trait on subculturing.

Genetic Changes

Few reports of mutation in culture have been verified by using cloned materials and breeding tests. Most of the well documented examples are of dominant, single-gene mutation that were selected for by resistance to some toxic substance, *e.g.*, antimetabolities, aminoacids and herbicides. There is also evidence for high incidence of stable changes that persists in plants regenerated from cultured materials and are sometimes transmitted to progeny of the regenerated plants (Larkin and Scowcroft, 1981). These changes often affect complex traits that are probably multigenic, such as growth, vegetative anatomy and disease resistance.

Some of this variation probably reflects changes in chromosomal constitution such as polyploidy, aneuploidy and rearrangements. Of 55 species studied in detail, only 11 retained the same chromosome number in culture as in parent plant (Bayliss, 1980); and even when chromosome number is conserved, there may be a high degree of chromosomal rearrangements. The extent and type of chromosomal variation depends on genotype and the conditions of culture. There is also a strong effect of development state. Tissues with a high degree of organizations, such as shoot meristems, usually give rise to plants with normal chromosome numbers, whereas callus cultures from the same plant may not.

Epigenetic Changes

The best documented example of epigenetic change in culture is a type of heritable cellular variation known as cytokinin habituation (Meins, 1982). The sustained proliferation of pith parenchyma tissues of tobacco requires exogenous sources of auxin and a cell division factor such as the cytokinin, kinetin. On prolonged culture, or under specific inductive conditions, tissues sometimes lose their requirement for cytokinin. This change from cytokinin requiring to cytokinin habituated phenotype persists when cells are cloned. Unlike most classical mutations, cytokinin habituation is a gradual, progressive process, which is strongly influenced by physiological and developmental state of the cells. The induction of habituation is directed and occurs at high rates, 10^{-3} per cell generation. Although the habituated state is very stable in culture, habituated cells regularly revert to the cytokinin requiring state when induced to form the complete plants. Finally, cytokinin habituation results in the expression of a pre-existent potentiality specified by a single nuclear gene. Thus, it appears that cytokinin habituation of tobacco pith cells represents the expression of a normally silent gene.

Tobacco cells in culture differ in their competence for habituation. Pith tissue can be induced to habituate by a brief treatment with cytokinin. Under the same conditions, leaf tissues remain cytokinin requiring. By using plant regeneration test outlined above, evidence has been obtained that changes in competence for cytokinin habituation, like the habituation process itself, have an epigenetic basis. In coleus, micropropagation may induce epigenetic changes in leaf variegation (Marcotrigianno, et al., 1990).

Mechanisms Causing Somaclonal Variation

Apart from karyotypic abnormalities, there is little hard evidence to support of favour any one of several possible mechanisms to account for somaclonal variation. However, a consideration of these possibilities may indicate the type of experiments required to elucidate the phenomenon.

Chromosomal Aberrations

It has been known for sometime that chromosomal abnormalities occur at a high frequency in tissue cultures (Bebeli et al., 1990; Bulk et al., 1990; Evans, 1988; Park and Walton, 1990).

Among regenerated plants and their progeny the incidence of gross chromosomal aberration is greatly reduced relative to cell in tissue culture. In the analysis of regenerants, chromosomal aberrations which preclude morphogenesis will not be observed. Where somaclonal progenies are analysed, those aberrations which cause sterility will also be eliminated. Thus, morphogenesis and meiosis are selection processes which eliminate many chromosomal abnormalities and strongly favour euploidy. In Haworthia, a comparison of chromosome number of regenerants with the tissue culture lines from which they were derived indicated a reduction in an aneuploid frequency in the regenerants.

Chromosome Number Changes

Aneuploidy is not uncommon in cultured cells and somaclones (D'Amato, 1985). Monosomics and trisomics can be useful in genetic analysis. We have used to good advantage monosomics recovered from culture in the mapping of traits to particular wheat chromosome (Brettell *et. al.*, 1986). Among 192 Plants regenerated from four different wheat cultivars, 29% were aneuploids.

Chromosome Structure Rearrangement

Cultured cells and regenerated somaclones can display cytologically observable structural changes in their chromosomes (D' Amato, 1985). Examples include many species and a range of observations, such as segment inversions; deletions (telocentrics, smaller chromosomes); altered banding patterns with different stains; meiotic multivalents (interchanges, bridges, and fragments); and isochromosomes. Inter changes may prove valuable for alien gene interrogation. We have recently obtained evidence of a duplicated segment on the alpha arm of chromosome 4A in a wheat C.V. Banks somaclone (Ba B2). The proximal "reduced height" gene (Rht I) was duplicated, as evidenced from the short hetrozygotes and very short homozygotes. However, the alcohol dehydrogenase gene (Adh-1A), which is distal was not duplicated, as indicated by isozyme pattern. Furthermore, duplication appears to be on the same chromosome arm since this arm fails to pair in meiosis in heterozygotes while pairing normally in mutant homozygotes. The duplicated segment is not large enough to be apparent cytologically as longer arm (Larkin et al., 1989).

Single Gene Mutations

Cell culture also induces point mutations. There are now two examples in maize of single base changes in the alcohol dehydrogenase 1 locus, resulting in an altered electrophoretic mobility in one case and a null mutation in the other (Dennis *et. al.*, 1987). The rate of detected nucleotide substitution in this gene in culture was 2 in 218 cultured embryos producing 1382 examined somaclones. A chlorsulfuron-resistant tobacco mutant isolated from culture was due to resistant acetolactate synthase and therefore a mutation in the coding region of the corresponding gene. Several monogenic mutations were recovered in R_2 of tomato, four of which were shown to be allelic to known, recessive, single gene mutants (Bulk et al., 1990).

DNA Amplification

The number of copies of a particular sequence in the genome may vary after cell culture. Examples of dramatic deamplification of repetitive sequence have been found in potato, *Triticale, wheat* and *Hordeum spontaneum.* Amplification has been detected in somaclones for the enolpyruvyl-shikimic acid phosphate (EPSP) synthase gene in carrot (conferring glyphosate resistance) and for glutamine synthase gene (conferring resistance to the herbicide phosphinothricin). Lapitan et at. (1988) described the amplification of a particular sequence on the short arm of chromosome 7R in a Triticale somaclone. Such gene copy number changes can alter the phenotype without altering the gene responsible.

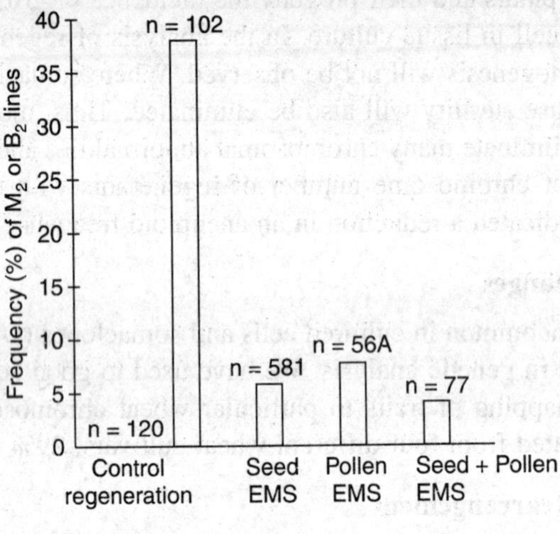

Fig. 9. Frequency of EMS induced mutation.

Altered Expression of Multigene Families

An increasing number of plant gene are proving to be encode on multigene families usually in clusters in one or more position in the genome. Somaclones have been readily found with altered patterns of products from such families. For example the gliadin storage proteins of wheat, hordeins of wild barley, the beta-amylase of wheat and the esterases of potato. Often these mutants express proteins not present in the donor plants. There may be heritably repressed genes in these families that are exposed following culture. Such rearrangements that alter control over those gene families have the potential to give access to a hidden germplasm.

Transposable Elements

The potential involvement of transposable elements in somaclonal variations has been demonstrated by Peschke et al., (1987). They used a line of maize with no active Ac (activator) element but with homologous sequences. After culture and regeneration, 3.7% of the somaclones showed Ac activity in appropriate test crosses.

A number of somaclonal variants in other systems behave as if they were caused by altered location and activity of a transposition. Two examples are, a tobacco mutation associated with the sulphur locus that causes a very high frequency of leaf (chlorophyll) spotting and flower (pigment) spotting and an alfalfa mutation to white flowers, which is unstable and reverts to colour at a high frequency, especially in vitro (Groose and Bingham, 1986).

Mutants can result from insertion of the element into either the structural gene or the regulator gene sequences. Several transposable elements have been isolated and characterized from mutants regenerated by the maize Ac - Ds system. By virtue of their movement transposable elements can inactivate structural genes, alter gene regulation, possibly reactivate silenced genes, and can generate duplications and developmental shock can induce the transposition of mobile elements.

Frequency of Somaclonal Variation

The frequency of somaclonal variation is often very much higher than chemically induced or radiation induced mutation. Thus, 2 point mutations at the Adh-1 locus were found in 1382 somaclones (from 218 cultured embryos); 1 B amalyse mutant was found in 90 wheat somaclones (from 30 cultured embryos); 1 mutant with a deamplified ribosomal sequence was found in 192 *Triticale* somaclones; and 5 mutants carrying group 4 translocations were found in 950 wheat somaclones.

A direct comparison of frequencies between somaclonal and ethyl methanesulfonate, (EMS) induced mutation has been made by Gavazzi et al., (1987) for tomato. Screening for the same set of characters in progeny of the treated generation, somaclonal variation produced on order of magnitude greater numbers of variants (Fig. 9).

Fig. 9; Frequency of EMS induced mutation and somaclonal mutation in tomato (data from Gavazzi et al. 1987). Mutation was assessed in 20-30 progeny from each line, for six classes of mutation (R2, progeny of regenerated plants; M2, progeny of mutated plants) *n*, number of families in each class. EMS mutagenesis was applied to seed or pollen or both, as indicated.

Novak et. al., (1988) compared mutation frequencies in maize between gamma ray induced and culture-induced mutation and a combination of the two. They examined selfed and back cross progeny for chlorophyll and morphological variants and found somaclonal variation to be often greater. The difference varied from 2-to 70-fold. Although these authors suggest that the spectrum of chlorophyll and morphological deviants is similar between gamma-induced and somaclonal variation, the wide range of differences in frequency (2-70x) indicates instead a range of different genetic consequences of the two treatments. A more valid comparison of the spectrum of mutation require examination of specific genes rather than broad classes of deviation.

Tomato, like maize, is a species in which there is a long history of mutational work. Evans (1986) has appraised extensive somaclonal and mutational research and concludes that there is a different spectrum of mutation.

In chemical or radiation-induced mutation, the frequency of dominant mutation is about 1% of the total number of mutations. Among heritable somaclonal mutations the relative frequency of dominant or codominant mutations is presently well over 10%. Given this and the very high absolute frequencies, it may be an attractive option for mutation in some circumstances.

Another distinction of somaclonal variation is the occurrence of apparent homozygous mutations. We are now aware of nine independent reports of putative homozygous (pure breeding) mutations arising directly from culture: in rice, two dwarfs; in wheat, a full-awned mutant, a white grained mutant and a gliadin mutant; in tomatoes, a jointless-pedicel mutant and a potato leaf mutant; in *Brassica juncea,* a yellow seeded mutant and in lettuce, a vigorous seedling mutant. In some of these cases test crosses established that these mutations were nuclear and not cytoplasmic in control. By whatever mechanisms this will eventually be explained, there seem to prospects for recovering recessive mutaitons in crops where self fertility or vegetative propagation precludes screening an M_2 population.

However, Bulk *et al.,* (1990) compared somaclonal variation to that of variation induced by treatment of seeds with ethyl methane sulphonate (EMS). The results showed that the mutant frequencies were higher after EMS treatment than those generated through tissue culture. With respect to the mutant spectrum, no clear differences were observed between the spectra obtained after EMS treatment and those after culture. However, tissue culture gave rise to polyploid plant, whereas no ploidy variants occurred after EMS treatment.

Applications of Somaclonal Variation to Plant Breeding

Creating the Variation

Despite the occurrence of literature documenting the occurrence of somaclonal variation, few experiments have rigorously explored the factors which might regulate the amount of variation produced. Manipulating the degree of variation would, however, be important to the efficient exploitation of somaclonal variation in a breeding programme. Where additional genetic variability is required it may be desirable to enhance the level of somaclonal variation. Alternatively, clonal uniformity may be essential where tissue culture is used for preservation of germplasm or for propagation as in the nursery industry, or where particular genotype in a breeding programme need to be cloned rapidly. In this case the occurrence of somaclonal variation has to be minimized which affect the frequency of cytogenetic alterations (Bulk et al., 1990). Somaclonal variation can occur from any explant but the type of explant can affect the level of phenotypic variability. In regenerants of potato, suspension culture protoplasts led to more variability than did mesophyll protoplast, but microspores gave even greater variation. However, the latter was probably a consequence of the heterozygous state of the donor plant being fully expressed in doubled haploids, rather than due to somaclonal variation alone. Length of time in callus culture has been shown to increase the frequency of cytogenetic abnormalities (Bulk et al., 1990). This is consistent with the sequential occurence of mutations in a rice callus culture. The initial event causing variation may occur very early in culture.

Comparisons of different culture treatments on the degree of variability have been difficult, since they required large number of experimental units and frequently lack methods of precisely monitoring genetic changes. Orton (1983) reviews experimental approaches to the study of somaclonal variation and suggest methods for the detection and measurement of the variation.

Exploiting the Variation

The exploitation of somaclonal variation in plant breeding has several aspects. The ability of the mechanism to cause variation which is independent of chromosome rearrangements is expected to be useful in supplementing existing genetic variation. In sexually propagated plant species, somaclonal variation may duplicate variation already available in germplasm pool, but generating the variation **de novo** in tissue culture confers the advantage that the background genotype can be an already adapted cultivar. In asexually propagated species, somaclonal variation may offer an essentially new means for varietal improvement, since hybridization is frequently difficult or impossible in these crops. Again, locally adapted cultivars with established agronomic and quality characteristics can be used as starting material.

Research associated with plant breeding may also be able to exploit some of the manifestations of somaclonal variation. It may offer a means to create isogenic lines rapidly and conveniently, possibly avoiding the linkage drag commonly associated with conventional methods of producing such material.

It may also exploit chromosome loss during culture for production of genetic stocks such as monosomics and disomics with a common background. The various uses of these include development of substitution lines or monosomic analysis. This type of material has proven very useful in assigning gene to specific chromosomes in polyploid species such as wheat and tobacco. In such species this is difficult using conventional procedure of homeologous genomes.

Selection of Variants

Selection of new genotypes generated during tissue culture can be expected to follow the same principles as selection in prevailing plant breeding programmes, but with the added benefit that some characters may rcspond to selection during the culture phase. Using planned selection strategies, a wide range of biochemical mutants have been successfully isolated in yeast and bacteria cultures, but the higher levels of organization present in plant may add complexity to the selection scheme for some characters, or totally preclude others from in vitro selection. Guidelines for establishing whether a plant function may be feasibly selected in culture are:

1. The function should be a cellular process, rather than one which is a property of higher levels of organization. For example, control of plant-water relations in the plant depends upon integration of diverse functions in the while plant; this integration is not expressed in culture.
2. The function should be expected to be expressed in culture; a function may satisfy criterion (1) but its expression may be specific to a tissue not present in culture. For example, the synthesis of seed storage proteins may not be expressed in culture. Morphological character which donot develop in culture also fit this category.
3. If genetic variation for the character already exist at the whole plant level, differential response between extreme genotypes in the proposed culture selection scheme should be demonstrated.

The combination of somaclonal variation and selection in culture offer a potentially very powerful route to plant improvement, since a very great number of possibly variant cells may be screened in little time with few resources. Examples of economic characters which could be expected to respond of selection in culture include tolerance to those herbicides which act at the cellular level, and disease resistance where the disease acts through a toxin. Other character such as salt tolerance, heat tolerance and amino acid composition have been widely postulated as amenable to selection in culture, probably because selection strategies are deceptively easy to devise. However, based on what is known about the biochemistry and physiology of these processes, they fail to satisfy the criteria listed and chances of success seem slight. Peterson and Krans (1990) formulated a new invitro selection technique to recover disease-resistant plants, namely Host-Pathogen Interaction system (HPIS) that exposes calli to a pathogen without physical contact between the organisms. The HPIS appears to provide an effective means for exposing calli to pathogen without complications associated with direct physical contact between the organisms.

The major benefit of this aspect of somaclonal variation is expected to be that of creating variation in adapted genotypes. Recognition of new genotypes at the whole plant level can only be as good as the selection schemes used, and problems in selection among somaclonal lines are expected to be no different to those already faced by plant breeders.

Selection Procedures

Variant cell lines can loosely be grouped into four categories

1. Lines with a selective advantage over wild type cells under certain conditions (*e.g.* resistant lines).

2. Lines visually distinguishable from the wild type.
3. Lines under a selective disadvantage compared to wild-type cells under certain conditions (e.g. Temperature sensitive or auxotrophic lines).
4. Lines indistinguishable form wild type except by chemical analysis. (e.g. many secondary metabolite producers).

The strategy used for selection depends on the category into which the required line falls. For the first two categories selection is conceptually straight forward, resistance or visual selection generally being adopted. For the other categories counter selection, total selection or some other indirect selection procedure is necessary. The principles of these different selection strategies are as follows.

Resistance Selection: (Fig. 10)

Where the variant phenotype is manifested as resistance to either an antimetabolite in the culture medium or advance culture conditions such as low temperature, variant cells can be selected directly as outlined in figure. Where protoplasts (Single Cells obtained by removal of the cell walls) are used as a source of cells, selections are usually performed on protoplast derived colonies rather than on protoplast themselves. The end result of this selection procedure is a cell aggregate, colony or callus mass, capable of withstanding the selective conditions, which can be isolated and bulked-up for further testing. This approach is not only applicable to the selection of resistant variants (Maliga, 1978), but also to the selection of other classes, such as lines capable of growing without added growth hormones.

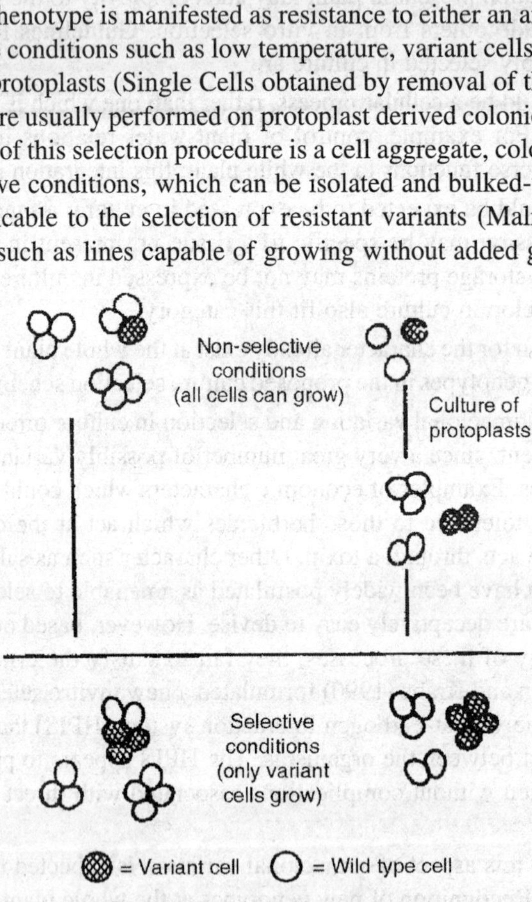

Fig. 10. Resistance selection using (left) callus or cell suspension, on (right) protoplast cultures.

Counter Selection

For many conditional lethal mutant, such as auxotrophs (mutants defective in the synthesis of an essential metabolite and hence requiring a supply of that metabolite for growth) or temperature sensitive mutants, there are no conditions in which the mutant cells have a selective advantage over the wild type, ruling out direct selection. In microbial cultures this does not provide a serious barrier, as replica-plating (the use of a stamp to transfer portions of colonies to non-permissive medium) permits rapid screening of large numbers of colonies. Unfortunately, the tight adhesion between plant cells prevents replica-plating by the usual methods, and it is necessary to seek indirect selection procedures. One approach, devised initially for selection of autotrophs in animal cell cultures (Puck and Kao, 1967), is theoretically applicable to the selection of any cell line which is inhibited in cell division (but does not die too quickly) under non-permissive conditions (*i.e.,* conditions which prevent growth of the variant, but not of the wild type cells). This procedure depends on the use of a counter selective agent, which kills only dividing cells, and is outlined in Fig. 11.

Puck and Kao (1967) used 5-bromo deoxy uridine, (BUdR), a thymidine analogue, as the counter selecting agent. The analogue becomes incorporated in place of thymidine into the DNA of dividing cells, which are then killed either spontaneously or on exposure to light. The cells are then washed and plated on permissive medium which permits growth of the auxotrophic cells.

A number of other counter selective agents have been used in plant cell cultures. A promising new counter selective agent has been suggested by Polacco (1979), who demonstrated that dividing soybean cells were selectively killed by arsenate. This provided the basis for a selection procedure used to obtain an isoleucine and valine-requiring cell line of *Datura innoxia.*

Visual Selection

Clearly, any phenotype which can be readily identified by appearence lends itself to visual screening of a large number of colonies, cell aggregate or even singly cells. This approach has been mostly used to obtain lines producing pigments, including chlorophyll, carotenoids, anthocyanin and other pigment classes.

Total Selection

For phenotypes not readily amenable to selection procedures total selection can be applied. This involves screening of a large number of individual colonies, applying appropriate tests to establish their pehnotypes. It is generally tedious and pains taking approach and only embarked upon when there is high level of confidence that the desired phenotype is represented in the cell population.

For the isolation of auxotrophs total selection is the alternative strategy to counter selection and may lose favour if highly effective counter selective procedures are developed. Identification of cell line producing secondary metabolites which cannot be visually distinguished also usually involves screening of a large number of colonies or callus pieces, but in this case selection is based on chemical analysis rather than growth tests.

Several analytical procedures which have been used include thin-layer chromatography to identify codeine-producing cell lines of *Papavar somniferum,* and high pressure liquid chromatography for ubiquinone-10 production by tobacco cells. There is certainly considerable scope for ingenuity in diversing other methods for detecting small quantities of individual secondary metabolites.

Other Selection Procedures

A comprehensive understanding of the physiological or biochemical basis of a particular phenotype, together with a little lateral thinking, can sometimes lead to a noble selection procedure for the character in question. A well documented example is the procedure devised by Muller and Grafe (1978) to isolate nitrate reductase deficient (NR-) line of *Nicotiana tabacum.* They used the fact that the enzyme, which converts nitrate to nitrite and is therefore central to the assimilation of nitrogen into

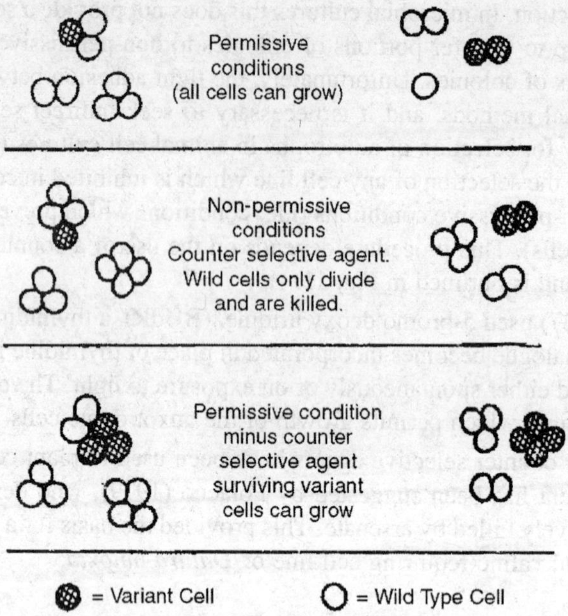

Fig. 11. Counter selection to obtain auxotrophs on temperature–sensitive mutants from (left) callus or cell suspension cultures, or (right) protoplast cultures.

organic molecule, will also recognize chlorate as a substrata. Chlorate is non-toxic but is converted by nitrate reductae to highly toxic chlorite. So, whereas one would expect counter selection or total selection to be necessary to obtain NR-line, relatively simple selection for resistance to chlorate can be used. NR lines are unable to convert chlorate to chlorite and so are not killed but are able to grow provided an organic nitrogen source is supplied.

Herbicide Tolerance (Table 14)

Creating crop genotypes with specific herbicide tolerance is becoming an attractive proposition for the following reasons:

1. Biochemical knowledge of herbicide action is increasing. Many herbicides act at specific biochemical steps, so that single-gene control of resistance is a feasible hypothesis. Postulated mechanism for resistance include impairement of the uptake system, altered sensitivity of a target molecule, or increased activity of existing degradation or detoxification enzymes.
2. Evidence for natural intraspecific genetic variation in herbicide resistance is accumulating. For example, naturally occuring atrazine-tolerant biotype have been found in a number of weed species, differential tolerance of wheat cultivars to metribuzin has been reported and resistance to diclotop -methyl has been recovered in natural population of *Lolium rigidum.*
3. Selection in culture may be possible, depending on the action of the herbicide. Specific knowledge of the herbicide action may increase the efficiency of the screen devised, but lack of knowledge would not preclude an exploratory 'Black-box' attempt. In fact, the recovery of tobacco mutants form culture enabled the identification of the biological process inhibited by the herbicides, Glean and Oust.
4. Incorporating specific tolerance into crops could improve the safety and efficiency of herbicide use.

Table 14. Herbicide Resistant Cell Lines

Herbicides	*Species*	*References*
Amitrole	**Nicotiana tabacum** (Tobacco)	Singer & Mc Daniel, 1984
Asulam	**Apium graveolens** (Celery)	Merrick & Collin, 1981.
Arazine	**Glycine max** (Soybean)	Zenk, 1974.
Chlorsulfuron	**Nicotiana tabacum**	Chaleff & Ray, 1984
	Triticum aestivum	Baillie et al., 1989.
2, 4 dichloro phenoxy	**Nicotiana sylvestris**	Zenk, 1874
aceticalid	**Citrus sinensis (Orange)**	Spiegel-Roye et al., 1983
Glyphosate	**Saccharum sp (Sugarcane)**	King & Maretzki, 1983.
Paraquat	**Nicotiana tabacum Lycopersicon**	Miller & Huges, 1980.
	peruvianum xL esculentum	Thomas & Pratt, 1982.
Picloram	**Nicotiana tabacum**	Chaleff & Persons, 1978
Isopropyl-N-Phenyl Carbamate	**Nicotiana tabacum**	Aviv & Galun. 1977
Sulfometuron methyl	**Nicotiana tabacum**	Chaleff & Roy. 1984
Trifluralin	**Triticum aestivum**	Baillie et al., 1989

Disease Resistance: (Table 15)

Breeding for disease resistance is likely to remain a major requirement in improving and maintaining plant productivity. New sources of resistance are particularly required where the pathogen is capable of genetic adaptation to specific resistance genes in the host. Somaclonal variation may therefore contribute to new host resistance and may be particularly exploited where selection against the pathogen or its toxin can be carried out in culture (Scowcroft et al., 1983). Even if the nature of the host-pathogen relationship means that in vitro selection is not feasible, screening at the seedling level is frequently possible and still permits relatively large numbers of regenerants or their progeny to be screened. Both types of selection strategy have been successfully used to isolate somaclonal variants for disease resistance.

Table 15. Disease Resistant Cell lines

Crop	*Disease*	*References*
Potato	Phytophthora infestans (late blight)	Shepard et al., 1980
	Alternaria solani (early blight)	Shepard et al., 1980
Maize	Drechslera maydis (leaf blight)	Shepard et al., 1980
Lucerne	Verticillium Wilt	Latunde-Data & Lucas 1983.
Banana	Fusariam Wilt	Hwang & Ko, 1990.
Tobacco	Phytophthora parasitica (black shank)	Daub & Jenns, 1989.
	Pseudomonas solanacearum (bacterial wilt)	Daub & Jenns, 1989.

Stress Tolerance: (Table 16)

Salt, water logging and drought, low and high temperatures, and mineral toxicity and deficiency are frequently cited as environmental stresses limitting higher yields and the expansion of crops into new or marginal areas. While interspecific differences in tolerance to those individual stresses may be conspicuous, intraspecific genetic variation is much less evident. In contrast to herbicide and disease resistance, a number of plant processes are probably directly involved in the response of plants to these stresses. For this reason, the physiological and biochemical basis of these stress responses is not yet well understood. Therefore, targets for genetic manipulation are difficult to define, and it may not be realistic of search for simple genetic solutions. Even to a single stress, alternative mechanism for tolerance may exist, some of which may depend on the whole plant and which may not be amenable to in vitro screening. Nevertheless, a few attempts have been made to use cell culture systems to screen for variants with increased tolerance to a particular stress.

Generally, the prospects for somaclonal variation to contribute to increased stress tolerance in crops are not immediately promising. Plant Physiologists have attempted to define morphological and physiological characters which confer resistance to particular stresses, but some have yet been very successful. Thus, although somaclonal variation may contribute to increased variability, selection of useful stress tolerance will depend upon a better understanding of whole plant and cellular responses to stress.

Table 16. Stress tolerant Cell Lines

Stress	*Species*	*Reference*
Temperature extremes	Antirrhinum majus	Melchers & Bergmann, 1959.
Freezing	Daucus carota (carrot)	Templeton-Somer et al., 1989.
Chilling	Nicotiana sylvestris	Dix, 1977.
	Capsicum annum (Sweet papper)	Dix street, 1976
Salt	Nicotiana tabacum	Watad et al., 1983.
	Glycine max (soybean)	Jia-Ping et al., 1981.
	Oryza sativa (rice)	Yano et al., 1982.
	Pennisetum americanum (Pearlmillet)	Rangan & Vasit, 1983.
	Triticum aestivum (wheat)	Xiao & Zhao, 1989.
	Solanum tuberosum (Potato)	Sabbah et al., 1989.
	Zea mays (maize)	Lupotto et al., 1989.
Aluminium	Lycopersicon esculentum (tomato)	Meredith, 1978.
	D.carota	Ojima & Ohira, 1983.
Manganese	D.carota	Ojima & Ohira, 1983.
Mercury	Petunia hybrida	Colijn et al., 1983.
Drought (Polyethylen glycol-induced water stress)	N.tabacum	Heyser & Nabors 1979.
	L.esculentum	Handa et al., 1983.

Qualtiy (Table 17)

Considerable effect has been expended on attempting to manipulate the amino acid composition of seed protein by in vitro selection for over production of a limiting amino acid (Maliga, 1984). The principle depends upon selecting for resistance to a toxic analogue of the target aminoacid. The analogue is owing to its incorporation into protein or to false feedback inhibition of one or more specific enzymes. In the mutants relaxation of the back inhibition allows over production of the target amino acid.

Table 17. Cell Lines with Improved Quality

Quality characters	*Crop*	*References*
Amylase Content & Gel Consistency	Rice	Sun et al., 1989
Glycine	Tobacco	Lawyer et al., 1980.
Lysine	Rice	Schaeffer & Sharpe, 1981
Methionine	Tobacco	Gonzale et al., 1984
Phenyl alanine	Tobacco	Flick et al., 1981
Proline	Tobacco	Dix et al., 1984.
Threonine	Barley	Bright et al., 1982.
Tryptophan	Carrot	Sung, 1979.
Valine	Tobacco	Vansh et al., 1982.

Field Performance

Somaclonal variation to complex characters is less easy to document than it is for more simply inherited characters. Where the environment contributes to the expression of a character significantly, replicated field trails are necessary. But seed or propagate supply is limiting in the early generation after regeneration, and segregation may complicate the analysis. However, for practical purposes, some indication of the potential of somaclonal variation to produce alteration in agronomic performance is important.

In rice, replicated glass house evaluation of lines derived form another culture showed significant differences for seed yield and protein yield components. Amongst the lines selected for increased protein content, one line maintained a higher percentage of seed protein concurrently with high seed size and seed number and is potentially useful germplasm. Field performance of this material is still necessary to confirm the value of this line.

In wheat heritable somaclonal variation has been reported in agronomic characters such as height and flowering time, (Larkin et al, 1984). Seed colour variants in mustard regenerates have also been reported; seed colour may have little agronomic significance but is often critical to consumer acceptance.

Recently, Zheng et at., (1989) observed variation in plant height, heading date, spike shape, awns, grain characteristics, glume colour, waxiness and α-amylase and prolamine content in wheat plantlets. Later Mohanand and Nabor (1990) regenerated somaclones from embryo having significant difference from their parents for plant height, spike length, grain/spike, and 100 grain weight. Arya et al., (1990) noticed variation among protoplast derived moth bean (*Vigna aconitifolia*) in two groups of characters. The first group includes, difference in seed colour, abortive seeds/pod and response to rot diseases. No difference were recorded in pollen stainability and meiotic behaviour. In the second group, analysis of variants showed significant difference for plant height, rachis length, length and breadth of leaflet, seeds/pod and seed weight.

Since somaclonal variation at various single-gene loci has been established, it seems reasonable to expect that loci affecting field performance can also be affected. Whether these changes can lead to increase in yield has been debated, but the potential benefit mean that the possibility should be properly explored.

Somaclonal Variaton and Alien Gene Introgression

A new option for achieving alien gene introgression has emerged from the observation that chromosome rearrangements occur in plants regenerated from tissue culture. The occurrence of chromosomal rearrangements during the tissue culture of interspecific hybrids and synthetic crop species, such as Triticale, are particularly compelling. Armstrong et al., (1983) found that the karyotypes of Triticale were quite unstable during culture. C-banding enabled identification of rye-wheat recombinant chromosomes.

In rye-grass, 4000 plants have been derived from callus initiated from a single immature triploid embryo (Ahloowalia, 1983). The triploid embryo was from a cross between diploid (*Lolium multiflorum* and tetraploid *L. perenne*). The regenerate plants showed wide variation in leaf shape and size, floral development, vigour and survival. Interestingly some plants displayed novel phenotypes. Cytological analysis of triploid ($2n$ = 21) and near triploids ($2n$ = 1920) revealed translocation and bridges and fragment which originated from chromosome fragmentation, deletions and inversion. In some plants penta–and hexavalent association were observed, which signifies that more than one reciprocal translocation had occurred during culture.

Larkin and Scowcraft (1981), and Ahloowalia (1983) recognized that the enhanced level of chromosome recombination in tissue culture might be employed to effect alien gene transfer. In crops such as wheat, wide-crossing programmes at CIMMYT (Centro Internaceonal de Mejoramiento de Maiz Y Trigo—International Centre for Maize and Wheat Improvement) and else-where are producing alien chromosome addition lines following hybridization with relatives such as *Aegilops, Agropyron, Elymus, Haynaldia* and *Hordeum.* These alien species carry genes for resistance to wheat pathogens and tolerance to environmental stress such as heat, salinity and metal toxicity. Research is being initiated to utilize tissue culture in an attempt to enhance the rate of transfer of specific alien genes from these wheat alien addition lines.

It is likely that the large translocations that are evident as multivalent at meiosis are the least useful for alien gene introgression, (Larkin *et al.,* 1989). Generally, only particular genes are desired and a cytologically apparent fragment will carry many hundreds of genes.

This somaclonal introgression approach, if indicated, may not be limited to aliens that can be crossed sexually to the crop of interest. Somatic hybridization is a method for gaining access to chromosomes from alien sources that are beyond the barriers of sexual crossabiltiy. Protoplasts from the two sources are fused and then cultured back to plants. It is wise to conceive of somatic hybridization as a means to temporarily access alien genes rather than as a means to construct weird (and almost certainly useless) hybrids. A recent example of this was the intergeneric interchanges observed in regenerates following fusion of *Nicotiana plumbaginifolia* and *Solanum tuberosum* (De Vries et al., 1987).

The direction of such limited gene exchange can be influenced by first gamma irradiating the donor protoplasts. This serves to inactivated them from further growth and promotes breakage of the genome. Somers et al., (1986) obtained immunological evidence of the transfer of the barley nitrate reductase gene to tobacco by this method. Similarly, Dudits et al, (1987) achieved transfer of methotrexate resistance from carrot to tobacco plants. It appears that taxonomic differences will be no barrier to such limited genetic exchange following somatic hybridization.

REFERENCES

Ahloowalia, B.S. (1983). Spectrum of Variation in somaclones of triploid rye grass. Crop Sci. 23:1141-1147.

Armstrong, K.C., C. Nakamura, and W.A. Keller, (1983). Karyotype instability in tissue culture regenerants of Triticale (x Triticosecale Wittmack) cv Welsh from 6 month old callus. Z. Pflan Zenzuchtg. 91:233-245.

Arya, I.D., S. Arya, D.V. Rao, and N.S. Shekhawat (1990). Variation amongst Protoplast derived moth bean Vigna aconitifolia plants. Euphytica. 47(1):33-38.

Aviv, D., and E. Galun. (1977). Isolation of tobacco protoplasts in the presence of isopropyl N-Phenylcarbamate and their culture and regeneration into plants. Z. Pflanzen physiol. 83:267-273.

Baillie, A.M.R., B.G. Rosswagel., and K.K. Kartha. (1989). Tissue culture regeneration and inviro selection in barley. Canadian Journal of plant science. 69(2): 545-546.

Barbier, M., and H.L. Dulieu (1980). Effects genetiques observes surdes plants de Tabac regenerees a partin de cotyledons par culture in Vitro. Ann. Amelior plantes. 30:321-344.

Bebeli, P. J., A. Karp and P. J. Kaltsikes (1990). Somaclonal variation from cultured immature embryos of sister lines of rye differing in heterochromatic content. Genome. 33(2): 177-183.

Brettell, R., M. Pallota, J.R. Gustafson, and R. Appels. (1986). Variation at the Nor loci in Triticale derived from tissue culture. Theor, Appl. Genet. 202:235-239.

Bright, S.W.J., B.J. Miflin, and S.E. Rognes. (1982). Threonine accumulation in the seeds of a barley mutant with an altered aspartate kinase. Biochem, Genet. 20:229-244.

Bulk, R.W., H.J.M. Loftler, W.H. Lindhout, and M. Koornneef (1990). Somaclonal variation in tomato: effect of explant source and a comparison with chemical mutagenesis. Theor. Appl. Genet. 806:817-825.

Chaleff, R.S., and M.F. Parsons (1978). Direct selection in vitro for herbicide resistant mutants of Nicotiana tabacum. Proc. Nat. Acad. Sci. USA. 75:5104-5107.

Chaleff, R.S., and T.B. Ray, (1984). Herbicide resistant mutants from tobacco cell cultures. Science. 233:1148-1151.

Colijn, C.M., A.J. Kool, and H.J.J. Nijkamp. (1983). An effective chemical mutagenesis procedure for petunia hybrida cell suspension cultures. Theor, appl. Genet. 55:101-106.

D' Amato, F. (1985). Cytogenetics of plant cell and tissue cultures and their regenerants. CRC Crit. Rev. Plant Sci. 3(1):73-112.

Daub. M.E., and A.E. Jenns. (1989). Field and green house analysis of variation for disease resistance in tobacco somaclones. Phytopathology. 79(5):600-605.

Dennis, E., R. Brettell, and W.J. Peacock. (1987). A tissue culture induced Adh I null mutant of maize results from a single base change. Mol. Gen. Genet. 210:181-183.

De Vries, S., M. Ferwerda, A. Looner, L. Pijnacker, and W. Feenstra. (1987). Chromosomes in somatic hybrids between Nicotiana plumbaginifolia and a monoploid potato. Theor, Appl. Genet. 75:170-176.

Dix, P.J. (1977). Chilling resistance is not transmitted sexually in plants regenerated from Nicotiana sylvestris cell lines. Z. Pflanzenphysiol. 84:223-226.

Dix, P.J., U.A. Me Lysaght, and R.S. Pearce. (1984). The potential of cell cultures for the production of salt tolerant cultivars. In efficiency in plant Breeding. Proc. 10th Congr. Eucarpia Eds. Lange, W., A.C. Zeven, and N.G. Hogenbooml. pp. 219-223. Pudoc, Wageningen.

Dix, P.J., and H.E. Street. (1976). Selection of plant cell lines with enhanced chilling resistance. Ann. Bot. 40:903-910.

Dudits, D., E. Maroy, T. Praznovszky, Z. Olah, J. Gyorgyey, and R. Cella. (1987). Transfer of resistance traits from carrot into tobacco by asymmatric somatic hybridization: regeneration of fertile palnts. Proc. Natl. Acad. Sci. USA 84:8438-8438.

Evans, D. (1986). Case histories of genetic variability in vitro: tomato. In Cell culture and Somatic Cell genetics in plants. Vol. 3. Plant regeneration and genetic Variability. Edited by I.K. Vasil. Academic Press, Orlando, FL. pp. 419-434.

Evans, D.A. (1988). Applications of somaclonal variation. In Biotechnology in agriculture (ed Mizrahi, A.), Newyork, USA; Alan R. Liss, Inc. 203-223.

Flick, C.E., R.A. Hensen, and D.A. Evans. (1981). Isolation, protoplast culture and plant regeneration of PFP resistant variants of N. Tabacum (Su/Su). Z.Pflanzen physiol. 103 ; 239-245.

Gavazzi, G., Tonelli, G.Todesco, E.Arreghini, F.Raffaldi, F.Vechhio, G.Barbuzzi, M.Biasini, and F.Sala. (1987). Somaclonal variation versus chemically induced mutagenesis in tomato (Lycopersicon esculentum L.). Theor. Appl. Genet. 74:733-738.

Gonzales, R.A., P.K. Das, and J.M. Widholm. (1984). Characterisation of cultured tobacco cell lines resistant to ethionine, a methinonine analogue. Plant physiol. 74:640-644.

Groose, R., and E.Bingham. (1986). An unstable anthocyanin mutation recovered from tissue culture of alfalfa (Medicago sativa). Plant Cell rep. 5:104-107.

Handa, A.K., R.A. Bressan, S.Handa and P.M. Hasegawa. (1983). Clonal variation for tolerance to polyethylene glycol induced water stress in cultured tomato cells. Plant physiol. 72:645-653.

Heyser, J.W., and M.W. Nabors (1979). Growth, water content and solute accumulation of two tobacco cell lines cultured on sodium chloride, dextran, and polyethylene glycol. Plant physiol. 68:429-435.

Hwang, S.C., and W.H. Ko. (1990). Tissue culture plantlets as a source of resistance to fusarial wilt of Cavendish banana. In Biological control of soilborne plant pathogens (ed Hornby, D) Walling ford, Oxon; U.K. CAB International. 345-353.

Jia-Ping, Z., E.J. Roth, W.Terzaghi, and K.G. Lark. (1981). Isolation of sodium dependent variants from haploid soybean cell cultures. Plant Cell report. 1:48-51.

King, J., and A.Maretzki. (1983). Isolation from sugarcane cultures of variants resistant to antimetabolites. Physiol. Plant. 58:457-463.

Lapitan, N., R.Sears, and B.Gill. (1988). Amplification of repeated DNA sequences in wheat x rye hybrids regenerated from tissue culture. Theor, Appl. Genet. 75:381-388.

Larkin, P. (1987). Somaclonal Variation: history, method and meaning. Iowa State J. Res. 61:393-434.

Larkin, P.J., P.M. Banks, R.Bhati. R.I.S. Brettell, P.A. Davies, W.R. Scow croft, L.H. Spindler, and G.J. Tanner, (1989). From somatic variation to variant plants: mechanisms and applications. Genome 31(2):705-711.

Larkin, P.J., S.A. Ryan, R.I.S. Brettell, and W.R.Scowcroft, (1984). Heritable somaclonal variation in wheat. Theor. appl. Genet. 67:443-445.

Larkin, P.J., and W.R. Scowcroft. (1981). Somaclonal Variation–a novel source of variability from cell cultures. Theor, appl. Genet. 60:197-214.

Latunde-Dada, A.O., and J.A. Lucas. (1983). Somaclonal variation and reaction to Verticillium with Medicago sativa L.plants regenerated from protoplasts. Plant Sci. Lett. 32: 205-211.

Lupotto, E., M.C., Lusardi and M. Mongodi. (1989). In vitro selection of maize (Zea mays) salt tolerant somaclones and plant regeneration. J. Genet. & Breed. 43(4):215-22.

Lawyer, A.L., M.B. Berlyn, and I.Zelitch. (1980). Isolation and Characterization of glycine hydroxamate resistant cell lines of Nicotiana tabacum. Plant Physiol. 66:334-314.

Lorz, H, and W.R. Scowcroft. (1983). Variability among plants and their progeny regenerated from protoplasts of Su/Su heterozygotes of Nicotiana tabacum. Theor, appl. Genet. 66:67-75.

Maliga, P. (1978). Resistance mutants and their use in genetic manipulation. In Frontiers of plant tissue culture, 1978 (Ed. Thorpe, T.A.), pp. 381-392. IAPTC, 1978.

Maliga, P. (1984). Isolation and characterization of mutants in plant cell culture. Ann. Rev. Plant Physiol. 35:519-542.

Marcotrigianno, M., T.H. Boyle, P.A. Morgan, and K.L. Ambach. (1990). Leaf color variants from coleus shoot cultures. Journal of American Society for Horticultural Science. 115(4): 681-686.

Meins, F.Jr. (1982). Habituation of cultured plant cells. In Molecular Biology of plant Tumors. (Eds. Schell. J., and G. Kahl), PP. 3-31. Academic Press, Newyork.

Melchers, G., and L. Bergamann. (1959). Untersuchungen an kulturen von haploiden Gwewden von Antirrhinum majus. Ber. Dtsch. Bot. Ges. 78:21-29.

Meredith, C.P. (1978). Selection and Characterization of aluminium resistant variants form tomato cell cultures. Plant Sci. Lett. 12: 25-34.

Merrick, M.M.A., and H.A. Collin. (1981). Selection of asulam resistance in tissue cultures of celery. Plant Sci. Lett. 20: 292-296.

Miller, O.K., and K.W. Hughes. (1980). Selection of paraquat-resistant variants of tobacco from cell cultures. In Vitro. 16:1085-1091.

Mohanand, A.S., and M.W. Nabors. (1990). Somaclonal Variant Plants of wheat derived from mature embryo explants of three genotypes. Plant Cell reports. 8(9):558-560.

Muller, A.J., and R. Grafe. (1978). Isolation and characterization of cell lines of Nicotiana tabacum lacking nitrate reductase. Molec. gen. Genet: 161:67-76.

Novak, F., S. Daskolov, H. Brunner. M. Nesticky, R. Afza, M. Dolezelova, A. Herichova, and T. Hermelin. (1988). Somatic embryogenesis in maize and comparison of genetic variability induced by gamma irradiation and tissue culture techniques. Plant breed. 101:66-79.

Ojima, K., and K. Ohira. (1983). Characterisation of aluminium and manganese tolerant cell lines selected from carrot cell cultures. Plant Cell Physiol. 24:787-797.

Orton, T.J. (1983). Experimental approaches to the study of somaclonal variation. Plant mol. Biol. Rep. 1:67-76.

Park, C.H., and P.D. Walton (1990). Morphology, cytology and fertility of plant regenerated from immature inflorescence cultures of Elymus canadensis. J. Genet & Breed. 44(3):231-239.

Peschke, V., R. Phillips, and B. Gengenbach. (1987). Discovery of transposable element activity among progeny of tissue culture derived plants. Science (Washington, D.C.), 238:804-807.

Peterson, M. and J.V. Krans. (1990). Evaluation of a new invitro cell selection technique. Crop. Sci. 30(1): 226-229.

Polacco, J.C. (1979). Arsenate as a potential negative selection agent for deficiency variants in cultured plant cells. Planta. 146: 155-160.

Puck, T.T., and F.T. Kao. (1967). Genetics of somatic mammallian cells. V. Treatment with 5-bromodeoxyuridine and visible light for isolation of nutritionally deficient mutants. Proc. Nat. Acad. Sci. 58:1227-1234.

Rangan, T.S., and I.K. Vasil. (1983). Sodium Chloride tolerant embryogenic cell lines of Pennisetum americanum (L.) K. Schum, Ann. Bot. 52:59-64.

Sabbah, S., and M. Tal. (1990). Development of callus and suspension cultures of potato resistant to NaCl and their response to stress. Plant cell, Tissue and Organ Culture. 21(2):119-128.

Schaeffer, G.W., and F.T. Jr. Sharpe. (1981). Lysine in seed protein from 5-aminoethyl-L-cysteine resistant anther-derived tissue cultures of rice. In vitro. 17:345-352.

Scowcroft. W.R., P.J. Larkin, and R.I.S. Brettell. (1983). Genetic variation from tissue culture. In the use to Tissue Culture and Protoplasts in Plant Pathology (Eds. Deverall. B.J., and J. Helgeson), pp. 139-162. Academic Press. Newyork.

Shepard, J.F., D. Bidney, and E. Shahin. (1980). Potato Protoplasts in crop improvement. Science. 28:17-24.

Singer, S.R. and C.N. Mc. Danie. (1984). Selection of amitrole tolerant tobacco calli and the expression of this tolerance in regenerated plant and progeny. Theor. appl. Genet. 67:427-432.

Somers, D., K. Narayana, A. Kleinhofs, S. Cooper, and E. Cocking, (1986). Immunological evidence for transfer of the barley nitrate reductase structural gene to Nicotiana tabacum by protoplast fusion. Mol. Gen. Genet. 204:206-301.

Spiegel-Roy, P., J. Kochba, and S.Saad. (1983). Selection for tolerance to 2,4-dichlorophenoxy acetic acid in ovular callus of orange (Citrus sinensis) Z. Pflanzen physiol. 109 : 41-48.

Sun, Z.X., Y.K. Luo, Y.K. Zhang, and Y.Y. Chnag. (1989). Somaclonal variation in grain quality of rice. Hereditas (Beijing). 11(5):5-8.

Sung, Z.R. (1979). Relationship of indole-3 acetic acid and tryptophan concentrations in normal and 5-methyl tryptophan resistant cell lines of wild carrot. Planta. 145: 339-345.

Templeton-Somers, K.M., W.R. Sharp, and R.M. Pfister. (1981). Selection of cold resistant cell lines of carrot. Z. Pflanzenphysiol. 103:139-148.

Thomas, B. R., and D. Pratt. (1982). Isolation of paraquat-tolerant mutants from tomato cell cultures. Theor. Appl. Genet. 63:169-176.

Vansh, R., D. Aviv, E.Galun. (1982). Valine resistant plant derived from mutated haploid and diploid protoplasts of Nicotina sylvestris and N.tabacum. Theor. appl. Genet. 64:51-58.

Watad, A.E. A., Reinhold, and H.R. Lerner. (1983). Comparison between a stable NaCl-selected Nicotiana Cell line and the wild type. K^+, Na^+ and proline pools as a function of salinity. Plant Physiol. 73: 624-629.

Xiao, H. L., and S.X. Zhao. (1989). Selection of wheat mutants tolerant to NaCl by tissue culture. Acta Agriculturas Nucleatae Sinica. 3(2):85-90.

Yano, S., M. Ogawa, and Y. Yamada. (1982). Plant formation from selected rice cells resistant to salts. In plant tissue Culture, 1982 (Ed. Fujiwara, A.), pp. 495–496. Japanese Association for plant Tissue Culture. Maruzen Co. Ltd., Tokyo.

Zenk, M.H. (1974). Haploids in pysiological and biochemical research. Haploids in higher plants. Advances and potential (Ed. Kasha, K.J.), pp. 339-354. University of Guelph.

Zheng, Q.C., Y.L. Zhu, and W.H. Chen. (1989). Plantlet regeneration from in vitro culture of young spikes of wheat and its variation. Acta Agriculture Nucleatae Sinca. 3(3):129-136.

CHAPTER 12

Anther Culture and its Role in Crop Improvement

Introduction

Haploid plants have the gametophytic number of chromosomes, that is, a single set of chromosomes in the sporophyte. Their significance for plant improvement and as a tool in various disciplines of plant science has been stressed since haploids are of great importance, especially studies on the induction of mutations and also for the production of homozygous plants they are needed in large numbers. Since conventional breeding methods are very laborious and not very efficient.

Guha and Maheswari (1964, 1966) first reported the direct development of haploid embryos from microspores of *Datura innoxia* by anther culture in vitro. Afterwards, many researchers have been interested in the induction of androgenetic haploids. The simplicity and ease with which haploids can be produced in large numbers by this technique has stimulated a great deal of activity in the production and utilization of haploids for fundamental and applied genetics. The another culture technique has been refined and extended to 171 species belonging to 60 genera and 26 families of angiosperms. Researchers in China began work in this field in 1970.

The first pollen drived plants in China were rice and wheat. Subsequently studies on another culture were also carried out on many species. Such as maize, rye, triticale, rubber tree, chinese tobacco, cabbage, cotton, soybean and pepper. Induction of pollen derived plants of many species was successfully achieved in China. Species from which pollen derived plants were first obtained in China :

Year	*Species*	*Reference*
1971	*Capsicum annum* L.	Wang et al., 1973.
	Triticale	Sun et al., 1973
	Triticum aestivum L.	Ouyang et al., 1973
1973	*Beta vulgaris* L.	Institute of Sugarbeet
	$(2n = 4x = 36)$	and Sugar crops, 1978.
1975	*Zea mays* L.	Institute of Genetics 1975
1977	*Hevea brasiliensis*	Chen et al., 1978
1978	*Sorghum vulgare* L.	Zhou, 1978
1979	*Beta vulgaris* L.	Shao, 1979 Chen et al.,
	$(2n = 2x = 18)$	1979
	Saccharum sinensis L.	
1981	*Vitis vinifera* L.	Zhou and Li, 1981

Anther Culture Technique

Anther culture refers to culture of single pollen grains or of the anther, containing the male gametophytes or microspores with the objective of producing monoploid plants. The technique for the excision and culture of anthers is relatively simple and efficient. Closed flower buds, which have anthers containing uninucleate microspores, are most suitable for the induction of androgenesis. The flower buds are excised, surface sterilized in batches of 25 with a 5% solution of commercially available bleach or a 1% calcium hypochlorite solution for 10 minutes and then washed twice with sterile, distilled water. However, the flower buds obtained from plants grown in green houses, if dissected carefully, need no sterilization. First an incision is made on one side of the flower bud and the stamens are gently taken out with a pair of fine forceps and collected in a sterile petridish. The filament from the stamen is then carefully removed and five anthers are normally transferred to ensure that they are not injured in any way. Damaged anthers should be discarded, as they often tend to produce callus from parts other than pollen. The anthers are cultured on agar solidified medium in glass tubes or small peridishes. They can also be grown in liquid media in Erlenmeyer flasks and kept on a slow rotary shaker. The cultures are incubated ata 24-27 °C and exposed to light of about 2000 °lx for a 14 h day. Although unopened flowers with their sepals and petals touching each other have been observed to be suitable for androgenesis by excised anther culture, this is only true of young plants. However, older plants, especially towards the end of lowering frequently produce small buds which although appearing to be at a suitable stage, contain anthers with a heterogeneous mixture of microspores and young pollen. In such anthers the number of abnormal pollen also increases and androgenesis is not only delayed but also considerably reduced. Therefore, it is important that anthers should be taken from relatively young plants which have been grown under good light conditions (Dunwell and Perry, 1973).

Depending upon the plant species it takes about 3-8 weeks for pollen plantlets to emerge from the anthers. The plantlets when approximately 5 cm tall can be removed from the culture and freed from agar by gently washing with running tap water. The plantlets are then transferred to small pots containing autoclaved soil. To reduce sudden shock and to prevent desiccation, it is advisable to cover these plantlets with glass beakers and to keep them in a wellwet, humid green house. After one week the beakers are removed, and after further two weeks, when the potted plantlets have grown to a sufficient size, they are again transferred into larger pots where they eventually develop to maturity and flower, (Fig. 12).

Haploids can be diplodized to produce homozygous plants by colchicine treatment. The plantlets while still attached to the anther are treated for 24-48 h with 0.5% colchicine solution (Kasperbawer and Collins 1972, Burk et al. 1972) washed thoroughly and replanted.

Schematic representation of the culture of excised anthers and the development of haploid plants directly by embryo formation or through haploid callus :

Flower bud

Excised anthers

Anther culture

Androgenesis

Irrespective of the early events in the division of microspore nucleus there are two modes of androgenesis, the direct and indirect androgenesis.

1. **Direct androgenesis.** The microspore behaves like a zygote and undergoes various stages of embrogeny stimulating those in vivo, as in *Atropa, Datura,* and *Nicotiana.* The embryos mostly at the globular stage are released form the exine and develop further.

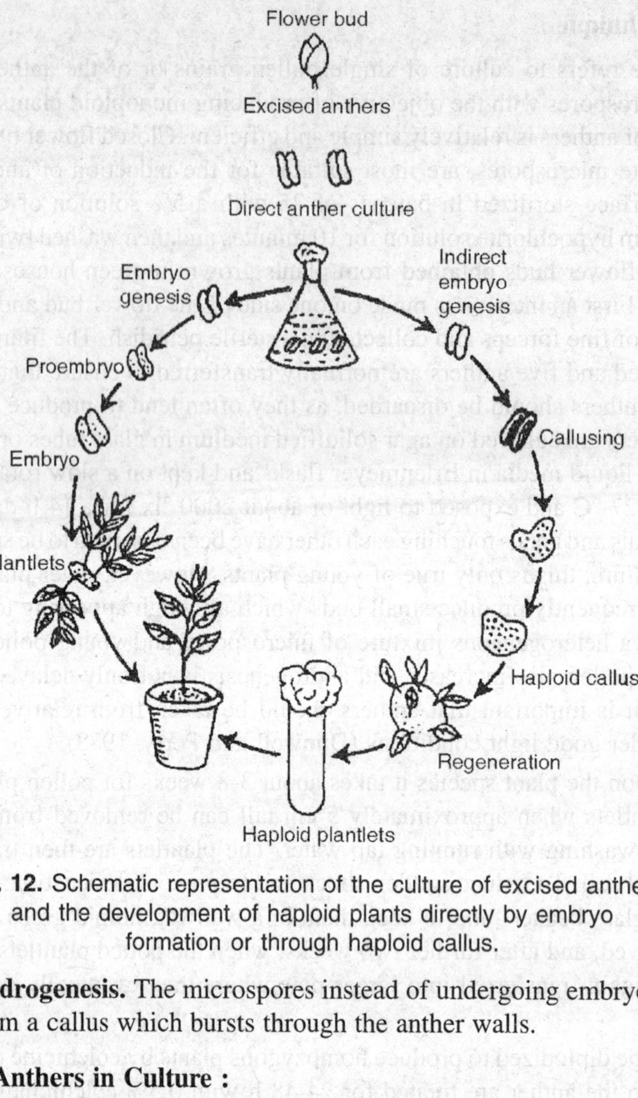

Fig. 12. Schematic representation of the culture of excised anthers and the development of haploid plants directly by embryo formation or through haploid callus.

2. **Indirect androgenesis.** The microspores instead of undergoing embryogenesis divide a few times to form a callus which bursts through the anther walls.

Factors Influencing Anthers in Culture :

1. Genotype of donor plants.
2. Stage of microspore development
3. Culture medium
4. Precultural and cultural conditions
5. Planting date of donar material
6. Conditioning factor or anther factor
7. Effect of anther orientation in culture
8. Light and temperature regime
9. Chemical Hybridization Agents (CHA)
10. Thermal shocks.

1. **Genotype of Donor Plants.** The genotype of the donar plants plays an important role in determining the frequency of pollen plant induction. Ouyang et al ; (1983) found that the induction frequencies of pollen calli and green pollen plantlets were with heritable characteristics, quantitatively, controlled by many genes. Foroughi-wher et al; (1982) reported the large differences among genotypes and predicted that genes from highly responsive genotypes might be transferred into breeding materials, which are reluctant in anther culture.

Androgenic plant regeneration from 2 of 31 sugarcane clones cultured was reported by Chen et al., (1979) while success from 2–10 clone cultures was reported by Fitch and Moore (1983).

Anther culture in corn is very genotype dependent. Migo et al. (1981) reported that of 159 genotypes tested in 1975, only 9 responded.

Petolino and Thompson (1987) examined maize anther culture response frequency in a diallel set of crosses. General (GCA) and specific (SCA) combining ability mean squares were calculated and determined to be highly significant. A significant GCA mean square indicates than an additive genetic component influences anther culture response. A significant SCA mean square implies that dominant or dominant epistatic genes also effect anther culture response. These results indicate that it should be possible to breed for improved anther culture responsiveness through several cycles of selection.

In barley, Dunwell *et al.* (1987) reported that, genotype is one of the most important factors that affects the efficiency of anther culture. He used the two most responsive spring cultivars Dissa and Sabarlis for anther culture and found that Dissa had a significantly higher level of callus production than the sabarlis, but the overall frequency of green and albino production was higher in sabarlis.

Foroughi-wher *et al.,* (1982) reported that although the involvement of cytoplasmic factors in anther culture was reported in reciprocal F_1 hybrids of Dissa and Aramir, no significant difference was observed between reciprocal F_1 hybrids of Dissa and Sabarlis in percentage of responding anthers green plant production and total plant production.

Growth environments of donor plants also present remarkable influence on the production of pollen callus or embryoid. After a series of studies, Heberle-Bors (1982) concluded that growth conditions of donor plants, particularly of light intensity and temperature affected the frequency of both anomalous pollen in vivo and pollen embryoids in anther culture as well as, their sexual balance. The effects become pronounced five weeks prior to anthesis. They believed that a short duration of about eight hours, and low temperature of 15-18 °C or even lower would be appropriate for growing tobacco plants for anther culture.

Grunewaldt (1975) reported that anthers from field grown barley plants had higher rates of pollen callus production and subsequent plant regeneration than those from green house plants.

Huang et al., (1982) with sabarlis barley, it was found that the donar plants grown during the first half of the year when day light was increasing gave the best yields of green pollen plants, while the anthers from plants grown during the second half of the year under conditions of decreasing day length produced fewer calli with poorer regenerative potential even if higher pressure sodium lights were used to supplement natural day light when it was less than 16 hrs.

Wen et al., (1991) reported that out of 7 cultivars namely in white, Tx 403-TSB, DDY, Sommer milo, *Tx* 2779, Brawley and Spur Federal, cultured callus induction frequency was higher only in 2 cultivars namely in white and T*x* 403-TSB.

2. **Stage of Microspore Development.** Although the pollen derived calli could be induced in culture, anthers were inoculated at different stages from pollen mother cell to mature pollen (He and Ouyang, 1984). For maximum production of pollen plants, the appropriate stage of microspore

development varies with plant species. As summarized by Sunderland (1982), anthers of some species (post mitotic for example tobacco and datura) are most productive when cultured at the first pollen mitosis or at later stages whereas those of others (premitotic for example barley, wheat and rice) are more productive when cultured at uninucleate microspore stage.

Members of the genus *Saccharum* flowers during the shortening days of autumn and early winter when minimum temperatures may be dropping to levels inhibitory or lethal to pollen. Plant vigour, pollen viabilitty and androgenic capacity declines as the flowering season progresses. Clones which produce anthers early in the flowering season have greater androgenic capacity (Fetch and Moore, 1983).

In maize, all investigators who have reported success with corn anther culture agree that the uninucleate microspore is the most responsive stage for culture, some have shown a preference for the mid-uninucleate stage while others prefer to the late uninucleate stage. The mid nucleate stage is described as the developmental stage of the microspore when the nucleus is situated in the centre of the cell before the formation of vacuole. The late uninucleate stage is described as the print in time when the nucleus has migrated to a position next to the cell wall opposite to the germination pore with the well formed vacuole.

In barley, it was shown that mid-unicellular stage is optimum either for pretreatment of tillers or spikes or for the culture of anthers directly taken from fresh spikes. The miduninucleate microspores contain well developed exine, and vacuole but nucleus small and still in G1 of the cell cycle.

3. **Culture Medium.** The culture medium is the principal factor controlling the induction and development of intact plantlets. Sucrose may not only regulate the osmotic pressure of the medium, but is also the best carbohydrate source. Nine percent sucrose is optimum for anther culture of wheat and 12 per cent for maize, rape seed and barley.

Several culture media developed by Chinese scientists are now widely used in China and abroad. Potato II medium (Chuang *et al.*, 1978) contains 10 per cent aqueous potato extract, half of macroelements of Whites (WH) medium and the Fe salt, and thiamine levels of MS (Murashige and Skoog) medium. The frequency of callus was much higher in this medium than in others. Nitsch (N) medium (Chu, *et al.*, 1985) has a low ammonium ion concentration and the optional ratio of ammonium nitrogen to nitrate nitrogen of efficiency of anther culture of rice and other cereals.

In barley, it is proved that potato medium was superior. Low salt medium seems more suitable for pollen callus formation than a high salt medium (Kao, 1981). Exogenises hormones in the medium usually play an important role in anther culture. A number of growth substances and their combinations have been used in barley anther culture.

Rout and Sharma (1991) reported that callus induction and green plant regeneration at high frequencies from interspecific hybrids *Oryza stiva* L. X *O.rufipogon* Griff were achieved by coordinating the growth regulators in the induction medium. The highest frequency was obtained on potato medium supplemented with 1 mg/1 2,4-D 2 mg/1 NAA (Naphthalein acetic acid) and 1 mg/1 kinetin.

4. **Precultural and Cultural Conditions.** In many cases, multi-cellular pollen grains (MPG's) are incduced by excision alone or under suboptional precultural and cultural conditions. Appropriate precultural and cultural conditions, however, are not only necessary for further development of MPG's but are beneficial for induction of MPG's.

Temperature stress pretreatment (mainly low temerature) has long been used to improve anther culture efficiency. In many genotypes of barley (Huang and Sunderland, 1982), and rice (Chen et al., 1980) low temperature pre treatment was considered essential for production of a large quantity of pollen calli.

The cultural temperature also plays an important role in the production of pollen plants. However, the optimum cultural temperature varies among geotypes. In some *Brassica* species, the high temperature (over 30 °C) is essential for production of pollen embryoids whereas in barley the temperature over 30 °C at an early stage of culture, kills almost all pollens. The wheat genotypes were classified into three categories; high temperature type (most productive when cultured at 30°C or higher) inter mediate type (28 °C) and low temperature type (26 °C). The response is highly sensitive to cultural temperature and 2 °C difference may lead to considerable variation in the production of pollen calli and plants. A response to temperature was found to be of a heritable character, controlled by multigenes (Ouyang *et al.,* 1983).

Chen and Chen (1983) demonstrated that development of pollen embryoids in rubber tree, sugarcane and rape seed might be divided into several stages. Different culture conditions were required for the various stages. Based on their results, Chen and Chen established the culture procedures for anther culture of rubber tree and rapeseed, which greatly increased the induction frequency of embryoid.

5. **Planting Date of Donar Material.** In maize significant differences were observed for planting date with the most responsive planting date, June 20, producing embryoids and calli with a frequency of 11-28% while the poorest responding planting date May 24 responded at a rate of 2.43%.

6. **Conditioning Factor or Anther Factor.** In barley, it is supposed that a discrete conditioning factor (CF) or anther factor (AF) leaches from the anthers floating in liquid medium which is essential to androgenesis in the initial stage of anther culture with cultivar sabarlis, the culture of anthers in a conditioned medium resulted in a tremendous increase of anther response and pollen callus production. The conditioning effect is highly reproducible.

7. **Effect of Anther Orientation in Culture.** In barley, a high percentage of responding anther was obtained (60% and 87%) respectively in two experiments, when the anthers were placed on edge, *i.e.,* with a single locule being in contact with the medium. Pollen embryos or calli developed only in the upper locule of anthers.

8. **Light and Temperature Regime.** Generally, barley anther cultures are carried out in darkness at 25 °C illumination reduced callus yield in barley. Pollen calli from illuminated culture were less vigorous and readily turned brown. Temperature and light influence productivity.

9. **Chemical Hybridization Agents (CHA).** Gametocides such an Ethrel have been known to cause male sterility (Rowell and Miller, 1971) and induce extra mitotic divisions and micronucler formation. To increase the frequency of pollen embyos in wheat CVS-Kolibri and Janus, the plants are treated with 4000 ppm Ethrel, the anthers, showed 4-6 nucleate pollen, with fewer starch grains than the controls.

In 1987, Picard et al., sprayed plants with fentidazone potassium chemical hybridization agent - (CHA), obtained significant improvement in another culture, and 10 to 20 fold increase in the production of dihaploid lines. Like wise Schmid and Keller (1986) observed positive effect of the gametocide and the androgenic process was significantly faster.

10. **Thermal Shocks.** The thermal shocks given to plants can alter the mode of division of the microspore nucleus. The stimulatory effect of thermal shocks for anther culture has been successfully extended to Datura, Tomato, *Atropa belladonna* and Tobacco to enhance androgenesis in anthers and isolated pollen.

Both cold treatment and brief exposure to higher temperature have been reported to stimulate repeated division of pollen (Table 18). The pretreatment of *Brassica hirta* at 4 °C in sucrose solution enhanced the intensity of callusing as well as the frequency of embryogenic anthers (Bajaj and Mhapatra, 1990). The temperature shocks appear to cause dissolution of microtubules and dislodging of the spindle which causes abnormal division of the microspore nucleus.

Table 18 Effect of various pretreatments of the floral buds on the androgenic anthers in Brassica hirta (Bajaj and Mahapatra, 1990).

Pretreatment condition	*Duraton of treatment*	*No. of anthers cultured*	*No. of ambryo-genic anthers*	*Frequency of ambryogenic anthers (%)*
1. Control (no treatment)		262	1	1.28
2. 4 °C in water	7 days	340	6	1.76
3. 4 °C in sucrose solution (0.4)(m)	3 days	432	14	3.24

Characteristics of Pollen Derived Plants : (Table 19)

1. Stability of Homozygous Pollen Plants

Homozygous diploids can be produced by doubling the chromosomes of haploids, thereby producing stable strains. In wheat (Ouyang et al., 1973) three characters of pollen derived lines (H_3) and the parental lines of a wheat cross was analysed. The coefficient of variation of all three characters of pollen derived lines (H_3) is less than or approximately equal to that of the parents. From this it is predicted that characteristics of pollen derived plants are uniform and similar to stable inbred varieties. Their haploids do not segregate.

Table 19. Analysis of characteristics of pollen derived lines (H_3) and their parents in Kotung 58 × 7479:

Character	*Kotung 58* $n = 20$	*7479* $n = 20$	*Kotung 58 × 7479* (H_3, $n=34$)
Plant height cm CV%	70.00 ± 3.69 4.24	95.15 ± 4.25 4.47	96.00 ± 3.00 4.06
Earlength CV %	8.42 ± 0.40 4.75	8.63 ± 0.39 4.52	8.54 ± 0.37 4.34
Seed weight (1000) seeds (g) CV%	50.90 ± 3.15 6.19	45.02 ± 3.54 7.86	55.22 ± 2.69 4.87

H_3—Indicates the third generation of pollen derived lines from the culture of F_1 hybrids.

C.V.—Coefficient of variation. Genetically uniform plants can be recovered by anther culture.

2. The Diversity of Phenotypes of Pollen Plants (H1) from F_1 Hybrids

As F_1 hybrids are often heterozygous for several genes and as recombination occurs during meiosis, plants derived from pollen grains of F_1 hybrids may represent different genetic recombinants. This phenomenon has been observed in experiments with rice and wheat as unique recombinants have been recovered.

3. Expression of Recessive Characters

The frequency of recessive characters derived from a pollen plant population are much higher than within a conventional F_2 population. Since a haploid plant only contains one set of chromosomes, dominant characters cannot mask recessive characters as in diploids, so recessive characters are expressed in high frequency. For example, in a wheat cross between sonora 62 with red grain color and Hongtu with white grain color, the red grain color is a dominant character. The segregation ratio of grain colour between F_2 and H_2 population was compared. The frequency of white seeds was much higher in an H_2 population than in F_2 population. In the cross of sonara 62 × Hongtu, (Table 20) there were 413 plants in the F_2 generation. Plants with red grain numbered 313, and there were 100 plants with white grain. The segregation ratio is 3.1 : 1. But in H_2 plant lines, the ratio of red to white grains was 0.8 : 1. The former is not significantly different from a 3 : 1 ratio while the latter is consistent with a ratio of 1 : 1.

Table 20. Comparison of the segregation ratio of grain colour between F_2 and H_2 populations

Expression of colour	*Sonara 62 × Hongtu*	
	F_2 plants	*H_2 lines (haploid)*
Red	313	14
White	100	18
Total	413	32
Red : White ratio	3.1:1	0.8 : 1

Strategies for Efficient Production and Use of Haploids and Polyhaploids:

The following are offered as possible points to consider in achieving efficient production of haploid plants:

1. Donar plants should be grown under optimum conditions with specific consideration to photoperiod, light intensity, temperature and nutrition.
2. An adequate number of genotypes should be sampled as some genotypes respond in culture while others will not produce any haploid plants. Breeding for improved anther culture should be considered.
3. Induction of abnormal pollen division is often promoted by thermic shock, thus a hot or cold pretreatment should be tested. stage of pollen development is important to culture successfully.
4. Organic nitrogen, on activated charcoal and high sucrose may be important additives to simple salts medium to improve culture response (Padmanabhan 1990).

The Possibility and Advantages of using doubled haploids in Crop Improvement

To use pollen haploids in crop improvement, the characteristics of anther culture and haploidy like pollen plants must be understood. Anther culture can solve some problems in genetic studies and plant breeding. The advantages of using pollen plants in crop improvement include,

1. Rapid achievement of homozygosity
2. Gametoclonal variation
3. Plant level gamete analysis and
4. Creating recombinants as well as variants and new forms.

1. Rapid Achievement of Homozygosity. It is well knowns that after chromosome doubling, homozygous doubled haploids might be rapidly obtained. They are stable and may be used directly in plant breeding and genetical studies. In some cases, the haploid breeding may reduce the time needed for developing a new variety by three to four generations. For example, it only took five years to develop the varieties of rice Hua Yu No. 1 and No. 2 from the initial hybridization to release of the new varieties (Hu and Hao, 1980). Three kinds of breeding methods of anther culture, bulk and pedigree were used to develop the four varieties. Among three methods, the duration needed for an anther culture is the shortest–only four generations. The development of pollen derived winter wheat Jinghua No. 1 is another example. In this case the evaluation of lines was carried out in H_2, whereas, by conventional method it would be done in F_6.

Development of Jinghua No. 1

Year	*Description*	*Generation*	*Year consumed*
1976	Hybridization Lorrin 18 × (5238-036) × Hong liang No. 4		
1977	Anther culture		
1978	Chromosome doubling	F_1	
1978	Evaluation of lines	H_2	
1979		H_2	6 years
1971	Trial test, multiplication of		
1980	trial test, Regional seeds test, performance test multiplication of seeds	H_3	
1981	Trial test, Regional test,	H_4	
1982	performance test multiplication of seeds Nomination of Jinhua No. 1	H_5	

2. Gametoclonal Variation. Variability of chromosome number and structure in regenerated plant cells in vitro is a common phenomenon, existing also in doubled pollen haploid plants. Recently it has been recognized that this method of introducing genetic changes could be used to develop new variants for plant breeding and genetics. The technology of introducing genetic variation by using cell culture has been termed, somaclonal and gametoclonal variation (Evans et al. 1984). In wheat, rice corn, tobacco, about 90 per cent of the pollen plants derived through anther culture are haploid, or homozygous diploid, whereas the residual to percent are heteroploids and ancuploids (Hu et. al. 1982).

Through anther culture of inbred wheat cultivars we have directly obtained many aneuploid plants in wheat, including nullisomics, monosomics, spontaneous aberrations in chromosomes and telocentrics (Hu, 1983). These results illustrate genetic variants may be produced without the involvement of hybridization.

Oono (1975) reported that pollen derived rice plants sometimes showed homozygous and sometimes heterozygous mutations. If all the variation preexised in the microspores (Pollen) all mutants would be homozygous. The heterozygous mutants must originate from the process of both gamete formation and regeneration of pollen plants. Experimental data also indicated that the frequency of aneuploid plants developed from a variety without the involvement of tissue culture is much lower than that of pollen derived plants (Han Hu 1984). As such the variation observed must have occurred during the culture phase. In wheat the first culture duration that is the in vitro de-differentiation stage of pollen is an important period. During this period, prolonged culture duration and a high cultural temperature are the different factors, influencing the chromosome doubling and fertility of pollen derived plants.

Phenotypic variations have been found recently in pollen derived plants in tobacco (De paepe et al., (1981), in rice (Oono 1975 ; Wakasa 1982), in pearl millet (Bai Dang Ha and Pernes, 1982). According to the experiments with several crop species mentioned above, the following variations have been observed based on evaluation of doubled haploid plants, *viz.*, changes in chromosome number or structure, single gene nuclear mutations, and several less defined genetic changes.

3. Plant Level Gamete Analysis. Since pollen grains from F_1 hybrids are heterozygous, different gene combinations of both parents of a cross occur in every grain. If pollen grains of F_1 hybrids are induced into pollen plants by anther culture, the haploid plants (H_1) show various phenotypes of both parents. Diversity of pollen progeny has ben observed in experiments with rice, wheat and tobacco. (Hu, 1984).

The possession of haploid and doubled haploid plants on which both dominant and recessive characters would be expressed, considerably shortens the process of selection for desirable characters. Various types of recombinants of hybrid (F_1) gamete could be fully expressed at homozygous plant level. It is significantly evident when distant hybrids were used for anther culture. Recently anther culture with three F_1 hybrids of hexaploid triticale and common wheat was carried out (Wang and Hu, 1985) more than 900 green plants were obtained and 588 of them were cytologically analyzed. The F_1 hybrids between hexaploid triticale and common wheat have the chromosome constitution of AABBDR. There are DR mixed genome, two sets of univalent DR, each of which at meiosis will distribute into daughter cells in one of the seven possible ways, and thus result in diversified chromosome constitutions. The gametes formed by meiosis are expected to contain 14 to 28 chromosomes. Through culture of anthers from F_1 hybrids, pollen plants with chromosome numbers ranging from 17 to 27, and the corresponding diploid after spontaneous chromosome doubling were obtained. Giemsa banding analysis of 12 pollen plants derived from the Rosner × Kedong 58 F_1 hybrid showed that they comprised in different numerical constitutions, of rye chromosome with wheat chromosomes of the D genome that is the chromosome of mixed or coexisted in these pollen derived plants. Theoritically, the types of pollen derived plants were the same as the types of chromosome composition formed by F_1 hybrid meiosis.

4. Creating Recombinant Variants and New Forms. The following figure shows that the most common system is to use F_1 hybrids as the parental material for haploid production, thereby fixing the new products of recombination between the parental genotypes at the earliest possible opportunity.

Chromosome breaks and reunions are likely to occur during anther culture. Anther culture can, be used to enhance the frequency of the exchange required in sexual hybrids for introgressing desirable alien genes, that is, the variants might be obtained, also new substitution, addition and translocation lines can be obtained. The research system in our laboratory is being established for transferring alien genes from *Triticale, Agropyron* into wheat cultivars through anther culture, and producing the relevant alien chromosome number variants simultaneously.

Anther culture, might be taken as an approach to create new forms, as the case shown in A233 ($2n$ = 48) and A116 ($2n$ = 16) which are relatively stable forms containing three pairs of rye chromosomes resistant to powdery mildew and setting seeds normally.

F$_1$ Doubled haploid system

	Variety A AAbb	×	Variety B aaBB	
		↓ F$_1$ AaBb		Haploidisation
Haploids	AB	Ab	aB	ab (Anther culture)
Doubled Haploids	AABB	AAbb	aaBB	aabb chromosome doubling
New variety (New recombinants)		↓ AABB		Field testing and selection

Recently, cytological analysis on aneuploid of pollen derived rice was carried out by Chu et. al., (1985). Rice aneuploids were obtained from 1715 pollen clones having a mean frequency of 10.2 percent in anther culture. Among the aneuploids obtained, the frequency of primary trisomics ranged from 5.4 percent to 6.7 percent, tetrasomics from 1.1 to 1.7 percent, monosomics from 0.9 to 1.3 percent, nullisomics from 0.5 to 1 per cent and double trisomics from 0.5 to 0.7 percent. The chromosome complement of aneuploids were identified by pachytene analysis of the absolute length of the extra chromosomes. Pollen clonal aneuploids showed a different range of variation in agronomic characters when compared to dihaploids from the same origin.

In anther culture of wheat and *Triticale* combined with irradiation, many variants and mutants were observed by Zheng et. al., (1984) at the institute for application of atomic energy. By irradiating anthers with gamma rays (100 kR) prior to inoculation, it was found that phenotypic frequency of haploids could be increased up to 55 percent whereas only 14.3 percent in control where anthers were not irradiated. The variations related to many characteristics such as plant height, spike shape, and grain shape and colour and size.

The gametoclonal variations were also observed in apple cultivar (Wu et al., 1984).

Application of Pollen Derived Doubled Haploids in Crop Improvement

Anther derived haploids have been used in basic as well as applied research. In basic research, protoplasts have been isolated from haploied plants and used in mutagenesis and selection schemes. In applied research, doubled haploids have been regarded as a possible means of rapidly producing true breeding lines. Theoretical treatment identified three important aspects to consider when deciding whether to use haploids in recurrent selection breeding programme.

1. The length of the selection cycle using double haploids versus segregating diploids.
2. The length of the selection cycle using doubled haploids versus segregating diploids and the number of fertile doubled haploids that can be produced and evaluated per cycle of selection.
3. The type of gene action (additive or dominant) involved in the determination of the character being selected and whether the character is influenced by environment. If the dominance affects phenotype more, the more the haploid system should be.

The benefits of doubled haploids for plant improvement include the rapid achievement of homozygosity and in consequence the rapid incorporation of new genes into breeding materials and the increase of selection efficiency.

Chinese scientists have extensively used this technique in plant breeding such as :

1. Developing and releasing many new varieties of rice, wheat, pepper and maize inbred lines of pollen plants.
2. Transferring desirable alien chromosome or gene into cultivars.
3. Establishment of haploid and dihaploid cell lines of pollen plants in maize and wheat.
4. Developing asexual line of rubber and popular of pollen plants.
5. Developing mutants resistant to diseases in tobacco, wheat and other crops.
6. Integration of haploid plant production with conventional breeding.
7. Production of pollen derived aneuploid and heteroploid plants.

Developing and Release of New Varieties

Haploid breeding techniques usually involve only one cycle of meiotic recombination. Many agronomically important traits like the yield are polygenically controlled. In such cases, one cycle of recombination is insufficient. If linkages between genes are present, then not all the variation potentially available in a cross will be released. For overcoming these disadvantages, Zhan (1980) and Hu (1984) developed a method comprising one cycle of anther culture followed by sexual hybridization between different genotypes of anther derived plants and another cycle of anther culture of the selected sexual hybrids. By using this method some polygenic traits have been successfully improved. An example is the release of the cold resistant rice variety named Hua Han Zao. A method combining anther culture with composite crossing has been developed to break linkages between genes. A winter wheat variety, Jinghua No. 1 with desirable characteristics has been released through this procedure (Han Hu 1984).

Rice

Eighty rice varieties and strains have been developed through anther culture in China. In the early period around 1975, some rice varieties derived from pollen were released. For example, Hau Yu No. 1, Xin Xiu and Huattan Zao. At present, a standard anther culture procedure is used, to produce double haploid lines from F_1 sexual crosses. Certain institutions, such as Crop institute of Chinese Academy of Agricultural Sciences (Beijing) using anther culture technique has released a batch of rice varieties. For example, Zhonghua No. 8 and No. 9 and Zhonghua 10.

Three kinds of breeding methods, anther culture, bulk and pedigree were used to develop those varieties. Among them the duration needed for anther culture is less, *i.e.,* only 4 generations. These varieties have high yield (7500 kg/ha) and resistance to bacterial leaf blight, blast, yellow dwarf, common dwarf and brown plant hoppers.

One line-shan-Hua 369, derived from hybrid rice Shan You No. 2 showed a yield increase from 4.4 to 11.4 percent and possessed desirable characters with early maturity, high protein content from 10 to 12 per cent, cold tolerance at the seedling stage and uniformity at the heading and maturing stages. It has been tried out and applied in the field production.

High quality rice variety Guan 18 has been produced by anther culture of hybrid rice. It has plant height of 78 cm, panicle length 20-61 cm, average number of grains per panicle 84, seed setting percentage 77.2, thousand grain weight 25 g, the grain shape is thin and long, resistance to blast. The yield is 5250-6000 kg/ha.

Research Highlights

1. Hybridization

Dihaploids of Indica × Japanica crosses were successfully raised through anther culture in rice. Androgenic embryoids were developed from five cross combinations exhibiting variations. These were transferred to field conditions for evaluation.

An evaluation of anther derived lines of ponni under field conditions indicated a considerable reduction in the duration (12.15 days) and stature of the plants (79-81 cm). The parent was 122 cm tall. The yield was less than that of the parent.

A comparison of 7 anther derived cultures of cross vaigai × Co 40 indicated the supremacy of lines 433A-R5 and 433-R6 in registering higher fields of 5.40 and 5.32 t/ha against 5.22 t/ha in the case of check, *i.e.*, Co. 40. The culture 4.33.R_1 was found to be resistant to blast and moderately resistant to rice tungro virus and leaf blight.

Two rice hybrids viz. Zs 97A/IR50 (115 days) and Er-Jiu-nan-IA/Co. 40 (145 days developed at Coimbatore have recorded an estimated grain yield of 4 t/ha respectively.

Wheat

Upto June 1982, Chinese Scientists had developed more than 20 new wheat varieties and strains through anther culture. Spring wheat Huapei No. 1 has several desirable characters, including vigorous tillering cold resistance, early maturity and good grain quality. The new winter wheat Jinghua No. 1, is the best example of anther culture's, potential in plant breeding programmes. Jinghua No. 2 and No. 3 have also been developed by anther culture. Jinghua No. 3 has good grain quality and a high protein content of 17 per cent.

In 1985, the first doubled haploid variety Florin obtained by anther culture was registered in France. F_1 hybrid anthers of winter type wheat were used as donor tissue for development of new strain Jingdan 2288 by anther culture. This new variety has large spikes and plenty of grains, vigorous, tillering, stripe rust and powdery mildew resistance, short stem and resistance to lodging.

Pepper

The sweet pepper variety Hiu Hua No. 3 was directly selected from pollen derived plants. This variety possesses desirable characters with short stature, early maturity (7-10 days earlier than the control) desirable fruit colour and size, good quality, resistant to disease and tolerant to heat. It's yield can be from 10 to 58 per cent higher than that of the control.

Maize Inbred Lines

Since 1975, many maize pollen derived plants have been obtained in China. More than 100 pure lines have been obtained form 30 different combinations.

Wu et al (1983) reported the development of promising elite doubled haploid named Qun Hua. The line was obtained by culturing anthers of F_1 hybrid 525 × C103. The maturity duration of Qun Hua is 103 days. The leaf shape and angle are similar to C103. The tassel shape and ear type are most like 525. The resistance to drought is moderate and has cold tolerance and early vigour. The lines resistance to *Helminthosporium turcicum* and *H. maydis* are similar to C103. The line was uniform for all agronomic traits. In the 2nd year Qun Hua was crossed with elite lines. Yield trials were conducted with controls Qundan 105 and Qundan 16 and yield data was tested. This has 10.56% increased yield over the control. The hybrid Qu × Qun Hua was named as Hua Yu No. 1.

2. Transferring Desirable Alien Chromosome or Gene into Cultivars

Because of chromosomal instability during anther culture, haploidy has the potential of rapid transfer of alien chromosome or gene to cultivars in wide crossing programmes. Breeding of high yielding rice varieties with high resistance to blast in rice growing region of North China has traditionally been done by backcross, which is time consuming. However, the successful introduction of the resistant gene P_1-Z in Toride No. 2 into a high yielding chinese variety Jinxi 17 through hybridization and anther culture was made. Two new cultivars Zhonghua No. 8 and Zhongua No. 9 with high yield potential, superior quality and good resistance to rice blast were developed. It took only 2 years through anther culture.

3. Developing Mutants Resistance to Disease in Tobacco, Wheat and Other Crops

It is important in crop improvement to screen mutants with resistance to disease. The selection of mutants which are easily evaluated at haploid level, can be carried out by test tube methods. Now tobacco mutant resistant to black shank disease has been obtained. Lines resistant to wheat scab (*Fusarium graminearum*) were obtained from a susceptible variety. Similar screening techniques were adopted in rice to screen for blast resistance.

4. Developing Asexual Line of Rubber Tree and Popular of Pollen Plants

Pollen plants can improve varieties of rubber tree. Chen *et al.* (1984) have obtained a pollen rubber tree taller than 6 metres, by asexual propagation. Several clones of pollen rubber trees were tasted. A peculiar merit of pollen plant in breeding is manifested in popular. Lu *et al.* (1984) have obtained six-pollen derived popular stocks. Comprising 150 plants at the age of eight to nine years, some of them were observed to be disease resistant.

5. Establishing Haploid and Diploid Cell Lines of Pollen Plants in Maize and Wheat

Through anther culture and tissue culture, both haploid and diploid somatic cell lines have been established in maize and in wheat. The corn cell lines have been maintained for more than 7 years, with ability of differentiation into green plants. These cell lines may represent a very useful system for both cytogenetical and molecular studies and for screening mutants with desirable characters.

6. Integration of Haploid Plant Production with Conventional Breeding

Six strains of rice have been selected from pollen plants derived from F_1 anther culture of the hybrid rice vriety Shan You No. 2. They are named as Shan, Hau, No. 7706, 78-1, 791, 792. Researchers have attempted to use these strains to replace the F_1 hybrid variety and save the trouble of producing hybrid seed. Using the conventional crossing method, Zhang (1980) combined anther culture with multiple crosses and selected the variety Hua Han Zao and other new strains such as strain 77001. In this way the difficulty of seletion using multiple crosses was overcome. Therefore, the anther culture has not only increased the efficiency of selection using multiple crosses, but also significantly reduced the time needed to develop new varieties.

7. Production of Pollen Derived Aneuploid and Heteroploid Plants

(*a*) Variability of pollen derived plants. Cultured cells of both animal and plant tissue are characterized by instability of chromosome number and structure. The phenomenon is expressed in regenerated pollen derived plants. The heteroploids are obtained through anther culture due to chromosome variations. These heteroploids are used for genetic investigations and plant breeding.

(*b*) The origin of pollen derived heteroploid plants. There are various ways in which pollen derived heteroploids might be produced (tables 21 & 22). They are meiosis, endomitosis,

Table 21

Species from which pollen derived plants first obtained in China

Year	*Species*	*Reference*
1971	*Capsicum annum*	Wang et al., (1973)
	Triticale	Sun et al., (1973)
	Triticum aestivum	Ouyang et al., (1973)
1973	*Beta vulgaris* ($2n = 4x = 36$)	Institute of sugar beet and sugar crops (1978)
1975	*Zea mays*	Institute of Genetics (1975)
1977	*Hevea brasiliensis*	Chen et al., (1978)
1978	*Sorghum vulgare*	Zhou, (1978)
1979	*Saccharum sinensis*	Chen et al, (1979)
	Linum usitatissimum	Sun, (1979)
1980	*Glycine max*	Vin et al., (1980)
1981	*Vitis vinefera*	Zhou and Li, (1981)

Table 22

Some Agricultural and tree species in which haploid callus (C) embryos (E) or plants (P) have been obtained by the culture of excised anthers/pollen

Crop species	*Nature of Haploids*	*Reference*
Arachis species	E,P	Bajaj et al., (1981)
Beta vulgaris	C,P	Atanassove (1973)
Brassica hirta	C,E,P	Bajaj and Mohapatra (1990)
Brassica juncea	C,E,P	George and Rao (1982)
Brassica napus	C,E,P	Swanson et al., (1987)
Brassica oleracea	C,E,P	Orton and Browers (1985)
Cajanus cajan	C,E	Bajaj et al., (1980 *b*)
Capsicum annum	C,P	Xuo et al., (1973)
Cicer arietinum	C,E	Bajaj and Gosal (1987)
Gossypium species	C,E	Bajaj (1982)
Helianthus annus	C,P	Bohorova et al., (1985)
Hordeum vulgare	C,E,P	Clapham (1973)
Coffea arabica	C,E,P	Xuo et al., (1973)
Cyamopsis tetragonoloba	E	Bajaj and Gosal (1988)
Hevea brasiliensis	C,E,P	Chen et al., (1979)
Oryza sativa	C,E,P	Mercy and Zapata (1987)
Pennistum americanum	C,P	Bai-Dang Ha and Pernes (1982)
Phaseolus aerecis	E,C	Bajaj and Singh (1980)
Solanum tuberosum	C,E,P	Wang et al, (1973)
Triticum aestivum	C,E,P	Wang et al., (1973)
Zea mays	C,E,P	Petolino and Thompson (1987)

spindle fusion, nuclear fusion and endo reduplication. The above abnormal phenomena are occuring during early sages of pollen development in vitro. If the initial uninucleate microspore is not a haploid because of some meiotic abnormality, the derived embryo and pollen derived plants will express this abnormal chromosome constitution. Fusion of nucleus during early nuclear divisions of the pollen grain is also possible and callus with doubled chromosome number may be formed in this way and nonhaploid or spontaneous doubled haploids may be produced. In some species *e.g.*, N. tabacum, the initial division of the vegetative nucleus is preceded by degeneration of the cytoplasmic organelles. As a result endomitosis may lead to a diploid restitution nucleus. Thus, doubled dihaploids are produced. Ultimately heteroploids are produced.

Conclusion

Distinct features of doubled haploids and the access to them, now allow to use them in genetical studies and breeding both gamete formation and regeneration of pollen plants. Through anther culture these different genotypes of gametes might be fixed in a short time. Thus, by regeneration of doubled haploids it is possible to create variants and recombinants. Anther culture provides a valuable material for breeding programmes.

REFERENCES

Atanassove, A. (1973). Anther culture of tobacco and sugar beet as a tool for producing haploid plants and callus of different ploidy levels. Genet. Sci. 6:501-507.

Bajaj, Y.P.S. (1982). Survival of anther and ovule-derived cotton callus frozen in liquid nitrogen. Curr. Sci. 51:139-140.

Bajaj, Y.P.S. (1990). Haploids in Crop Improvement I In : Biotechnology in Agriculture and Forestry. 12. Springerverlag Berlin Heidelberg. New York.

Bajaj, Y.P.S. and S.S. Gosal, (1987). Pollen embryogenesis and chromosomal variation in cultured anthers of chickpea. Int. Chickpea Newslett.

Bajaj, Y.P.S. and S.S. Gosal (1988). Establishment of tissue cultures and the regeneration of plants in quar (Cyamopsis tetragonoloba (L) Taub.) an industrial legume crop. SABRAO. J. 20:145-149.

Bajaj, Y.P.S., and D.Mahapatra (1990). Pollen embryogenesis and chromosomal variation in anther culture of Brassica hirta (Sinapis alba L.).

Bajaj, Y.P.S., H.Singh and Gosal, S.S. (1980). Haploid embryogenesis in anther culture of pigeon pea Cajanus cajan, Theor. Appl. Genet. 58:157-159.

Bajaj, Y.P.S., A.K. Ram, K.S. Labra, and *H.Singh (1981).* Regeneration of genetically variable plants from the anther derived callus of Arachis hypogaea and A.villosa, Plant sci. Lett. 23:35-39.

Bohorova, N., A. Atanassor, and J. Georgieva-Todorova (1985). In vitro organogenesis, androgenesis, and embryo culture in the genus Helianthus L. 2. pflanzon zucht 95:34-44.

Bai Dang Ha, D. and J. Pernes (1982). Androgenesis in pearl millet. I. Analysis of plants obtianed from microspore culture. Z. pflan Zenphysiol. 108. 317-27.

Burk, L.G., G.R. Gwynn. and J.R., Chaplin, (1972). Diploidized haploids from aseptically cultured anthers of Nicotiana tabacum. A colchicine method applicable to plant breeding J. Hered 63:355-360.

Chen, Y, R.F. Wang, W.Z. Tian, Q. X. Zuo, S.W. Zheng, D.Y. Lu and P.G. hang (1980). Studies on pollen culture in vitro and induciton of plantlets in Oryza sativa subsp. Keng. Acta Genet. Sin. 7(1): 46-54.

Chen, Z.H., F.Chen, T.C., Chien, C. Wang, S., Chang, Hsu, H.OU, Y. HO., and T. Lu, (1978). Induction of pollen plants of Hevea brasiliensis. Muell-Arg. Acta Genet. Sin. 5:107.

Chen, Z.H., C.F. Quian, M. Qin, C.H. Weng, C.J. Suo, F.Z. Chen, and Z.T. Deng, (1979). The Induction of pollen plants of sugarcane. Annual report of the institute of Genetics. Academia Sinica, p. 92-93.

Chuan, C.C., T.W. Onyang H. Chia, S.M. Chou and C.K. Ching. (1978). A set of potato media for wheat anther culture. Proceedigns of symposium on plant tissue culture Science Press, Peking pp. 51-56.

Chu, Q.R., Z.H. Zhang and Y.H. Gao. (1985). Cytogenetical analysis on aneuploids obtained from pollen clones of rice (Oryza sativa L.) Theor, Appl. Genet.

Clapham, D. (1973). Haploid Hordeum plants from anthers in vitro Z.pflanzenzucht. 62:305-310.

Depeape, R.D. Bleton and F.Geangbe. (1981). Basis and extent of genetic variability among doubled haploid plants obtained by pollen culture in Nicotiana sylvestris. Theor, Appl. Genet. 59: 77-184.

Dunwell, J.M., R.J. Francis and W.Powell (1987). Anther culture of Hordeum vulgare L: a genetic study of microspore callus production and differentiation. Theor. Appl. Genet. 74: 60-64.

Dunwell, J.M. and M.E. Perry, (1973). The influence of in vitro growth conditions of Nicotiana tabacum plants on the invitro embryogenic potential of their anthers. John Innes Annu. Rep. No. 64.

Evans, D.A., W.R. Sharp, and H.P. Medina Filho. (1984). Somacloal and gametoclonal variation. Amer. J. Bot. 71(6): 759-74.

Fitch, M.M. and P.H. Moore (1983). Haploid production from anther culture of Saccharum spontaneum L.Z. plan zenphysiol. 109: 197-206.

Foroughi-Wehr, B., W. Friedto and G. Wenzel (1982). On the genetic improvement of androgenetic haploid formation in Hordeum vulgare. L. Theor. Appl. Genet. 62: 233-239.

George, L. and P.S. Rao (1982). In vitro induction of pollen embryos and plantlets in Brassica juncea through anther culture Plant Sci. Lett. 26: 111-116.

Guha, S.R. and S.D. Maheshwari, (1964). Nature London 204:497.

Guha, S.R. and S.D. Maheshwari, 1966. Cell division and differentiation of embryos in the pollen grains of Datura in vitro. Nature, London 221:97-98.

Grunewaldt. J. (1975). Observations on anther callus from Hordeum vulgare L. Barley genetics III, Porc. 3rd Int. Barley. Genetic Symp, Garching, pp. 367-373.

Han Hu. Z.Y.X., J.K. Jing and X.Z. Wang (1982). Production of aneuploids and heteroploids of polien derived plants. In: A. Fujiwara (ed), Plant tissue culture in 1982. The Japanese Association for plant Tissue culture, pp. 421-24.

Han Hu. Z.Y.X. (1983). Genetic stabiltiy ad variability of pollen derived plants. In: S.K. Sen and K.L. Giles (eds.), Plant Cell Cultue in Crop Improvement. Pleanum Press. New York and London pp. 145-57.

Han Hu., Z.Y.Z. and S.Hao. (1980). The present status of investigation of plant tissue and cell culture in China In: F Sala, B. Parasi, R. Cella and Cifrri (eds.), plant cell culture-Results and perspectives Biometrical Press Elsevier/North Holland, pp. 89-104.

Han Hu Z.Y.Z. (1984). Crop Improvement by anther culture In V.L. Chopra, B.C. Joshi, R.P. Sharma and H.C. Bensal (eds.) Genetics New Frontiers. Proceedigns of the XV International Congress of Genetics. Vol. VV. Oxford and IBH. Publishing Co., New Delhi pp. 77-84.

Heberle-Bors, E. (1982). On the time of embryonic pollen grain induction during sexual development of Nicotiana tabacum L. Plants Planta 156: 402-496.

He, D.K. and J.W. Ouyang. (1984). Response of wheat anthers at different development stages to culture in vitro (in Chinese). Hereditas, (Beijing) 6(2): 6-9.

Huang, B. and N. Sunderland (1982). Temperature sress pretreatment in barley anther culture Ann. Bot 49: 77-88.

Huang, Q.E. H.Y. Yang and C.Zhou (1982). Embryological observations on ovary culture of unpollinated young flowers in Hordeum vulgare. L. Acta Botsin 24: 295-300.

Institute of Genetics, Ist Division, 3rd Laboratory, Academic, Sinica. (1977). Investigation on some induction factors and genetic expression of wheat pollen plants. In : collection of works of haploid breeding, pp. 65-81, Science Press, Beijing.

Institute of sugarbeet and sugar crops, breeding laboratory, Heilungkiang province (1978). Induction of pollen plants in sugar beet. In Proceedings of symposium on plant tissue culture (Hu Han, ed.) p.303 Science Press Beijing.

Kasperbawer, M.J. and G.B.Collins (1972). Reconstitution of diploids from leaf tissue of anther derived haploids in tobacco. Crop sci. 12;98-101.

Lu, Z.H., Y.X.Liu and P.G.Zhang (1984). Study on the variation of the chromosome number of popular plants obtained by anther culture in vitro, Intern.Sym.on Genetic manipulation in crops Abstracts 17.

Migo S.H.,C.S.Kuo, Y.L.Kwei, A.T.Sun, S.Y.Ku, W.L.Lu, and Y.Y. Wang (1981). Induction of pollen plants of maize and observations on their progeny. Proc. symp.plant tissue culture, Beijing 1978. Pitman, Boston, pp. 23-24.

Oono, K. (1975). Production of haploid plants of rice. (Oryza sativa) by anther culture and their use for breeding. Bull. Nat. Inst. agrl. Sci., D. 26: 139-222.

Ouyang, J.W., H. Hu., C.C. Chuan and C.C. Tseng (1973). Induction of pollen plants from anthers of Triticum aestivum L. cultured in vitro, Sci., Sin. 16: 79-95.

Ouyang, J., S.M. Zhou and S.E. Jia (1983). The response of anther clture to culture temperature in Triticum aestivum, Theor. Appl. Genet. 66:101-109.

Orton, T.J. and M.A. Browers (1985). Anther culture of Triticale Crop Sci. 16: 120-121.

Padmanabhan. C. (1990). Anther clture and its application in crop improvement. Summer Institute on Biotechnology and innovative breeding for crop improvement. Tamil Nadu Agricultural University, Coimbatore.

Petolino, J.F. and S.A. Thompson (1987). Genetic analysis of anther culture response in maize. Theor. Appl. Genet 74:284-286.

Picard, E., C. Hairs, S. Gregoire, T.H. Phan. and J.P. Meunier (1987). Significant improvement of androgenetic haploid and doubled haploid induction from wheat plants treated with a chemical hybridization agent. Theor. Appl. Genet. 74: 289-297.

Reinert. J. and Y.P. S. Bajaj (1990). Applied and fundamental aspects of palnt cell tissue and organ culture. Narosa Publishing House.

Rout, J.R. and N.P. Sharma (1991). Anther callus induction and green plant regeneration at high frequencies from an interspecific rice hybrid Oryza sativa L. O. rufipogon Griff. Euphytica 54: 155-159.

Rowell, P.L. and D.G. Miller (1971). Induction of male sterility in wheat with 2-chloretnylphosphoric acid etherl, Crop. Sci. 11: 629-631.

Schmid, J. and E.R. Keller (1986). Effect of Gametocide on the induction of haploids in Triticum aestivum In: Horn. W. Jensen C.J. Odenbach W. Schieder O (eds) Genetic manipulation in plant breeding Berlin, pp. 347-349.

Sharp, W.R., L.S. Caldas and O.J. Crocomo (1973). Studies on the induction of Coffea arabica callus from both somatic and microsporogenous tissues and subsequent embryoid and plantlet formation, Am. J. Bot. 60: 13.

Sun, H.T. (1979). Preliminary report on anther culture of flax Kexue Tong bao 24: 948-950.

Sun, C.S., Wang, C.C. and Chu, C.C. (1973). Cytological studies on the androgenesis of triticale, Acta Bot. Sin. 15: 165-173.

Sundarland, N. (1982). Induction of growth in the culture of pollen In: M.M. Yeoman and D.E.S. Truman (eds) Differentiation in vitro. Cambridge Unviersity Press Cambridge, pp. 1-24.

Swanson, E.B., M.P. Coumans, S.C. Wu., T.L. Barsby and W.D. Beversdorf (1987). Efficient isolation of microspores and the production of microspore derived embryos from Brassica napus. Plant cell Rep. 6: 94-97.

Wang, C.C., C.C. Chu, C.S. Sun., S.H. Wu, K.C. Yin. and C.Hsu. (1973). The androgenesis in wheat (Triticum aestivum) anthers cultured in vitro. Sci. Sinica 16: 218-22.

Wang, X.Z. and H.Hu. (1985). The chromosome constitution of plants derived from pollen of hexaploid triticale *x* common wheat F_1 hybrids, Theor, Appl. Genet. 70: 92-96.

Wakasa, K. (1982). Application of tissue culture of plant breeding–Method improvement and mutant production. Bull. Nat. Inst. Agr. Sci. D. 3: 121-200.

Wen, F.S., E.L. Sorensen, F.L. Banett and G.H. Liang (1991). Callus Induction and Plant Regeneration from anther and Inflorescence culture of sorghum. Euphytica 52: 177-181.

Wu, J.Y., D.Q. Huang and D.S. Han (1984). Induction of haploid plants from apple anthers and their cytological and histological study. Intern. Symp. on Genetic manipulation crops, Abstracts 4.

Xu. Z.H., B. Huang and N. Sunderland (1981). Culture of anthers in conditioned media. J. Exp. Bot. 32: 767-778.

Yin, K.C. S.C. Li, Z. Xu, L. Chen, Z.Y. Zhu, and F.Y. Bi. (1980). A study of anther culture of Glycine max. kexue Tongbao 25: 976.

Zhang P.G. (1980). Application of anther culture technique to rice breeding. symposium celebrating the 20th Anniversary of the International Rice Research Institute, Los Baunos, S. Philippines.

Zheng, Q.C., X. C. Pan and C.Y. Chen. (1984). The character inheritance and the use of the pollen plants derived from the hybrid rice (Oryza sativa subsp. Hsien) Intern. Symp. on genetic manipulation in crops abstracts 11.

Zhou, C.J. and P.F. Li, (1981). Induction of pollen plants of grape (Vitis vinifera L.). Acta Bot. Sin. 23: 79-81.

Zhou, W.B. (1978). A preliminary study on the anther culture of sorghum. Acta Genet Sin. 5(A): 337-338.

CHAPTER 13

Asexual Propagation of Relevance to Mutation Breeding

Irradiation of bulbs, tubers, rhizomes, cutting, grafts, other plant parts or whole plants, all having buds with multicellular apices composed of a number of fairly autonomous cell layers, automatically leads to the formation of chimeras. Irradiating such buds produces small-sectored mericlinal chimeras, and the chances of recovering the induced mutations from these are relatively low. This is the main obstacle in mutation breeding, especially in species in which there is no ontogenetically young stage of buds to be irradiated.

The ideal situation is for a plant (part) to originate from a single cell, in vivo or in vitro, or from one or a few vegetative daughter cells of a mutated cell. In that case chimera formation is avoided.

Adventitious bud formation from one cell : general aspects

An important method for the commercial plant breeder is the adventitious bud technique, which makes use of the phenomenon that (the apices of) adventitious buds, *e.g.*, on detached leaves, ultimately may originate from only one cell, often of epidermal origin (Broertjes et al., 1968 ; Haccius and Hausner, 1974).

Adventitious buds are formed exclusively by cells at the lowest part of the petiole near the cut end, as has also been found to be the case in other plants, such as Dendrophthoe (Nag and Johri, 1970). In experiments with Saintpaulia and other species, the basal 5 mm of the petiole was always cut off after completion of irradiation and prior to planting. In this way, epidermal cells situated higher on the petiole, which had no previous stimulus to divide, were activated to do so. Consequently, all such epidermal cells underwent the radiations treatment in a resting (non-dividing) stage of development, either G_1 or G_2 of the mitotic cell cycle (Broertjes, 1972 c).

An important phenomenon is the fact that the overwhelming majority of the adventitious mutants thus produced and observed in Saintpaulia, Achimenes, Streptocarpus, Kalanchoe, Begonia and other species, were solid and non-chimeric. Generally, low percentage of the mutants may be chimeric. This can be explained by the occurence of spontaneous mutations or by genetic instability during the outgrowth of the apex of an adventitious shoot.

Some of the chimeras used in regeneration experiments through adventitious shoots, are not reproduced true-to- type. Examples are the "pin wheel flowering" African violet cultivars used by Lineberger and Druckenbrod (1985) and the cultivar Desert Dawn, described by Eyerdom (1981). Other "chimeras", however, could be propagated true-to-type, such as the seedling cultivar Valencia, also described by Eyerdom (1981).

Adventitious bud formation on leaves

Many plants can be propagated by different types of adventitious plantlet formation on detached leaves. Broertjes et.al. (1968) listed over 350 of such species, belonging to various plant families, which are on record in the literature ; species, belonging to various plant families. For example, Begonia, various Gesnerianeae and peperomia, all being propagated on a commercial scale using leaves or parts of leaves. Many, however, are not listed, including a number of species in economically very important plant families, such as Gramineae (cereals) and Solananceae (potato, tomato). In Kalanchoe, the cultivar Annettee produces fewer plantlets than the cultivar Josine. Simiiar differences were found using various *Chrysanthemum* cultivars. One should therefore not conclude that a given species cannot produce adventitious shoots if only one cultivar has been tested.

Broertjes and Leffring (1972) compared physiologicaly old leaves, fully-grown leaves and young leaves of Kalanchoe. Although all leaves rooted readily, the young leaves produced adventitous plantlets at the base of the petiole sooner than the older leaves, and in much greater number. In *Chrysanthemum* more buds were formed on callus either on the upper parts of roots or at the base of the petiole in light than in darkness. In Streptocarpus, however, old leaves produced more but smaller plantlets in comparison with young leaves (Brown, 1971). The petiole also seems to play a part : leaves of the Kalanchoe cultivar Annette rarely produced plantlets when the petiole was cut off, whereas cv. Josine produced some plantlets on leaves at all three developmental stages, although many more grew from those with a petiole.

The size of the leaf as well as of the leaf part is an important factor for the number of adventitious buds formed in monocotyledons. In Ornithogalum, leaves of approximately 20 cm, in length were compared with leaf tops, leaf bases and mid-leaf pieces, all approximately 10 cm long. Whole leaves produced twice as many adventitious bulblets as any of the leaf pieces. Basal parts produced the lowest number of bulblets, but those produced were larger (Broertjes, 1972 a).

Adventitious bud formation on (artificially made) bulb scales

In *Lilium*, solid mutants can be obtained by irradiating bulb scales immediately after scaling the bulbs or, even better, by scaling after irradiation of the bulbs. Since clonal propagation of selected mutants, especially during the early years, is slow, it takes approximately 5 years before enough plants are available and enough experience of the mutants has been gained to justify commercial exploitation of a mutant.

Adventitious bulblets are obtained in some other bulb crops from scooping the bulbs, a traditional way of vegetative propagation of, for instance, hyacinth. Preliminary results with Amaryllis show that apparently solid mutants are produced on scale pieces of irradiated bulbs, but also a certain percentage of chimeras was observed.

Significance of mutation breeding

A convincing demonstration of the value of the adventitious bud technique for mutation breeding in vegetatively propagated ornamental plants has been found in Streptocarpus, using the cultivar constant Nymph. Irradiated or colchicine treated half-leaves produced a strikingly high percentage of mutants or polyploids, respectively. Approximately 95% of the adventitious mutants produced after irradiation with the optimum X-ray dose of 30 Gy were solid ; after treatment of half-leaves for 6 h with 0.1% colchicine, over 30% of the plantlets were non-cytochimeral tetraploids. In total, 857 mutants were obtained, five of which were sold commercially within 3 years of the start of the experiment (Broertjes, 1969).

Similar results were obtained with Achimenes. Though fewer mutants were produced than in Streptocarpus, irradiation of leaves of the cultivar paul Arnold with 30 Gy X-rays produced a number of solid mutants. Within slightly more than 3 years of irradiation, several mutants were being commercially exploited, namely 'springtime' and 'Early Arnold', both of which flower approximately 3 weeks earlier than 'Paul Arnold', and other mutants.

Hundreds of solid mutants of *Ipomoea batatas* (sweet potato) have been produced by irradiation of detached leaves.

Several of them were promising with respect to various characters, such as dry matter and protein content, skin and flesh colour (**B**-carotene) of the tuberous roots, type of plant, and yield (Love, 1968 ; Wilson, 1973).

In Begonia too, hundreds of solid mutants have been obtained by the irradiation of detached leaves from a few, triploid hybrids.

Peperomia can also be propagated easily from detached leaves (Bigot and Brulfert, 1968). The detached leaves of several members of the Liliaceae, for example, Endymion, Hyacinthus, Muscari, Ornithogalum and Scilla, readily produces adventitious bulblets. However, even following an optimum X-ray dose, the mutation frequency in these species is very low as a consequence of their limited heterozygosity.

In Lilium hundreds of solid mutants have been obtained after irradiation of bulb scales immediately after scaling, or after irradiation of bulbs just before scaling. Several promising mutants or tetraploids (obtained after colchicine treatment of bulb scales) are being propagated for further testing.

The adventitious bud technique was applied in an experiment with *Nicotiana alata.* Solid mutants were obtained, again after irradiation of detached leaves. By using this method, mutant individuals of a self-compatible clone were obtained, which were characterized by the same genetic background as the stock plant, and were consequently free from inbreeding effects (De Nettancourt et al., 1971).

Adventitious buds originating from more than one cell

Buds on stem and roots often develop endogenously from more than one cell, probably from a restricted number. In *Chrysanthemum* the shoots had originated from two or more cells which may or may not have derived from different apical layers. In Poinsettia (*Euphorbia pulcherrima*) much larger sectors and even complete (periclinal) mutants were obtained by adventitious bud formation after removal of the terminal bud as well as all lateral buds on the irradiated plants (Love, 1972).

Large sectored and apparently solid mutants of potato can be obtained in various ways. Ferwerda (1965) irradiated single-eye pieces and among the mutants obtained, a relatively high percentage (30%) seemed to be solid. Dormant eyes or tubers irradiated directly after harvest, de-eyed tuber pieces and very small, prematurely harvested tubers (3-5 mm) respectively showed gradually declining chimerism.

The best way known to obtain solid (non-chimeral) mutants in vivo is to de-eye tuber parts shortly (1-2h) before irradiation, to apply relatively high doses and dose rates, and to use adventitious sprouts which develop 3 months after irradiation ; any shoot developing earlier is discarded.

The use of callus, in general, may be another valuable method for reducing chimerism. *Chrysanthemum* leaves can be rooted easily ; they produce callus at the base of the petiole or on the upper part of roots on which one or a few adventitious shoots are occasionally formed. The majority of the mutants obtained in *Chrysanthemum* were chimeras, although several (approximately 20%) apparently solid mutants were observed as well.

Buds in ontogenetically young stages of development

A situation similar to adventitious bud formation of stems, roots or callus is found when existing bud meristems are irradiated in an ontogenetically very young stage of development. In this case, chimerism can also be restricted or sometimes even avoided.

The optimum moment for irradiating Dahlia tubers is immediately after harvest, when no visible buds are present, whereas after winter storage the tubers carry several relatively large sprouts. The irradiation of freshly harvested tubers results in chimeras as well as apparently solid mutants roughly 50% of each. Numerous apparently solid mutants have been commercialized and grown on a relatively large scale. None showed any sign of chimerism, indirectly proving that they must have been solid.

In Iris, the best irradiation time is immediately after lifting, when the growing point is at the youngest possible stage of development (Hekstra and Broertjes, 1968).

The production of solid polyploids

What has been stated about mutated cells also applies to a certain extent to polyploid cells, which result, of instance, from colchicine treatment of a multicellular apex. The generally slow growing, polyploid cells often lose the competition with the faster growing, original (diploid) cells. Colchicine treatment in combination with the adventitious bud technique, in which freshly detached leaves were used, resulted in a very high percentage of solid, non-cyto-chimeral polyploids.

In Vitro techniques

In vitro techniques include aseptic regeneration of (adventitious) shoots and bulblets from explants, callus, cells and protoplasts on an artificial medium. In vitro techniques have gained practical application in a rapidly increasing number of crops for fast true-to-type and virus free vegetative multiplication.

In vitro methods include the production of solid mutants, the application of chemical mutagens for the production of autodiploids and autotetraploids from haploids and di(ha)ploids, respectively, either spontaneously or by colchicine treatment.

From the mutation breeder's point of view, in vitro regeneration of adventitious shoots on plant organs or plant parts also seems very useful. Explants are expected to originate from one or a few genetically identical cells only. They may develop on leaf-disc explants, small leaves and epidermal strips, but explants of petioles, peduncles, pedicels, bulb scales and capitula have also demonstrated the ability to produce adventitious shoots in a range of crops.

What has been said about decapitated seedlings and detached cotyledons is also appropriate for endosperms, poolen, zygotes and embryos. The heterozygosity of vegetatively propagated plants is generally large, which causes the above mentioned material to be unsuitable for use in mutation breeding. In true-breeding botanical species, it may perhaps be used in conjunction with self fertilization to avoid loss of mutations in loci L-I and L-III and unfavourable segregation in M_2 because of chimerism in the loci L-II of the treated materials. The mutation breeder should investigate the crop of interest with regard to its possibilities for vegetative propagation both in vivo and in vitro. He should preferably look for methods in which, (*a*) as few cells as possible take part in the formation of the apex of an adventitious shoot (higher mutation frequency, wider mutation spectrum, easier selection and easier clonal propagation) and (*b*) a direct regeneration takes place (fast propagation, maintenance of genetic integrity).

It is a fact that most mutational events in vegetatively propagated plants result from chromosome aberrations, rather than from single gene mutations. In other wolds, more than one character is in general mutated. In solid mutants all mutated characters are expressed, including the unfavourable ones, whereas in a periclinal chimera, where for instance the loci L-I is mutated, only some of the mutated characters are expressed (Broertjes et. al., 1976) and "true-to-type" vegetative propagation, using in vivo or in vitro adventitous bud techniques, will in most cases be impossible.

Another possibility to obtain mutants is a method in which plants are produced in vitro from undifferentiated callus. Compared with a cell in a well-organized apex, a mutated cell in a mass of callus seems to have a larger change of surviving competition with other cells. The majority of the mutants produced by this method will be solid, especially if shoots are regretted from a repeated sub-culture of irradiated callus, since the number of cells from which adventitous shoots originate in vitro seems to be restricted. A high ratio of non-chimeric to chimeric plants can also be obtained if (*a*) the number of initial cells of an adventitious bud is very small and (*b*) the ratio between the total area of mutated and non-mutated callus tissue to that of the border between the two, is large (Broertjes and Keen, 1980).

REFERENCES

Bigot, C. and J. Brulfert. (1968). Dynamics of the natural appearance of neoformations of roots and buds from isolated leaves of peperomia qriseo-argentea.

C.R.A.cad. Sci. Ser. D, 267 : 1568 - 1588.

Broertjes, C. (1969). Mutation breeding in Streptocarpus. Euphytica, 18:333-339.

Broertjes, C. (1972 a). Improvement of vegetatively propagated plants by ionizing radiation. In : Induced mulations and plant Improvement. IAEA, Vienna, pp : 293-299.

Broertjes, C. (1972b). Mutation breeding of Achimences. Euphytica, 21:48-63.

Broertjes. C. (1972c). Use in plant breeding of Acute, chronic or Fractionated Doses of X-rays or Fast neutrons as Illustrated with leaves of Saint-paulea. Thesis, Centr. Agric. Publ. Doc., Wageninger, Agric. Res. Rep. 776, 74 pp.

Broertjes, C., B. Haccius and S. Weidlich. (1968). Adventitious bud formation on isolated leaves and its significance for mutation breeding. Euphytica, 17 : 321-344.

Broertjes, C and A.Keen (1980). Adventitious shoots : do they develop from one cell ? Euphytica, 29:73-87.

Broertjes, C., S.Roset and G.S. Bokelmann. (1976). Mutation breeding of Chrysanthemum morifolium Ram. Using in vivo and in vitro adventitious bud techniques. Euphytica, 25 : 11-19.

Brown, A.G. (1971). The effect of supplementary light on Streptocarpus 'Constant Nymph'. John Innes Inst. Annu. Rep., 62 : 50-51.

De Nettancourt, D., P. Dijkhuis, A.J.G. Van Gastel and C. Broertjes. (1971). The combined use of leaf irradiation and of the adventitious bud technique for inducing and detecting polyploidy, market mutations and self compatibility in clonal populations of Nicotiana alata Link and Otto. Euphytica, 20 : 508- 521.

Eyerdom, H. (1981). Flower colour sports and variations in Saintpaulia hydbrids, Afr. Violet Maq : 34 (4) : 32-37.

Ferwerda. F.P. 1965. Mutagenic effects of X-rays and EMS on potato sprouts. In : The use of Induced Mutations in plant Breeding. Rep. Meet. FAO/IAEA, Rome, 1964. Pergamon Press, Oxford, pp. 687-690.

Haccius, B. and G.Hausner. (1974). Adventivknospen and nicht-zygotische Embryonen-Grundlagen and Anwendung. Vortr. 22. Vortragstaq. Ges. Arzneipflanzenforsch., Tubingen, pp. 1-10.

Hekstra, G. and C. Broertjes. (1968). Mutation breeding in bulbous Iris. Euphytica, 17 : 345-351.

Lineberger, R.D. and M. Druckenbrod. (1985). Chimeral nature of the pinwheet flowering African violets (Sanint paulia ; Gesneriaceae). Am. J. Bot., 72 : 1204-1212.

Love, J.E. (1968). Mutation breeding in the sweet potato by means of fast-neutron irradiation. IAEA - SM - 121/34, IAEA, Vienna, pp. 331-336.

Love, J.E. (1972). Somatic mutation induction in poinsettia and sweet potato. In : J. Constantin (Editor), Mutation Breeding Workshop, Knoxville, Fenn. University of Tennessee, Knoxville, Tenn.

Neg, K.K. and B.M. Johri. (1970). Effects of Cytokinins and injury on the formation of shoots by leaves of Dendrophthoe falcata. Planta, 90 : 360-364.

Wilson, L.A. (1973). Stimulation of adventitious bud production in detatched sweet potato leaves by high levels of nitrogen supply. Euphytica, 22 : 324-326.

CHAPTER 14

Mutation Breeding for Root and Tuber Crops

Root and tuber crops are grown on some 50×10^6 hectares throughout the world, the total production amounting to about 6×10^8 metric tonnes. From the point of view of breeding, several root and tuber crops are rather difficult to improve, and so it is worthwhile to consider carefully the possibilities of applying mutation breeding.

Cassava

Cassava, also known as manioc or tapioca, is an important source of food in the low land tropics. Selection programmes aim mainly at high yields, better nutritional quality, low hydrogen cyanide content, resistance to viruses and insects, and storability.

Cassava and its wild relatives show an enormous genetic variation and there are a large number of cultivars. The crop is normally outcrossing but can also be propagated vegetatively. Interspecific hybridization is also used. Bud mutations have been reported to occur for many years, as has the use of colchicine in attempts to produce polyploids.

Moh (1963) was the first to use radiation on this crop. He irradiated stem cuttings (nodes) with up 40 Gy of gamma rays (4-5 Gy/min), subjecting 150 nodes to each treatment. The (lethal dose) LD_{50}, *i.e.*, the dose of radiation at which 50% of the treated material does not survive treatment, was approximately 30 Gy. Abraham (1970) reported that maximum sprouting of buds from irradiated stem cuttings was obtained at doses of less than 15 Gy, whereas at 50 Gy no sprouts were produced at all. Moh and Alan (1973) compared the irradiation of pollen with that of seeds and stems. The advantage of pollen irradiation is that no chimeras are formed. It is hard to distinguish between the variability caused by outcrossing and that resulting from mutation induction. The characters which are used to distinguish them are early root maturity and low hydrogen cyanide content which are rarely found in natural cultivars. From Taiwan (Anonymous, 1978), it was reported that ten mutants with better resistance to *Xanthomonas manihotis* had been induced by ethyl methane sulphonate (EMS) treatment.

It must be concluded that mutation breeding has played only a minor role so far in the improvement of Cassava.

Garlic and shallots

Garlic and other food alliums like the onion, shallot and leek are among the most popular vegetables in many countries.

During its early domestication garlic became completely vegetatively propagated by cloves or by inflorescence bulbils. Sometimes modern cultivars do produce flowers, but they never set seed.

Despite this they differ in many of their characters, including age at maturity, bulb size, clove size and number and scale colour, but it is not known whether this variation originated during the period when garlic reproduced sexually or has arisen through bud mutation since it became vegetatively propagated and sterile. Since the causes of the sterility are also unknown, the only way to obtain further genetic variability is by mutation, either spontaneous or artificially induced.

Skylar (1973) reported that cloves irradiated with gamma rays or treated with chemicals resulted in mutants in $_{v}$M1, *i.e.*, the first cycle of vegetative multiplication after mutagenic treatments. The optimum radiation dose seems to be below 10 Gy, though how much depends upon the cultivar used.

Novak et. al. (1984) and others have referred induction of onion mutants in invitro systems. These systems may speed up the rate of vegetative propagation and diminish problems related to chimera formation and selection, especially for characters that are not directly visible.

Potato

The potato, an important food crop, especially in the northern temperate zone, ranks high in terms of total production of energy for human consumption.

Our knowledge of potato genetics is very limited because inheritance is tetrasomic ($2n = 4x = 48$) and the chromosomes are small. Many cultivars show poor flowering and low fertility. Crossing work with primitive cultivars or wild relatives is further hampered by crossing barriers and by differences in ploidy level. Other stumbling blocks are the high degree of heterozygosity and the low heritability of many important characters.

Spontaneous mutations in vivo

Several cases of spontaneous mutations in potato indicated by bud variations. As early as 1907, Cramer referred to some cases in which bud variations had led to the production of cultivars of practical value, for example, cv. White fortyfold with its white tubers, obtained from cv. Purple fortyfold.

According to Krantz (1951), 15% of the certified seed (small tubers) production of commercial varieties in the U.S.A in 1951 came from bud sports. These sports displayed a change in the skin of the tuber, *i.e.,* towards either a russet structure or a different, more attractive colour. The total observed frequency of spontaneous mutations appears to be very low. Heiken (1960) found spontaneous aberration rates varying from 1.5×10^{-3} (for so called botters in certain cultivars) to 1.2×10^{-5} (for certain foliage mutants).

Induced mutations

Early reports : Johnson (1937) found increased tuber formation and increased weight per hill and per tuber after exposure to 15 Gy of X-rays. Demidovic (1934) who claimed that storage of true potato seed for 5 years induces more mutations than any other method. Stanton and Sinclair (1951) treated their material with phosphorus-32 and described morphological changes in the leaves. They also reported that the apical area of shoots displayed a higher radiosensitivity than other plant parts.

The starting materials

At first, all mutagenic treatments of potato were performed on whole tubers or tuber halves. In the latter case one half was mostly used as a control. Dormant as well as sprouting material is treated. The use of dormant tubers normally leads to a lower proportion of chimeric plants, but sprouting tubers may produce higher mutation frequencies (Heiken, 1960). The treatment of smaller structures, for example, young tubers, tuber parts, single eye pieces, cuttings or adventitiously developed sprouts, is preferable. Unrooted cuttings, the basal area of which is inserted into a piece of lead during irradiation is preferable, for example, if tuber production of a certain clone is low and if it takes too much time before the next tuber generation is available. Another positive point of the method is the high uniformity of the starting material.

The most recent development is the rather extensive use of in vitro methods in mutation breeding of potato. Very effective and relatively simple is the use of explants in vitro. Sonnion *et al.* (1986) grown buds on a solid medium and irradiated the resulting plante lets which were subsequently propagated by making single-node microcuttings.

In addition to irradiation of vegetative parts, potato seed and pollen have also been treated for different purposes. Practically, all mutation work has been performed with the common, tetraploid potato. Diploid or dihaploid material derived from the common potato has also been used for mutation breeding. An experiment involving 150 *x* - irradiated small dihaploid tubers of cv., desire was published by Van Harten and Bouter (1973). In this investigation all emerging sprouts were dissected when about 3 cm, long and planted individually over a period of 8 months. Amongst the latest developing sprouts especially, the majority were of adventitious origin and produced relatively high frequencies of non-chimeric mutants.

Mutagenic treatments

Over the years many different mutagenic agents have been applied. In the beginning mostly X-rays, 32_p and, to a smaller extent, gamma rays were used, in a wide range of doses, dose rates and concentrations. More recently neutrons have also been administered and, in a still increasing number of cases, different chemicals.

For X-rays and gamma rays, doses range from 4-100 Gy for treatment of vegetative parts in vivo, 20-30 Gy being the doses preferred by most workers. Doses of 80-100 Gy are usually lethal. The dosages used vary from about 10 to 30 Gy depending on the kind of explant material used.

Ferwerda (1965) and Umaerus (1966) were the first to apply ethyl methane sulphonate (EMS) solutions, which were dropped on eyes at concentrations of, for example, 1.5% EMS has remained the most often used chemical agent although several others, like ethylene imine (EI), diethyl methancsulphonate (DMS), N-nitroso-methyl urethane (NMU), hydroxylamine have also been used on different occasions.

Types of induced mutations

Most cases in which the induction of mutations have been proved definitely refer to genetic changes of morphological characteristics, for example, size, shape or colour of the foliage or of the subterranean parts.

The work of Heiken (1960) has confirmed that most mutations manifest themselves as periclinal chimeras and often show pleiotropic effects. Heiken reported an increased pigmentation in the corolla after mutagenic treatment. The induction of mutations for earliness, increased resistance to different diseases and increased starch content of the tubers was reported by Solomko.

Kuklimura and Takemata (1975) studied induced variation in starch content, stem height and tuber yield after treating dormant tubers with ^{60}Co gamma rays. Doses of 40-160 Gy were given. Mutations have been induced by applying X-rays at a dose of 10 Gy to detached leaves of a monohaploid potato clone. From the irradiated leaves, adventitious plantlets were produced in vitro on leaf explants. This is one of the first examples of a successful combination of mutation breeding and in vitro methods.

Several characteristics of economic importance, like cooking quality, tuber flesh colour, eye depth, under water weight, stolon length, resistance to phytophthora and leaf roll virus, could be improved using mutagenic treatment. Resistance to phytophthora was induced in a few clones of cv., bintje, which remained significantly more resistant during 6 successive years of testing in infested field plots, without showing any aberration of the Bintje phenotype. This fact demonstrates that desirable mutations are not always accompanied by pleiotropic effects or, in any case, that such effects can be eliminated by further selection without losing the desired genetic change itself.

The commercial mutant of potato

Name of new cultivar	*Place and date of release*	*Conditions and date of mutagenic treatment (parent cultivar)*	*Main improved attributes of cultivar*
Konkei, No. 45	Japan, 1973	30 Gy X-rays, 1962 (Benimaru)	Skin colour : red-yellowish white

Sweet Potato

Sweet potato is grown on a large scale in tropical areas, especially in Africa. Considerable quantities are also produced under more temperate conditions, especially in Japan and on a smaller scale in China, Taiwan, U.S.A. and New Zealand.

Tubers are used for human consumption, boiled or baked, but can also be used as fodder. Several products flour, starch, glucose syrup and alcohol can be manufactured from the tubers. Leaves are eaten as vegetables and fresh vines are given to cattle as fodder. Plants are normally propagated by means of stem cuttings 30 – 50 cm long. In temperature regions, sprouts or slips from small tubers, which are pulled from nursery beds and transplanted at a length of 20 – 30 cm, are often used. The cuttings are very easy to root. Propagation by single—leaf cuttings is also possible.

From the point of view of breeding, the sweet potato is a very difficult crop. Flowering is uncommon, especially in temperate regions, and in all climates seed set is low. Both incompatibility and sterility barriers are known to exist. The hexaploid condition ($2n = 6x = 90$) makes genetic studies very complicated. It has been shown that many important characters are inherited quantitatively.

Most cultivars of sweet potato are highly heterozygous and variable. In the past, improved cultivars were obtained by selection among existing clones in which spontaneous mutations occur quite frequently. Moreover, chance seedlings occur in nature. At present, breeding work is carried out with seedlings from open pollination as well as with controlled crossing. Breeding work in many countries aims mainly at producing high yielding, early, well-shaped, shallow rooting types with good storage and cooking qualities, palatability, and resistance to viruses and pests. Attractive skin colours, high vitamin content and frost resistance are also favoured.

Miller (1935) was also one of the first authors to work on X-ray induced mutations in this crop. Some differences in tuber colour and taste between treated and non-treated material were observed. Love *et.al.* (1977) did not find yield improvement but noted skin colour and flesh colour mutations after exposure of rooted cuttings to fast neutrons. Changes in flesh colour occurred more often than those in skin colour. Miu et. al. (1973) reported improved tolerance in the important native cultivar Red-flesh-tail, from Taiwan, after treatment with 50 Gy of gamma rays. Plants with darker leaves and higher yields were also obtained. Treatment with 32_p were performed by pouring a diluted solution into a hole (2 cm^3) made in a tuber. One practical importance is that favourable induced mutants were entered into a crossing programme from which further selections were made. A resulting line with higher starch yield and a high general level of resistance to pests and diseases was used as parental material in further breeding work.

Yams and various other (tropical) root and tuber crops

Yams, *Dioscorea* spp. are mainly herbaceous climbing plants, grown for their edible tubers. More recently the genus *Dioscorea* has become important because of its medicinal value, *i.e.*, the presence of steroids such as diosgenin, a precursor or cortisone from which steroidal hormones like progesterone can be prepared.

Breeding work mainly aims at high yield, shallow rooting, palatability, lack of toxicity, resistance to necrosis and virus diseases, long life cycle and improved shelf life of the tubers.

Abraham (1970) reported that, by treating tubers with gamma rays, high yielding mutants were obtained, Koo and Cuevas Ruiz (1974) irradiated the aerial tubers of several varieties after storage for 1–2 months at room temperature. The dose used was 20 Gy of gamma rays given at a rate of 10 Gy/min. The first shoots to appear were removed, forcing new shots to develop; these latter shoots were allowed to grow and were observed. The results showed that solid mutant shoots appeared as early as the second generation.

Yams, which are normally propagated from tubers or tuber pieces, can also be multiplied from vine cutting and detached leaves, which are far more attractive for mutation breeding. Tissue culture, for instance from nodal segments has become another method of vegetative propagation of Yams.

Mutagenic treatment for increasing diosgenin content have been performed on tubers. Tubers were irradiated with 150 Gy X-rays. In vM_2 and vM_3 a remarkable increase in diosgenin content was observed. Seeds were irradiated with gamma rays (25-125 Gy) or treated with NMU (Nitroso methyl urea) (0.01-0. 05%) or DES (Diethyl sulphate) (0.10-0.40%). Chemical mutagens were found to lead to higher mutation rates than did gamma rays. vM_2 and vM_3 = vegetative mutant 2 and 3 generation ; 1 Gy = 100 rads.

Abraham et.al., (1976) reported on the mutagenic potential of ginger and turmeric.

REFERENCES

Abraham, A., (1970). Breeding work on tapioca (Cassava) and few other tropical tuber crops. In : D.P. Plucknett (Editor), Tropical Root and Tuber crops Tomorrow, Vol. 1. University of Hawaii, Honolulu, HI, PP. 76-78.

Abraham, S., S. K. Abraham, and G. Radhamony. (1976). Mutagenic potential of the condiments, ginger and turmeric. Cytologia (Japan), 41 : 591-595.

Anonymous, (1978). Taiwan Sugar Research Institute : Annu. Rep., 1976-77, p. 65.

Cramer, P.J.S., (1907). Kritische Ubersicht der bekannten Falle Von Knospenvariation, 4. Natuurk. Verh. Holl. Maatsch. Wet. Haarlem, 474 pp.

Demidovic, A., (1934). A new possibility in breeding, Semenmvodstvo (Seed growing), 4 : 8-10 (in Serbo croatian).

Ferwerda, F.P., (1965). Mutagenic effects of X-rays and ehtyl methane sulphonate on potato sprouts. In : The use of Induced Mutations in Plant Breeding. FAO/IAEA Meeting, Rome, 1964. Radiat. Bot., 5 (Supl. 687-690).

Heiken, A., (1960). Spontaneous and X-ray Induced Somatic Aberrations in Solanum tuberosum L. Almqvist and Wiksell, Stockholm, 125 pp.

Johnson, E.K., (1937). Tuberization of the colorado wild potato as affected by X-radiation. Plant physiol. 12 : 547-551.

Koo, F.K.S. and J. Cuevas Ruiz, (1974). Exploratory induction of solid mutations in yams by gamma irradiation. In : A.M. Srb (Editor), Genes, Enzymes and populations. Plenum publishing, New York, pp. 331-336.

Krantz, F.A., (1951). Potato breeding in the United States. Z. Pflanzenzucht. 29 : 388-393.

Kukimura, H. and T.Takemata. (1975). Induced quantitative variation by gamma-rays and ethleneimine in tuber bearing plants. Gamma Field Symp., 14, 25-36.

Love, J.E., T.P. Hernandez and M. Mahmood. (1977). Mutation studies with the centennial sweet potato. Hort Science, 12 : 405 (abstr).

Miller, J.C., (1935). Further studies of mutations of the porto Rico Sweet Potato. Proc. Am. Soc. Hortic. Sci., 32 : 460-455.

Miu. T.S., C.Y. Yu and C.S. Chang. (1973). An inquiry into assessment of cold tolerance in groups of irradiation induced lines of sweet potato. Tai-wan Nung Yeh Chi K'an (Taiwan Agri.c.o), 9 (1) : 28-40 (in Chinese).

Noh, C.C., (1963). Radiosensibilidad de las especies de plant tropicales, Carica papaya, Manihot dulcis Y swietenia humilis. Turrialba, 13 : 180-181.

Moh, C.C. and J.J. Alan. (1973). Methods for inducing mutations in cassava and the possible uses of the mutant. In : Induced Mutations in vegetatively propagated plants. IAEA, Vienna, pp. 67-75.

Novak. F.J., L.Havel and J. Dolezel. (1984). In vitro breeding system of Allium. In : A. Fujiwanas (Editor), plant Tissue culture, Proc. 5th Int. Congress, Tokyo, Japan, 1982, pp. 767-768.

Skylar, G.E.,(1973). The induction of Somatic chemical and radiation mutants in garlic. Biofiz. Issled. Rast. Geterozis, 1973, 90-95 (in Russian) : Plant Breed. Abstr. 46 : No. 2899.

Sonnino, A., G. Ancora and C. Locardi. (1986). In vitro mutation breeding of potato : the use of propagation by microcuttings. In : Proc. Int. Symp, on Nuclear Techniques and In-Vitro culture for plant Improvement, Vienna, 1985. IAEA, Vienna, pp. 385-394.

Stantion, W.R. and W.K. Sinclari. (1951). Effect of high concentration of phosphours 32 on growth of potato. Nature (London), 167 : 234-235.

Umaerus, M., (1966). Somati aberrations in Solanum tuberosum induce by ethyl methane sulfonrate and X-irradiation. Z. pflanzenzucht., 55 : 238-245.

Van Harten. A.M. and H.Bouter. (1973). Dihaploid potatoes in mutation breeding : Some preliminary results. Euphytica, 22 : 1-7.

CHAPTER 15

Genetic Manipulation and Gene Transfers in Plants and Animals

Introduction

Techniques for the isolation and characterisation of DNA fragments encoding the genetic information of many proteins of agriculture and medical importance have developed over the past decade. Rapid advances can be expected in the spectrum of plants modified by recombinant DNA technology and cell regeneration. Molecular processes of gene regulation and cell development are being explained. Many practical applications are evidence, including the improvement of nutritional value and processing qualities of crop plant products. Commercial opportunities have been already demonstrated by the development of plants that are resistant to herbicides, insects and plant pathogens. Expressions of foreign gene in transgenic plants was first reported by Murai et al. (1983) in which the transferred gene expressed the synthesis of the seed protein phacscolin in sunflower callus. In 1986, Fraley et al provided the detailed information about genetic information of higher plants. Lamb and Beachy (1989) reported that the field animal transgenesis is rapidly growing in many laboratories as a model for human disease ; the first animal to be patented being a transgenic mouse. Innovative programmes in domestic animals have just started using subtle approaches to achieve resistance to disease, controlled reproduction, increased growth and production, altered growth of wool and composition of milk, as well as the development of transgenic protein production units. The progress in transgenesis in fish was some what slow to start but may well be the first species to claim economic success in production (Church 1989).

Genetic Manipulation of Higher Plants

In vitro manipulation of higher plants include in vitro culture techniques *e.g.*, in vitro multiplication, pathogen elimination, in vitro selection, haploid production, plant regeneration from protoplasts. Variation and change in higher plants is normally achieved by sexual means; namely crossing and selection. Alternatively, genetic modification and variability can be now created by in vitro techniques and genetic engineering. The new techniques should however not be considered to replace or to compete with conventional plant breeding, but rather to complement plant breeders work, (Austin et al., 1986). Several possibilities are given below. On how to achieve new combinations of the genetic information of plant cells and how to introduce defined genes into a receipient cell.

There are two different types of gene transfer systems in plants.

1. Vector Dependent Gene Transfer

Agrobacterium

DNA viruses & RNA viruses

2. Direct Gene Transfer

DNA uptake into protoplasts

Microinjection of DNA into Zygotic/Somatic Cells

Pollen transformation

Injection of DNA into plants.

1. Vector Dependent Gene Transfer

Agrobacterium tumefaciens has been observed to transform only dicotyledonous species and a few monocots in the lilly family, (De cleene and De Lay 1976). In practice, to date a much broader range of transgenic plants have been produced with *A. tumefaciens* mediated transformation (Table 23). Since it can transform regenerable tissues with relative case, white protoplasts of most species are still difficult or impossible to regenerate. The difficulties of DNA transfer and selection for transformed colonies during recovery and growth have proven to make low frequency regeneration from protoplasts a major problem. Only transgenic tobacco and carrot plants have been produced via protoplasts. Free DNA delivery to regenerable protoplasts may be the eventual key to transformation of important crops such as corn, wheat or rice, but because of the practical case of *Agrobacterium tumefaciens* mediated transformation, it is likely to remain the method of choice for routine production of transgenic plants for species within its host range.

Surface sterilized leaf discs or other axenic explants are infected with the appropriate strain of *A. tumefaciens* carrying the vector of choice, and cocultured on regeneration medium for 2-3 days. During this time the virulence genes in the bacteria are induced, the bacteria bind to the plant cells around the wounded edge of the explant, and the gene transfer process occurs. After the transformation has occured the explants are transferred to regeneration selection medium. This contains 500 mg/ml carbenicillin to kill the bacteria and the appropriate antibiotic to inhibit untransformed plant cells, usually kanamycin. During the next 3 weeks the transformed cells grow into callus or differentiate into shoots in a organogenesis. Between 3 and 6 weeks, the shoots develop enough to remove them from the explant and induce rooting in preparation for transfer to soil.

Table 23. Transgenic Plants Produced with A. tumefaciens Mediated Transformation

Species	Reference
Nicotiana plumbaginifolia	Horsch et al, 1984
Petunia	Horsch et al, 1985
Tobacco	De Block et al, 1984
Tomato	Mcormick et al, 1986
Arabidopsis thaliana	Eloyd et al 1986
Medicago sativa	Deak et al, 1986

Characteristics of Agrobacterium

A. tumefaciens causes crown gall disease by transferring a defined segment of DNA (T-DNA) from its tumour inducing (Ti) plasmid into the nuclear genome of cells in an infected wound on many dicotyledonous plants (Nester et al 1994). The T-DNA contains genes that encode enzymes that catalyze the synthesis of phytohormones IAA (Schroder et al, 1984) and an auxin and cytokinin. This over production of phytohormones is responsible for the proliferation of wound callus into a gall which can harbor a population of the bacteria. Other T-DNA genes cause synthesis of unique matabolities called opines cannot be utilized by many other micro organisms or plants. One strain of *A. tumefaciens* (A 208) that produce snapaline synthase, an enzyme that produces nopaline from Arginine and alpha ketoglutarate.

The movement of the T-DNA is mediated by genes in another region of the Ti-plasmid. The virulence genes are not themselves transferred with T-DNA but act in trans to cause the transfer (Framond *et al* 1982). The virulence genes are induced by phenolic compounds present in wounded plant cells) (Stachel *et al,* 1986). The T-DNA is defined by a 25 base sequence, the border sequence, which is present as a direct repeat outside the ends of the T-DNA (Barker et al, 1984).

These properties of *Agrobacterium* DNA transfer system are invaluable for developing a powerful vector system for plant transformation. It is possible to remove all of the original T-DNA which was responsible for the disease and replace it with any DNA, as long as the border sequence is maintained, (Zambrysi *el al*, 1983). Once the T-DNA region of the Ti plasmid was identified, it seemed very likely that foreign genes inserted within this region would be transferred to the host genome alongwith the T-DNA.

A shuttle vector (Fraley *et al* 1983) that can replicate in *E coli* where recombinant manipulations are easily handled and then transferred into *A. tumefaciens* in preparation for transfer into plants. There are two basic modes of maintenance of the shuttle vectors in *Agrobacterium*, either by integration into the Ti plasmid by recombination at a region of DNA homology (Fraley *et al.,* 1983) or by autonomous replication in trans to the Ti plasmid (An *et al* 1984). The former type of vector is referred to as a *cis* or *integrating* while the latter is called a *trans* or *binary* vector. In most cases they accomplish the same goals of shuttling genes for *E.coli* to *A.tumefaciens* in a T-DNA package that can then be transferred to plants.

Selectable markers

Since only a small portion of cells are transformed by *A.tumefaciens,* it is important to have a selectable marker to recognize the transformed cells and supress the growth of wild type ceils. The first marker to be used was the neomycin phospho transferase gene from *E.coli*. This gene has been successfully used in many systems including mammalian cells, fungi and *Dicotyostellium.* In solanacious plants such as petunia, tobacco and tomato it provided an excellent, marker.

Leaf disc transformation

The simple and powerful interface between *A.tumefaciens* and plant cells is illustrated by the leaf disc transformation system, (Horsch et al., 1985) where gene transfer, selection and regeneration are coupled together in an efficient process. Tobacco is an excellent host for *A.tumefaciens* and also responds exceedingly well in culture. While the technique is more easily practiced in tobacco, it has been applied to number of other species, lettuce, celery, popular flax, *Brassica napus,* sunflower and cotton.

Inheritance

Progeny of transgenic plants produced using this process always have inherited the T-DNAs found in the parent plant. Progeny usually inherit the T-DNA according to Mendelian rules for dominant genes. In cases where there is only one T-DNA, salfed progeny show a 3:1 ratio for kanamycin or napaline resistance, while backcrossed progeny show a 1:1 ratio.

DNA Viruses and RNA Viruses

In addition to bacterial vector systems viral vectors based on DNA as well as RNA viruses have been developed for plant transformation, too. Cauliflower mosaic virus has been used successfully to replicate and to express a foreign gene in a dicotyledonous and this virus can be considered useful as a plant gene vector providing a self replicating system (Brisson and Hohn, 1986). Similarly, wheat Dwarf virus, a member of Gemini viruses which infects several Gramineae Species has been cloned and introduced into wheat protoplasts. It has been shown that the cloned virus DNA replicates in wheat as well as maize protoplasts and the replacement of viral coat protein has led to the expression of newly introduced, foreign gene after transfection of wheat and maize cells. The process of plant transfection by wheat Dwarf Virus (WDV) is strictly dependent on leaf hopper transmission or the genetically engineered virus DNA has to be introduced directly into protoplasts.

2. Direct Gene Transfer

DNA Uptake Into Protoplast

Direct gene transfer to plant cells mediated by DNA uptake into protoplasts and it has been shown that the transferred DNA is stably integrated and transmitted to the progeny in a Mendelian inheritance pattern (Hain *et al.,* 1985). This transformation method, applying chemical treatment or electric pulses (Shillito *et al.,* 1985) to stimulate DNA uptake, is a very general one and also has been used to successfully transform *Triticum monococcum* (Lorz *et al.* 1985), *Lolium multiflorum* (Potrykus *et al.,* 1985), *Zea mays* (Fromm et al., 1986) and *Oryza sativa* (Uchimiya *et al.,* 1986) protoplasts. In these experiments plasmids have been used which contained a selectable marker chimeric gene with the coding region of the Aminoglycoside phosphotransferase type II gene.

Micro Injection of DNA into Zygotic/Somatic Cells

Another direct way of transferring cloned genes into protoplats is the micro injection of DNA. The procedure is most efficient when DNA is injected directly into the nucleus of protoplasts (Crossway et al., 1986). As this method is very labour intensive and only a limited number of protoplasts can be treated this way, the application of the procedure is so far limited only to protoplast systems with high efficiency (Lawrence and Davies, 1995).

Micro Injection Techniques

Protoplasts with partially reformed cell walls can be attached to glass coverslips with lysine without damaging the cells (Steinbiss and Stabel, 1993). After one day in culture the cell walls of protoplasts are firm enough to provide support and protection yet they are easily penetrated with micropipettes. At the same time DNA synthesis has reinitiated in alfalfa (*Medicago sativa*) protoplasts. At the end of the experiment or the next day the micro injected cells can be segregated and transferred to optimal conditions for low density cell culture.

Mariotti et al., (1984) reported that alfalfa is difficult to infect with *A.tumefaciens* and oil seed rape exhibits a strong hyper sensitive response to the bacteria that transforments are difficult. So, transformation can be easily done with microinjection technique. The recovery of large numbers of transformants clonally and without selection by microinjection is very significant for many plant species. Proteins or their antibiotics can be introduced into specific cell compartments to study post synthesis processing, intracellular transport and influence on cell architechture or function.

Pollen Transformation

Pollen transformation has been studied intensively with petunia and Nicotiana and more recently also in maize (Ohts, 1986). DNA treated pollen grains which are pregerminated are used for pollination to transfer the foreign gene.

Injection of DNA Into Plants

DNA was injected directly into developing inflorescences. Previous work analyzing the effect of various drugs injected at different developmental stages of the rye germ cells had shown that the archesporial cells of rye are especially sensitive to exogeneously applied toxin about two weeks prior to metaphase I (Puer-tas et al., 1984). Plasmid DNA coding for kanamycin resistance has been injected into developing inflorescences of rye cv JNK in field and glasshouse grown plants. (Fig. 13). Flowering plants were crossed with each other and seeds from DNA treated and control plants were germinated in the presence of kanamycin. Transformants were normally developed from the seven green seedlings selected, two of them showed the presence of active gene product.

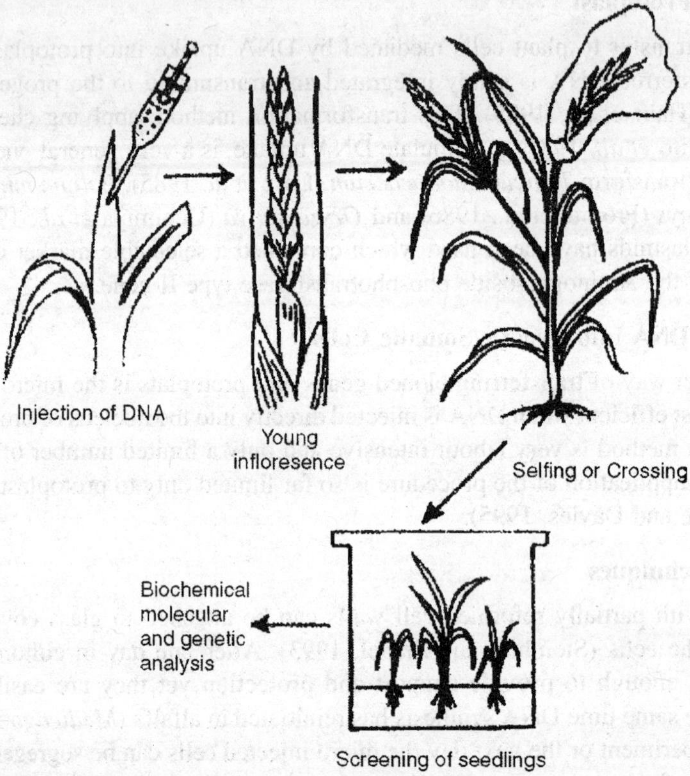

Fig. 13. Schematic summary of the transformation procedure of rye plants medicated by injection of isolated DNA into Developing infloresence.

Protocol for Routine High Transformation Frequencies

The following protocol gives routinely high frequencies of transgenic plants with leaf protoplasts of *N.tabacum.* (Shillito et al., 1985). Protoplasts are isolated as appropriate for subsequent culture (Potrykus and Shillito 1985). They are suspended in 0.4 M Mannitol containing 6 mM $MgCl_2$ and 0.1% MES buffer (Morpholino ethansulfonic acid) at pH 5.6 to a population density of 1.5×10^{6ml}. $MgCl_2$ is added to bring the electric resistance of the suspension to the range of 1.2 k ohm. This is followed by incubation at 45 °C for 5 minutes and cooling to room temperature on ice. The gene is added to a final concentration of 10 g/ml and carrier DNA to 40 g/ml followed by polyeth leneglycol (MW 6000) to a final concentration of 8%. For electroporation aliquots (one ml solution) are transferred to a chamber to an electroporator and three high voltage pulses applied at 10 second intervals to give 50% survival. Thereafer the protoplast suspension is transferred to petri dishes, a 10 fold volume of culture medium containing 0.6% agarose is added without prior washing of the protoplasts. The protoplasts are carefully distributed and the gel allowed to solidity. After incubation under optimal growing conditions for 7 days the protoplast agarose gel is cut into segments and transferred into a 10 fold volume of liquid culture medium containing if desired, a selective agent.

Factors Affecting High Transformation Frequencies in Direct Gene Transfer

(1) Host range
(2) Frequency of stable integrative transformation
(3) Reproducibility
(4) Transforming DNA
(5) Carrier DNA
(6) Heatshock treatment
(7) Electroporation
(8) Integration into the host genome
(9) Stability of the foreign gene
(10) Instability of the foreign gene

(1) Host Range

One of the major reasons to develop the biological vector independent gene transfer was the hope that this would enable gene transfer to such interesting plants as (*e.g.*) the cereals. This is demonstrated for *Lolium multiflorum* (Potrikus *et al.*, 1985), *Triticum monococcum* (Lorz et al., 1985) and *Zea mays* (Fromm *et al.*, 1986). The major problem with genetic engineering of cereals is in plant regeneration from protoplasts. There are several presentation on plant regeneration from rice protoplasts. Whereas it is still difficult task to repeat with other cereals what might be possible with rice (Yamada et al., 1986). With protoplasts from herbaceous dicots we have not seen any limitation so far with regard to plant species or protoplast type and we would therefore assume that direct gene transfer is possible with protoplasts.

(2) Frequency of Stable Integrative Transformation

The transformation frequency is related to the number of clones growing after the transformation treatment. The calculation is based on either screening without prior selection of a few hundred randomly picked protoplast derived clones ; or by relating the number of clones surviving selection to the number of clones growing in an unselected aliquot of the treatment. This gives precise figures on the frequency of stable integration of at least one functional copy of the foreign DNA). With this method of calculation ; high frequencies of routinely 1-2% are achieved with *N.tabacum* (Shillito *et al.*, 1985) and other partially optimized protoplast systems *e.g., Petunia hybrida, Hyoscyamus muticus.*

Nieotiana plumbaginifolia, N. tabacum cell lines - frequencies around 10% are obtained. Screening for integrating without function of the foreign gene yielded frequencies around 15%.

(3) Reproducibility

Direct gene transfer has produced transgenic clones and plants at frequencies ranging from 1-2% (Loerz et al., 1985).

(4) Transforming DNA

The effect of the transforming DNA on transformation frequency is four fold :

(*a*) There is a clear concentration effect with a linear range between 1 to 50 g/ml.

(*b*) Linearized molecules transform about 2 fold better than circular ones.

(*c*) The size and shape play a role in as for as small and large molecules transform better than medium sized ones.

(*d*) Stronger expression signals produce higher transformation frequencies under selective conditions than weaker ones.

(5) Carrier DNA

The presence of carrier DNA is not a prerequisite for even high transformation frequencies. But it improves and stabilizes transformation frequencies. There is a concentration effect which levels off. Also size plays an important role, yielding optimal effects with small molecules and toxic effects with large ones. There is not much difference with regard to the type of carrier DNA when comparing plant DNA, Calf thymus DNA, Salmon sperm DNA and lambda DNA (Shillito et al., 1985).

(6) Heat Shock Treatment

Heat shock treatment optimally 5 minutes at 45°C for tobacco leaf protoplasts may improve transformation frequency by 10 fold. It also may improve the integration pattern of the transforming DNA by reducing rearrangement and inactivation. Both effects are however not entirely consistent in different experiments and they may vary considerably with plant species and protoplast type.

(7) Electroporation

Electroporation describes the phenomenon of pore formation in membranes under the influence of short discharge pulses with high voltage. This is conveniently done with protoplasts by subjecting them in a cylindrical chamber between parallel electrodes, to controlled discharge pulses from a capacitor. High initial field strength (1400 volts) and very short decay constant (10 sec) in several repetitions balanced 50% survival, yield the highest effect on stable integrative transformation, which is upto 10 fold with *N. tabacum* leaf protoplasts. The initial field strength has to be adjusted to the size of the protoplast type uses. Electroporation is no prerequisite for even high frequencies direct gene transfer and its effect on transformation frequency may vary with plant species and protoplast type (Shillito *et al.*, 1985).

(8) Integration into the Host Genome

The foreign genes integrate, obviously via., illegitimate recombination, and therefore at random sites, into chromosomal DNA. There is no clear case of integration into either ch1 DNA or mt DNA known. In 77% of 220 primary transformants studied integration was at one single locus. The number of integrated functional copies can range from 1 to 30 or more. Also in the case of multiple copies these are preferentially at one locus. There is however also simultaneous integration at independent loci.

(9) Stability of the Foreign Gene

Stability determinates :

(*a*) sub-populations from one experiment yield the same transformation frequency when grown directly under selective conditions as when grown for 5-6 weeks without selection and thereafter transferred to selective conditions, (*b*) of 60 subclones regenerated from two primary transformants, 57 maintained the gene under non selective conditions,

(*c*) transgenic plants are neither chimeras nor mosaics,

(*d*) gene is regularly transmitted to sexual offspring,

(*e*) the inheritance of the genes follows strictly Mendelian rules and

(*f*) the gene has been maintained over several sexual generations.

(10) Instability of the Foreign Gene

Besides those clones with a high degree of stability there were others which showed considerable instability. A self cross of a homozygous transformant yielded a R_2 population of 422 resistant and 4 sensitive seedlings. Sixty further R_2 populations of the same plant produced by selfing, 60 flowers from primary, secondary and tertiary inflorescences all produced seedling populations containing 0.5% sensitive individuals. One resistant R_2 plant tested further produced R_3 populations with 0.5% sensitive individuals one resistant, R_2 plant tested further produced R_3 populations with 0.5 sensitives, one produced a population segregating 840 : 98 for resistant versus sensitive.

Practical Applications for Transgenic Plants

For practical modification of crops, DNA containing the desired information must be introduced into a cell that can subsequently be regenerated to yield a mature, fertile plant. As has been described, procedures are well established for the introduction of foreign genes and their expression in dicots. The major remaining barrier is the limited range of dicots that can be regenerated from transformable cells. In monocots, the regeneration barrier appears to be even greater and much research remains to be done before efficient expression of transferred genes in monocots is achieved. Gene transfer is done in plants in the following aspects.

— Development of herbicide resistant plants
— Development of resistance to pathogens
— Development of resistance to insects
— Improvement of nutritional and processing qualities of crop products especially proteins
— Alternation of tastes through the addition of proteinaceous sweetners.
— Development of resistance to heavy metals
— Cloning genes for Biosynthesis of stilbene type phytoalexins.
— Cloning the mRNA of Acc Synthae
— Engineering of a new flower colour variety of petunia.

Development of Resistance to Herbicides

Resistance to several different herbicides has now been engineered into a number of crop species. An early report was resistance to glyphosate, a broad spectrum herbicide in transgenic tobacco containing a modified gene from *Salmonella typhimuran* (Comai *et al.*, 1985). Glyphosate kills plants by interfering with 5 enol pyruvyl shikimate 3 phosphate (EP SP). Synthase involved in the schematic pathway; the approach overcomes this through the use of a mutant bacterial gene resistant to this block. A similar approach has been used to obtain tomato plants that are resistant to 0.854 kg ai/ha (Fillatti *et al.*, 1987).

Introduction of Human Alfa Interferon

Introduction of recombinant human Alfa interferon gene influences resistance to virus diseases. Several human infering are able to defense plants from virus diseases. This is obtained by the way of introducing this gene into potato cultivars. (Smirnov *et al.*, 1991).

Towards Better Understanding of Coat Protein (CP) Gene Mediated Protection

Powell et al., (1986) reported that transgenic plants expressing the tobacco mosaic virus (TMV) coat protein (CP) gene were protected against TMV infection. Ponz and Bruenzing (1986) reported that CP mediated protection were similar to those of classical cross protection where a mild strain of virus is used to protect a plant against infection caused by a more severe strain. Register et al., (1988) conducted experiments using transgenic tobacco plants that express the TMV CP gene have revealed insights to the mechanisms of protection. In these, plant protection functions at two levels viz.,

(1) at the initial site of infection and

(2) as a block to long distance spread.

Anderson *et al.,* (1989) observed the block to long distance spread when they recorded a delay in TMV movement into the upper leaves of TMV CP + transgenic plants as compared to controls. Wilson (1984) demonstrated the protection against long distance spread by which involved grafting of transgenic tobacco TMV CP + stem segments between lower and upper sections from a CP-plant. When the Lower CP - leaves were infected, movement of the virus through the CP + segment into the upper CP - section was inhibited. This mechanism is due to the involvement of inhibition movement of the virus into the vascular tissue. Register III and Beachy (1988) demonstrated that protoplasts derived from transgenic TMV CP + plants have protection at an early stage of virus infection.

Powell *et al.,* (1989) demonstrated that protection in transgenic TMV CP + tobacco plants was due to the presence of CP and not CPmRNA. This was supported when it was shown that at temperature above 30 °C, the level of TMV CP in transgenic tobacco plants sharply declines with a concurrent decline in protection levels against TMV. When the temperature was shifted back to 25 °C both CP and protection levels recovered. In transgenic tomato plants the decline in CP levels is not accompanied by a decline in protection. The above results showed that protection is linked to the accumulation of TMV CP.

Broad Spectrum Protection in the Tobacco Virus Group

Heterologous protection, *i.e.,* protection against a virus rather than the virus from which the CP gene is derived; within a virus group has only recently been described (Stark and Beachy, 1989). Such heterologous protection could be valuable because it would provide protection against several viruses by using a limited number of viral CP genes. Xanthi nc tobacco plants expressing the TMV CP gene were inoculated with TMV, (tomato mosaic virus) TOMV, or tobacco mild green mosaic virus (TMGMV). Following the inoculation the numbers of chlorotic or necrotic lesions on TMV CP + and TMV CP – plants were compared. (Nelson et al., 1987).

Table 24. Numbers of Necrotic Lesions on Tobacco plants

Virus concentration		*Necrotic lesions/plants*	
mg/ml	*TMV*	*TOMV*	*TMGMV*
0.1	"nc"748	"nc"748	"nc"748
0.25	577 4	123 1	274 0
	800 6	296 1	522 0

Lesions on *N.tabacum* Cv Xanthi and TMV coat protein expressors (line 748) after inoculation tobacco mosaic virus (TMV) tomato mosaic virus (TOMV) and tobacco mild green mosaic virus (TMGMV).

From the Table 24, the plants accumulating TMV CP show over 100 fold fewer necrotic lesions than controls at both inoculum concentrations for all three viruses tested. The protection appears to be equivalent against all three viruses. Similar experiments were carried out on Cv. Xanthi, a systemic host of these tobacco viruses.

Table 25. Percent Plants Showing Symptoms

(The percentage of tobacco plants showing systemic disease symptoms 14 days after inoculation with tomato mosaic virus (TOMV) and tobacco, mild green mosaic virus (TMGMV), tobacco plants expressing TMV coat protein CP + are lines 36.46 and 3404) (Abeles et al., 1986).

Virus concentration	*TOMV*		*TMGMV*	
g/ml	CP –	CP +	CP –	CP +
0.01	100	0	60	0
10	100	0	100	0
100	100	80	100	100

From the Table 25, it is shown that plants accumulating TMV CP exhibit a high level of protection against both TOMV and TMGMV infection similar to that seen against TMV. These results with TOMV agree with and expand upon those previously reported. (Nelson et al., 1988).

Broad Spectrum Protection Within the Potyvirus Group

Potyvirus are a large, widely distributed group economically important plant virus group and account for over one-fourth of all plant viruses, (Francki et al., 1985). The CP amino acid sequences are known for over 26 strains of 13 different potyviruses. Most potyviruses share strong sequence homologies in the carboxy terminal two-thirds of their CP's. With overall amino acid homology between different potyviruses averaging 55-60%. Due to the relatively low overall amino acid homology, a demonstration of heterologous protection would coroborate the suggestion that structural homology is important in CP – mediated protection. In order to test this hypothesis, a soybean mosaic virus (SMV) CP gene was cloned and expressed in transgenic tobacco plants (Eggen berger 1989).

SMV was chosen because it is a non-pathogen on Xanthi tobacco. Protection in this sytem would therefore demonstrate that the expression of the CP gene from a virus which does not normally infect a plant can confer protection against pathogenic viruses within the same virus group.

Table 26. Percentage of Plants Showing Disease Symptoms on Successive Days After Inoculation

CP — Coat protein gene

PVY — Potato virus Y

TEV — Tobacco etch virus

CP + — Plants are seedlings of transgenic line 1052 which express the (soybean mosaic virus) SMV CP gene.

CP–9 — Plants are a combination of those 1052 seedlings which do not express the SMVCP gene and a progeny of a plant transformed to harbor the intermediate plasmid.

Table 26 shows that plants accumulating high levels of SMVCP (Stark and Beachy, 1989) are protected against infection by two unrelated potyviruses, tobacco etch virus (TEV) and potato virus Y

(PVY). Protection span at least a 100 fold increase in virus inoculation concentration protection in this system is manifest as a delay in the development of systemic disease symptoms, with a concomitant decrease in virus disease severity in those CP + plants which eventually show symptoms.

Virus Inoculation	*Genotype*	*Days after inoculation*						
		4th	5th	6th	7th	8th	9th	14th
PVY 0.5 g/ml	CP+	0	0	0	5	5	5	80
	CP–	0	5	75	90	95	100	100
PVY 5 mg/ml	CP+	0	0	0	0	25	70	80
	CP–	0	10	90	100	100	100	100
PVY 50 g/ml	CP+	0	0	27	80	87	87	100
	CP–	14	93	100	100	100	100	100
TEV.05 g/ml	CP+	0	0	0	0	0	0	10
	CP–	0	10	50	55	55	55	55
TEV 5 g/ml	CP+	0	0	50	5	20	20	65
	CP–	5	85	100	100	100	100	100
TMV 50 g/ml	CP+	0	0	7	33	100	100	100
	CP–	93	100	100	100	100	100	100

Development of Resistance to Insects

Engineering Enhanced Natural Resistance to Insect Pests

Insects are both beneficial, *e.g.,* acting as pollinating agents, and adverse since they use plants as a major source of food. Thus, in spite of heavy use of costly chemical protective, treatments on average, crop production losses are of the order of 15.1 per cent world wide.

During the course of evolution flowering plants have developed an impressive array of protection mechanisms so that, despite the potential incessant attack by a wide range of insect pests, relatively little damage occurs although occasionally a crop may be devastated in nature. Protective mechanisms may be physical such as sticky exudates or thick culiclesor biochemical. In general, plants possess chemical protectants against insects. Since that time a large number of biochemical compounds have been suggested as protectants, but for the most part with little definitive evidence. Thus, alkaloids, pyrethrins rotenoids, juvenile hormone analogs, glucosinolates saponins, sesquiterpene lactones, non-protein amino acids, isothiocyanates, fluro-fatty acids, condensed tannins, phenoplice, enzyme inhibitors and lectins have all been implicated as playing a role in plant protection.

Plants are such a prolific source of secondary compounds that originally it was thought that these might serve no purpose, but, more recently, opinion has changed to the view that, if a secondary compound is produced in any quantity, it is likely to have at least one function. Seeds, fruits and storage organs which are particular target organs for insect attack, are a ready source of these protective chemicals (Ryan, 1973).

Although no generalisations concerning resistance mechanisms can be made at present, it would appear that any given plant possesses several, such mechanisms and that no single one is responsible for complete protection. These chemical protectants are usually produced in relatively large amounts upto 1% of the dry weight, and in some cases even greater. Many of them (*e.g.,* tannins) act against a variety of pest and disease organisms. On the other hand, others are more specific. But by no means exclusive, being targeted against several of the major incessant pests and it would appear that this strategy is an ecologically stable one, in contrast to the instability of exogenous chemical protection.

If an endogenous chemical protectant is identified as being a protein, it may then be possible to exploit this form of resistance since the gene encoding it could be isolated and engineered into a new host, thereby transferring a protective mechanism from one plant to a crop plant where natural resistance does not exist. This approach is exemplified by the Cowpea trypsin inhibitor which was the first resistance gene of plant origin to be successfully transferred to a different host plant, resulting in significantly enhanced levels of insect resistance. (Hilder *et al.,* 1987). Originally elevated levels of trypsin inhibitor were identified as playing a protective role in the seeds of a particular resistant variety of cowpea (*Vigna unguiculata*) towards its major storage pest, *Callosobruchus maculatus* (Gatehouse, 1979). Subsequent feeding trials demonstrated that when the inhibitor was added to artificial seeds at the physiological level present in the resistant variety, it was toxic to the developing larvae. Since this resistance was to be transferred to a suitably commercial agronomic background, detailed inheritance studies were carried out on the mechanism of inheritance of insect resistance, in relation to the presence of Cowpea Trypsin Inciciptor (CP TI) (Redden *et al.,* 1983).

These studies showed that :

(*a*) the expression of resistance is mainly determined by the maternal genotype.

(*b*) resistance may be inherited as a major gene effect, with resistance recessive, and with the presence of modifier genes of different effects in different crosses.

(*c*) presence of trypsin inhibitors are partly associated with expression of resistance, but uncorrelated in F_2 segregation and appear to be quantitatively inherited and variable between crosses.

Hilder et al., (1987) reported that the cowpea trypsin inhibitor (CPTI) is only part of the defence mechanisms of the cowpea. The gene when transferred to tobacco, enhanced the tobacco resistance to its herbivorous pests. The cowpea trypsin inhibitor fits the general characteristics expected of a natural resistance, mechanism in that it is active against a wide range of insect pest species including both Coleoptera ane Lepidoptera, at relatively high concentrations (Table 27). Ryan (1973) reported that

Table 27. Insect Pest Species 11 Tested for Efficacy of CPTI in Artificial Diets

COLEOPTERA	Anthonomus grandis (Boll weavil)
	Diabrotica undecimpunctata (Corn rootworm)
	Callosobruchus maculatus (Cowpea Seed Weevil)
	Tribolium confusum (Confused flour beetle)
	Costelytra zealandica (Grass grub)
LEPIDOPTERA	Heliothis virescens (Tobacco budworm)
	Heliothis zea (Corn earworm)
	Spodoptera lituralis (Army worm)
	Chilo partellus (Stem borer)
	Insect pest species tested for efficacy of CPTI in Transgenic tobacco plants.
LEPIDOPTERA	Heliothis virescens (Tobacco bud worm)
	Heliothis zea (Corn earworm)
	Spodoptera lituralis (Army worm)
	Manduca sexta (Tomato horn worm)
	Autographa gamma (Silver Y-moth)

inhibitors agains the four classes of proteases probably exists in plants and as nucleic acid and protein sequences data on these inhibitors accumulate. There are atleast a dozen different classes.

Ryan (1989) reported that all classes of inhibitors act as insect protectants, or that an equivalent inhibitor from two ralated plant species would necessarily act in the same way. For example, the protease inhibitor of tobacco appears to be ineffective against *Manduca sexta* while transgenic tobacco plants containing an equivalent inhibitor from potato are effective against Manduca sexta. Gate house and Boulter (1983) reported that CPTI way insecticidal against Callosobruchus maculatus, limabean trypsin inhibitor and the soybean kunitz trypsin inhibitor were relatively ineffective towards this pest. Vack *et al.,* (1987) reported that modified endotoxin genes from Bacillus *thuringiensis* was effective when transferred to several higher plants. In comparision with the Bt genes are more specific, although difference following major groups of insect pest, Diptera, Lepidoptera and Coleoptera.

Now-a-days, only single genes for insect resistance have been used in genetic engineered plants. However, using this technique, it is also possible to produce plants with several resistance genes. These genes can either be introduced into the crop palnt in the form of a 'multiple gene package' or they can be introduced singly and the transgenic plants containing different genes crossed with one another. There are many immediate advantages in producing such crops. Thus, resistance will be multi mechanistic thus more closely resembling that which occurs in nature.

Incorporation of Bt Gene

Bt gene from *Bacillus thuringensis* has been incorporated into plants, like tobacco and tomato varieties. These transgenic plants are highly resistant to lepidopteran insects (Khush, 1991).

Improvement of Nutritional and Processing Qualities of Crop Products Especially Proteins

Many opportunities exists for improving the nutritional and processing qualities of crop products especially seed protein. Legume seed proteins, such as phaseolin are typically low in the essential aminoacid methionine, and cereal seed protein such as the zeins in maize are low in lysine. Insertion of codons for the deficient amino acids into the coding regions of the seed protein genes should result in improvement of their nutritional value.

Alternation of Taster Through the Addition of Proteinaceous Sweetners

It includes alteration of taste through the addition of proteinaceous sweetners such as thaumatin (Eden et al., 1982) or monellin (Ogata et al., 1987). The expression of proteins in new locations *e.g.*, seed proteins in leaves, may be an important agronomic opportunity and as determinants of morphology become known, modification of plant structure raises many possibilities.

Development of Resistance to Heavy Metals

Heavy Metal Resistance in Transgenic Plants Expressing a Human Metallothionein Gene

Over the past years, a modern agricultural practices such as the excessive use of phosphatic fertilizers and sewage sludge have resulted in contamination of agricultural soils with heavy metals. The passive uptake of toxic metals moves them into the food chain. Consumption of such contaminated food and tobacco result in chronic exposure whcih poses a serious threat to human health, (Sherlock, 1984). In addition industrial activities, such as mining and smelting operations have produced large areas with Cu and Zn contaminated soils where climatic factors are otherwise favourable for crop production, (Petolino and Collins, 1984). The increasing levels of toxic metals in the soils, warrants the production and use of plant varieties capable of (*a*) heavy metals tolerance (*b*) sequestration of toxic metals in non-consumed plant parts. A logical approach to this problem is through the expression of genes coding heavy metal binding/sequestration proteins derived from vertabrates and fungi into transgenic plants. Heavy metals in vertebrates and fungi are detoxified by the metallo thioneins (MTS), which are low molecular weight cysteine-rich and heavy metal binding proteins. Their synthesis is regulated at the transcriptional level in response to stress. These proteins are however, not found in plants. Instead phytochelating (small peptides, which are not gene products) have been shown to sequester heavy metals (Grill et al., 1985). Despite the presence of these peptides plants are generally susceptible to enhanced levels of cd. In order to confer tolerance to toxic levels of heavy metals, we have introduced a chimeric human (MT) metallothionein gene into *Brassica napus* and *Nicotiana tabacum* cells on a disarmed Ti plasmid of *Agrobacterium tumefacines.* Our results show that growth of transgenic seedlings was unaffected by upto 0.1 mM $CdCl_2$ in the media, whereas, control seedlings showed severe inhibition of root and shoot growth. Inheritance studies on transformed seeds show that the tolerance phenotype segregated in a manner consistent with Mendelian inheritance.

Transformation selection of Regeneration of B. napus and N.tabacum

A metallothionein processed gene representing a full length copy of its corresponding mRNA was used (Varshney and Gedamis, 1984). The chimeric gene was constructed. For transformation, leaf discs of *N.tabacum* Cv. W38, and stem epidermal explants of B.napus Cv. westar were inoculated with *A.tumefaciens* cells containing the Cointegrate plasmid. The intermediary vector carried a chimeric neomycin phosphotrans ferase 11 gene for selection of transformed tissue on kanamycin and an intact nopaline synthase gene as a screenable marker (Sanders *et al.,* 1987). The transformed cells of B.napus and N.tabacum were selected and regenerated on media containing 0.1 g/ml and 0.25 g/ml of kanamycin, respectively.

Finally regenerated shoots and rooted on B5 media containing 0.1 g/ml kanamycin were screened for expression of nopaline synthase and neomycin phospho transferase II activities. A total of 5 transgenic plants of *B.napus* and 15 different transgenic plants of *N.tabacum* were regenerated. Finally only three plants of *N.tabacum* and two plants of *B.napus* that showed significantly higher levels of nopaline synthase and NPT 11, enzyme activities were selected for further analysis.

Integration and Expression of the Chimeric MT (metallo thionein) Gene in Transgenic Plants

Protein analysis was conducted on putative transformants to detect the presence of the MT protein, protein were labelled with (^{35}s) cysteine. The MT protein was detected in all five of the Kan^R resistant plants. In transformed *N.tabacum* and in *B.napus* a band was observed that co-integrated with MT band. This band was absent in non-transformed cells of *N.tabacum* and *B.napus.*

Heavy Metal Tolerance of Transgenic Plants : (Table 28)

Seed progeny from self pollinated transformants (s1) and control plants were germinated on MS (Murashige and Skoog) media containing 0.1 mM $CdCl_2$. The seedlings were then scored for root length and shoot length. The fresh weight of *N.tabacum* and *B.napus* was reduced to 83 and 49% respectively (Misra and Gedamu, 1989). Similarly, in each case the root length showed 85-84% reduction relative to the seedlings growing on control media. In contrast, the progeny of the selfed transformants segregated for root and shoot growth. A majority of the seedlings grew normally on media containing 0.1 mM $CaCl_2$, after 3 weeks on 0.1. mM $CaCl_2$ the average.

Table 28. Heavy Metal Tolerance of Transgenic B. napus and N. tabacum S.Progeny

Seed sample		*Media*	*Root length (m)*		
N.tabacum	1	– Cd	5.5	±	0.02
B.napus	2	+ Cd	0.91	±	0.18
	1	– Cd	4.01	±	0.5
	2	+ Cd	3.25	±	0.16
		– Cd	10.00	±	3.1
		+ Cd	1.7	±	0.03
		– Cd	9.0	±	0.5
		+ Cd	7.75	±	0.2

1 = Control, 2 = Transformed. (Data represented are means of 100 seed samples of *N.tabacum* and 15 Seeds of *B.napus* each), root length of Cd-tolerant *N.tabacum* Seedlings was 3.25 ± 0.16 cm compared to 0.9 ± 0.1 cm of control seedlings. In *B.napus* the average root length of Cd tolerant seedling was 7.75 ± 1.2 cm compared to 1.7 cm of control seedlings. Also, the fresh weight of Cd-tolerant seedlings of tobacco and *Brassica* remained unaffected whereas the controls were severely affected. (Misra and Gedamu, 1989).

A chisquare analysis conducted on segregating (S1) populations of *N.tabacum* demonstrated that the ratio of tolerant to susceptible plants was 3 : 1.

These transgenic plants show tolerance to toxic levels of Cadmium suggesting that the MT protein synthesised in *B.napus* and *N.tabacum* may be involved in heavy metal detoxification/sequestration. Also in nature, plants differ in their ability to grow in soils containing elevated levels of toxic metals. Natural selection on mine sites has resulted in ecotypes within species capable of growth in the presence of high concentrations of toxic ions (Gemmell, 1977). The Ti-plasmid mediated genetic transformation of MT gene in plants provides a valuable method of reclamation of waste lands and mine soils.

Cloning Genes for Biosynthesis of Stilbene Type Phytoalexins and Key Enzymes of Biosynthesis

The term stilbene describes a group of secondary plant products which shares a common diphenylethylene backbone and some of the examples are :

(1) Resveratrol

(2) Pinosylvin

(3) Hydrangeic acid

(4) Peterostilbene

These basic molecules can be modified in different ways. They are :

(*i*) methylation of hydroxyl groups (methyl ethers),

(*ii*) attachment of sugars to hydroxyl groups (glycosides),

(*iii*) polymerisation *e.g.*, (viniferings)

and a number of other substances may be divided from stilbene backbones by other or additional modifications (Hart, 1981).

Stilbenes have been detected in wide range of unrelated plant families. Often they are found in the heartwood and/or in the sapwood of trees, or in roots and leaves. Synthesis is correlated with stress in many cases, either by wounding or infection (Kemp and Burden, 1986). Most stilbenes are toxic, as highly reactive polyphenols they can inactivate proteins and other macromolecules by formation of bonds. It is thought that their induced formation in plants represents part of the general defence mechanisms against infecting organisms. Thus, stilbenes are considered as phytoalexins.

Key enzymes of biosynthesis of stilbenes are synthesis which catalyze formation of the stilbene backbone and resveratrol synthase (RS) from peanut cell cultures (*Arachis hypogaea*) is the best known example. Chloroplast DNAs and genomic clones have been characterised for different genes which are differentially induced under various stress conditions. One of the genes was introduced in tobacco. It is transcribed and found that whether it is increasing the resistance against pathogen attack in the new host.

Cloning the mRNA of ACC Synthase

Ethylene is a plant hormone that influences many aspects of plant growth and development. Ethylene production is augmented during several stages of plant growth including fruit ripening, seed germination leaf and flower senescence and abscission. It is also induced by a variety of external factors, including the application of toxins, wounding, anaerobiosis, viral infection, chilling injury, drought, Cd+ ions (Abeles, 1973).

The biosynthetic pathway of ethylene was elucidated by the pioneering work of SF Yang and his associates. It has been established that methionin is the biological precursor of ethylene in all higher plants (Yang and Hoffman, 1984). It is converted to ethylene via the following biosynthetic route.

Ethylene biosynthesis :

Methionin ———→ SAM (Adenosyl/methionin) ———→

———→ ACC (Aminocyclopropane–1–Carboxylic acid) ———→

ACC Synthase C_2H_4 (Ethylene)

The rate limiting step is the conversion of SAM to ACC, catalyzed by the enzyme ACC synthase. Induction of ethylene production by a variety of agents is due to de novo synthesis of the enzyme, (Bleecker et al., 1988). It has been partially purified from several sources, with wide discrepancies in its molecular weight. Isolation of the gene encoding ACC synthese has been difficult, in part because the enzyme is unstable and present in low abundance. The cDNA sequence to ACC synthase mRNA is isolated from Cucurbita pepo fruit, which requires partially purified enzyme. The genomic sequences to ACC synthase from cucurbita is isolated.

The cloning of ACC synthase mRNA has been difficult because traditional cloning approaches require that the protein be purified and at least partially sequenced. Unfortunately, ACC synthase has been difficult to purify to homogeneity because of its low abundance (*e.g.,* 0.000 p/o, (Percentage)) of the total protein in ripe tomatoes (Bleecker et al., 1986).

The experimental approach consists of six main steps :

(1) Partial purification of the ACC synthase from a highly induced tissue by conventional protein purification procedures.

(2) Use of the partialy purified enzyme to produce an antiserum to ACC synthase that inhibits enzyme activity.

(3) Purification of the crude antiserum by chromatography with total proteins from uninduced tissue.

(4) Immunoscreening cDNA libraries with the purified antiserum and isolation of putative cDNA clones to ACC synthase.

(5) Identification of the cDNA clones by immuno blotting analysis with antiserum released from putative cDNA clones and

(6) Finally, by expressing and recovering ACC synthase activity in *E.coli* and yeast using putative ACC synthase cDNA clones.

The cloning of the ACC synthase mRNA provides the potential to manipulate endogenous ethylene production and consequently to prevent senescence. This is of importance in agricultural industry, where billions of dollars each year are lost worldwide due to over ripening of fruits and vegetables during transportation from one location to another. It may be possible to reduce ethylene production by reducing expression of the ACC Synthase gene. This could potentially be accomplished by a variety of approaches, including gene disruption (Scherer and Davis. 1979) dominant negative mutation (Herskowits, 1987) expression of antibody genes (Carlson, 1988).

Engineering of a New Flower Colour Variety of Petunia

In most plant species flavonoids are the major components contributing to flower pigmentation. *Petunia hybrida* is one of the best analysed systems where the pathway of anthocyanin biosynthesis has been analysed both genetically and biochemically (De viaming et al., 1984). The basic colouration of the flower is mainly determined by the conversion of the three dihydroflavonols, dihydrokaemferol, dihydroquercetin and dihydromyricetin into pelargonidin cyanidnand delphinidin derivatives. The enzyme dehydroflavonol-4-reductase (DFR) plays a key role in anthocyanin production in *Petunia hybrida* as it accepts both dihydroquercetin and dihydromyricetin as a subtrate, but not dihydro kaempferol (Forkmann and Ruhnau, 1987). For this reason only blue delphinidin derivatives and red cyanidin derivatives are made by petunia vareities, but no pelargonidin pigments which would contribute a brick-red flower colour. DFR-enzymes from other species do not contain the same substrate specificity as the petunia DFR. In maize, DFR which is encoded by the A1 locus, converts dihydroquercetin into leuco pelargonidia (Reddy et al., 1987) but it also accepts dihydrokaemferol as a substrate, which is converted into leucopelargonidin. It has been demonstrated by feeding experiments that in *Petunia hybrida* leucopelargonidin, can be further processed into pelargonidin pigments. A C-DNA of the type 2 A1 gene of *Zea mays* was cloned between the viral 35 s-prometer and terminator and this construct was introduced by direct transfer into protoplasts of petunia line, R101, which accumulates dihydro-kaemferol (Schwars-sommer et al., 1987). Transgenic plants which expressed the maize A1-construct exhibited a brick-red flower colour which was due to the conversion of dichydrokaemp-ferol into pelargonidin-3 Glycosides (Fig. 14) (Meyer *et al.,* 1987).

Crosses of Transgenic A1 Plants with Commerical Varieties

When brick red transformant with small size flowers was crossed to a commercial line of large size flowers and intensive aganidin pigmentation, the F_1 plants displayed a cyanidin-red phenotype with only a minor proportion of pelargonidin. The F_2 generation from a selfed F_1 plant featured three different types of pigmentation: a cyanidin type, a pelargonidin type and a mixed type with a major proportion of cyanidin as already observed in the F_1 plants. These different colours were seen at variable intensities in large as well as small sized flowers, demonstrating the usefulness of this material for commercial breeders.

Fig. 14 The introduction of the A1-gene from *Zea mays* which encodes the enzyme dehydroflavonol-4-reductase (DFR) leads to the production of leucopelargonidin in *Petunia hybrida* which is processed into brick red pelargonidin glycosides.

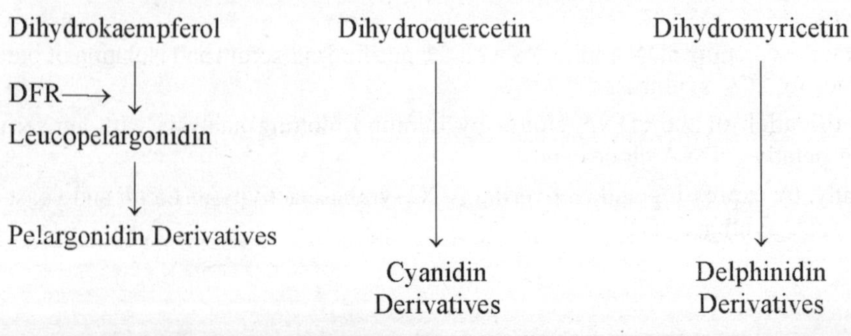

Transgenic Plants Display Strong Differences in A1 Expression and Pelargonidin Production

Among individual transformants which contain the A1 construct have three different phenotypes of flower colouration. Plants which showed pigmentation of the whole flowers were termed "Red", plants expressing the new pigment only in some floral cells were termed "Variegated" and plants which did not show any pelargonidin colouration were named "white".

Variability of A1-Expression

Two different examples were observed which illustrated the variability of A1-expresson. An F_2 plant which derived from the selfing of a red transformant that contained one integrated A1-Copy, developed a branch with whits flowers, where the A1-gene was obviously no longer active.

One variegated plant, R101-16, which contained several copies of the A1-Gene, developed branches with fully coloured flowers (named 16+) next to branches which did not show any A1-expression (named 16–). A branch of each type was cut and rooted to produce a red 16+ plant and a white 16-plant. An analysis showed that both plants contained an identifical pattern of integrated A1-copies. When plant 16– was selfed, the F_1 progeny consisted of white flowering plants. The progeny of a cross16 – X 16 + also was completely white, while selfing of plant 16+ and the cross 16+ × 16– approximately resulted, in a 1:1 ratio of white and red progeny. The data indicate that the A1-Copy which had been activated in plant 16+ is only expressed in the next generation, if plant 16+ is the female partner of the cross. If the gene is transmitted via pollen, it is inactivated again.

Transgenic Animals

In animal breeding modification of germ cells and improvement of the animal are the desired goals. The modification of germ cells by gene manipulation in vitro is possible and this process is given the name transgenesis. Transgenesis is not restricted to animals but be achieved in plants.

Gene Delivery Methods

Regardless of whether somatic or germ cells are to be modified, a method of introducing exogenous DNA into these cells is required.

- Methods of introducing exogenous DNA into Mammalian cells
- Use of viral vectors
- Calcium phosphate mediated DNA uptake
- Micro injection of eggs
- Fusion of DNA loaded membranous vesicles
- To target cells (*e.g.*, liposomes)
- Electroporation (DNA uptake mediated by electric current).

Use of Viral Vectors

A number of different virus vectors are available for cloning in mamalian cells. Each vector system has its advantages and disadvantages. To some extent this is a reflection of the different host ranges of the different viruses. The different viruses vectors are given below :

SV40 Vectors

Simian virus 40 (SV40) was the first eukaryote virus to be harnessed as a vector simply because it was the first such virus for which a complete nucleotide sequence was available. The genome of

SV40 containing very little non essential DNA. So, it is necessary to insert the foreign gene in place of essential viral genes in order not to exceed the size limits imposed by the viral protein coat.

Adenovirus vectors

The major disadvantage with SV40 and polyoma based vector is the small size of the genome. Since pakaging constraints are strict, only small gene inserts can be made. Adenoviruses are much larger than SV 40 and can accomodate much more extraneous DNA.

Vaccinia vectors

Vaccinia virus is used in a slightly different way. Exogenous genes are first cloned adjacent to a vaccinia promotor present in the middle of a vaccinia derived thymidine kinase gene which is carried on an E.coli plasmid. After manipulations are complete in E.coli, the recombinant plasmid and wild type vaccinia virus are co transfected into a Monkey cell line. (Fig. 15)

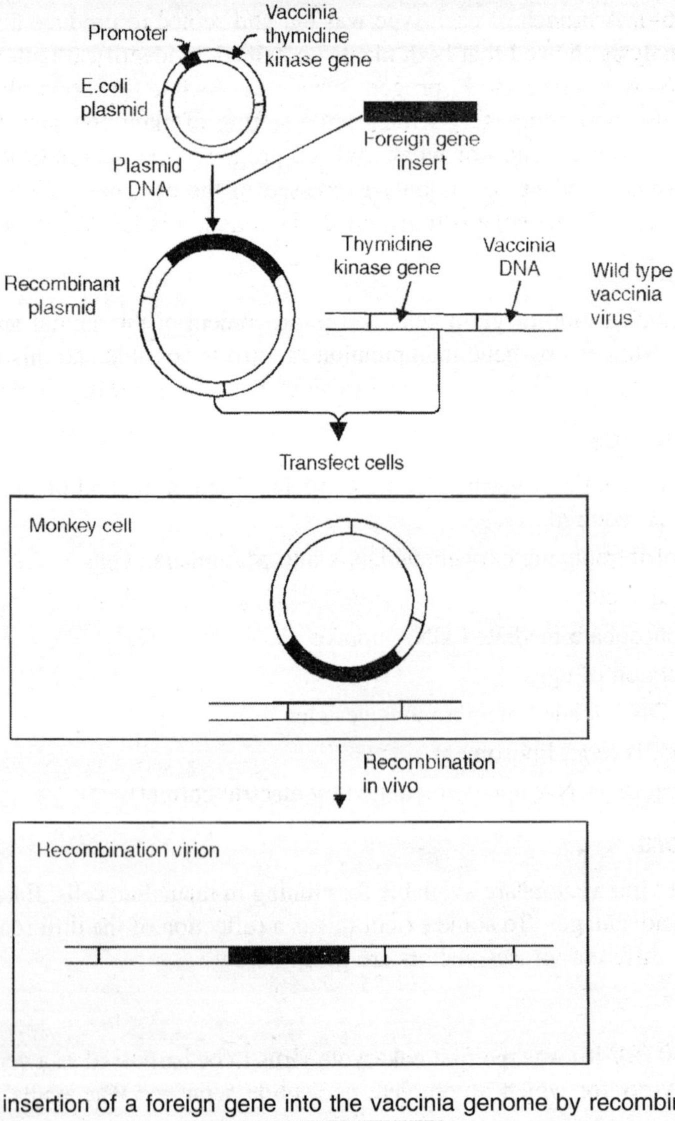

Fig. 15. The insertion of a foreign gene into the vaccinia genome by recombination in vivo.

BPV Vectors

Vectors based on bovine papilloma virus (BPV) are unique in that not only are they propagated as multicopy plasmids but they cause neoplastic transformation of the host cell.

Retrovirus Vectors

Retrovirus have single stranded RNA genomes and would appear to be unpromising candidates as vectors but the peculiarities of the retrovirus life cycle mean that these viruses provide perhaps the most promissing vectory system of all.

Calcium Phosphate Mediated DNA Up Take

Introduction of unpackaged DNA into animal cells in culture is known as transfection. Fragments of DNA can be phagocytosed by means of the calcium phosphate co-precipitation method and in at least a few of the transfected cells will become stably integrated at random in the genome.

Microinjection of Eggs (Table 29)

Microinjection is a widely used technique but is a laborious procedure as only one cell at a time can be injected, it has also a high failure rate. Even in the hands of experienced operators many of the eggs are damaged.

Table 29. Efficiency of production of transgenic animals by micro injection of a growth hormone gene. (Hammer et al., 1985)

Animal species	*Number of Ova injected*	*Number of off-spring*	*Number of transgenic off-spring*
Rabbit	1907	218	28
Sheep	1032	73	1
Pig	2035	192	20

Electroporation and Cell Fusion

They are not used much because little is known about the mechanics of the processes.

Production of Transgenic Animals

The aim of transgenesis is to produce animals with a heritable change in their genotype such that the benefits of gene manipulation can be passed to their offspring. With animals the technique used is the micro injection of DNA into fertilize eggs followed by implantation of the manipulated ova in a foster mother. For experimental purposes the mouse is a particularly appropriate choice of animal a superovulated mouse can yield upto 40 eggs; reimplantation is relatively easy and a mouse can carry upto 20 offspring.

Cloned plasmid DNA is injected into one of the pronuclei of the newly fertilised egg. The male pronucleus derived from the sperm is larger than the female pronucleus and so it is chosen for injection. The two pronuclei subsequently fuse to form the diploid zygote nucleus of the fertilized egg. The injected embryos are cultured in vitro. Until they have undergone a number of divisions and are then implanted in pseudo pregnant foster mothers. Between 3% and 40% of the animals, developing from these embryos contain copies of the exogenous DNA integrated into their chromosome. Most experiments on transgenesis have made use of the mouse metallothionein (MMT) gene promotor. MMT encodes a cysteine rich polypeptide which is involved in heavy homeostasis. By coupling the MMT promotor to a Herps simplex thymidine kinase (TK) gene it is possible to raise transgenic mice in which TK activity shows a tissue distribution similar to MMT and is inducible by cadmium ions. This experiment shows that it is possible for a foreign gene, in this case from a virus, to be integrated into the genome of the mouse (Particularly TK gene).

Another example is the coupling of the MMT promotor to a rat growth hormone gene. Transgenic mice carrying this gene construct grow significantly faster and larger than their litter mates. In some cases, the transgenic mice grow to almost double the weight of normal mice. In ideal transgenesis experiment, the promotor selected should be one which enables the normal cellular control patterns to be exerted on the gene of interest, *i.e.,* the gene should be switched on and off at the correct time and in the correct tissues. Success of this kind has been achieved with transgenic mice carrying the human insulin gene. Not only was gene expression restricted to the cells of pancreas but serum levels of human insulin were properly regulated by glucose and amino acids.

Expression of Foreign Genes in the Lactating Mammary Gland of Transgenic Animals

Our primary goal is the large scale manufacture of human proteins for eventual use in the pharmaceutical industry. The mammary gland of transgenic animals is an ideal vehicle for the large scale production of exogenous proteins. The protein would ideally be produced in the lactating mammary gland and be secreted into milk from which it could be purified. Milk is an ideal vehicle for the production of exogenous proteins because milk is an abundant body fluid, and its removal is simple. So, mouse models for the mammary expression of tissue plasminogen activator (TPA) and protein C were developed. TPA is a fibrinolytic protein (enzymatic protein) and has been shown to be efficacious in the treatment of myocardinal infection. Protein C acts invivo as an anticoagulant (Church, 1989).

Antifreeze Protein (AFP) Expression in Transgenic Salmon

In general, fresh water freezes close to 0 °C but the body fluids of most fishes (Teleotas) freeze at – 0.5 °C to – 0.8 °C owing to their salt content. This indicates that fish inhabiting fresh water are in no danger of freezing, but those living in sea water, where the salt concentration in approximately three times that of fish serum and the freezing point ranges from – 1.7 °C to – 2 °C can encounter waters which are a degree or more below their freezing temperatures. Such freezing conditions exists in polar regions and neighbouring waters for at least part of the year, and yet fishes native to these waters do not freeze and die. Scholander et al., (1957) studied the above phenomenon in number of marine fish inhabiting the coast of Northern Labrador. Devries et al., (1970) reported that proteinaceous macro molecules were responsible for the antifreeze activity in fish sera. Four distinct classes of antifreeze proteins (AFP) differing in carbohydrate content, amino acid composition, protein sequence and secondary structure have been isolated from the plasma of diverse marine teleotas (Devries et al., 1988). Winter flounder are protected from freezing in ice-laden sea water by high concentrations of antifreeze protein (AFP) that circulate in the blood. Atlantic Salmon do not have the benefit of protection from these macro molecular antifreezes and will freeze at temperatures below – 0.7 °C. The lack of AFP in salmon poses a severe limitation of their aquaculture in Seapens along the atlantic coast. In an effort to improve the freezing resistance of this species AFP gene from winter flounder have been transferred to salmon by micro injection of AFP genes into Salmon eggs.

Transgenesis in Large Animals

While transgenic mice are of interest of an experimental tool, the technique needs to be applied to farm animals. Working with large domestic animals is much difficult than with mice. The reasons are :

(1) They do not produce as many eggs *e.g.*, an ewe on average only gives three or four.

(2) Reimplantation of manipulated embryos is more difficult because sheep and cattle will not give birth to more than two offspring.

(3) The eggs of many domestic animals have such an opaque cytoplasm that it is impossible to see their pronuclei or nuclei without restoring to special techniques.

The enhanced growth of mice after the transfer of the human growth hormone gene is an effect that has been duplicated in other animals. For example, the pigs expressing high levels of growth harmone do not grow bigger, instead they develop anthritis–like symptoms and various other pathalogies. So, better control of gene expression is required or even a more appropriate gene. The results with human insulin expression in mice suggest that ultimately success will be achieved with animals of commercial value. For that certain histocompatibility genes might be selected carefully and could be used to enhance disease resistance (Church, 1989).

Conclusion

Transgenic plants visibly exhibiting desirable traits have been developed by plant gene transfer. Now financial support for genetic engineering of crop plants have been reinforced by the novel methods available and considerable increase in research activity in plant biotechnology can confidently be expected in the coming years. But the attainment of commercial goals will undoubtedly require expanded efforts towards understanding of basic processes of gene regulation and function in plant growth and development. As far as the animal transgenesis is concerned, transgenic biological production units for the synthesis of important peptides in milk has been accomplished in mice, goat and sheep. Transgenic animals have been generated to produce novel proteins by proteins by the incorporation of a fusion constructs containing a gene of interest.

REFERENCES

Abeles, F.B. (1973). Ethylene in Biotechnology. Academic Press, New York.

An, G.,B. Watson, S. Stachel, M. Gordon and E.Nester (1984). New Cloning Vehicles for transformation of higher Plants, EMBO. J., 4 : 277–284.

Anderson, E.J., D.M. Stark, R.S. Nelson, P.Powell, N.E.Tumer and R.N. Beachy (1989). Transgenic plants that express the coat protein genes of TMV of ALMV interface with disease development of some non-related viruses. Phytopathol.

Austin, R.B., R.B. Flavell, I.E. Henson and H.J.B. Lowe. (1986). "Molecular Biology and Crop Improvement". Cambridge University Press.

Barker, R., K.Idler, D.Thompson and J.Kemp. (1983). Nucleotide sequence of the T-DNA region from the Agrobacterium Plant. Mol. Biol., 2 : 335-350.

Bleecker, A.B., W.H. Kenyon, S.C. Somer Ville and H.Kenden. (1986). Use of monoclonal antibodies in the purification and Characterisation of 1-aninocyclopropane-1-carboxylate. Synthase, an enzyme in ethyleno bio synthesis. Proc. Natt. Acad. Sci. (USA), 83 : 7755.

Bleecker, A.B., G.Robinson and H.Kende (1988). Studies on the regulation of 1-aminocyclopropane-1-Carboxylate Synthase in tomato using monoclonal antibodies. Planta., 173 : 385.

Brisson, N. and T. Hohn. (1986). Plant virus Vectors : Cauliflower mosaic virus. Methods. Enzymol, 118 : 659-668.

Carlson, J.R. (1988). A new means of inducibly inactivating a Cellular protein. Mol. Cell. Biol., 8 : 2638.

Church., R.B. (1989). Transgenic Models in Medicine and Agriculture : UCLA Symposia on Molecular and Cellular Biology Vol. 116. John Wiley and Sons, Inc. Publication, New York.

Comai, L., D. Facciotti, W.R. Hiatt, G.Thompson, R.Rose and D.Stalker (1985). Expression in plants of a mutant or A gene from Salmonella typhimurum confers resistance to Glyphosate. Nature (Lond)., 317:714–744.

Crossway, A., J.V. Oakes, J.M. Irvine, B.Ward, V.O.K.Aauf and C.K.Shewmaker. (1986). Integration of foreign DNA following micro injection of tobacco mesophyll protoplasts. Mol. Gen. Genet. 202:179–185.

Deak, M., G.B.Kiss, C.Koncz and D.Dudits. (1986). Transformation of Medicago by Agrobacterium mediated gene transfer. Plant Cell Reports. 5 : 97–100.

De Block, M., L. Herrera–Estrella, M.Van Montagu, J.Schell and P.Zambrysti. (1984). Expression of foreign genes in regenerated plants and their progeny. EMBO.J., 3 : 1681.

De Cleens, M. and J.Delay. (1976). The host range of crown gall. Botanical. Rev. 42 : 389.

De Viaming, P., A.G.M. Gerats, H.Wiering, H.J.W. Wijaman, A.Cornu, E.Farey and D.Maizonnier. (1984). Petunia hybrida: a short description of the action of Al, genes their origin and their map location. Plant Molec. Biol. Rep. 2 : 21–42.

Eden, L., L.Heslinga and R.Klok. (1982). Cloning of *c* DNA encoding the Sweet. Tasting plant protein thaumatin and its expression E.Coli. Gene (Amst)., 18 : 1–12.

Eggenberger, A.L., D.M. Stark and R.N. Beachy. (1989). The nucleotide sequence of a Soybean mosaic virus coat protein coding region and its expression in E.Coli. Agorbacterium tumefaciens, and tobacco callus. J.Gen. Virol.

Fillatti, J., J.Kiser, R.Rose and L.Comcei. (1987). Efficient transfer of a glyphosate tolerance gene into tomato using a binary Agrobacterium tumefaciens vector. Bio. Technol., 5 : 726–730.

Forkmann, G. and B.Ruhnau. (1987). District substrate specificity of dihydro flavonol-Reductase from flowers of Petunia hybrida. Z. Naturforsch., 42 : 1146-1148.

Fraley, R.T., S.G. Rogers, R.B. Horsch, P.R. Sanders, S.P. Adams, M.L. Bittner, L.A. Brand, C.L. Fink, C.R. Galluppi, S.B. Goldberg, N.L. Hoffmann and S.C. Woo. (1983). Expression of bacterial genes in plant cells. Proc. Natl. Acad. Sci. USA., 80 : 4803-4807.

Francki, G.I.H., R.G., Milne and T.Hatta. (1985). Atlas of plant Viruses. Vol. I & II. C.R.C. Press. Inc. Boca Raton.

Fromm, M., L.P. Taylor and V.Walbot. (1985). Expression of genes transformed into monocot and dicot plant cells by electroporation. Proc. Natl. Acad. Sci. USA. 82 : 5824-5828.

Fromm, M., L.P. Taylor and V.Walbot. (1986). Stable transformation of maize after gene-transfer by electroporation. Nature., 319 : 791-793.

Gatehouse, A.M.R., J.A. Gatehouse, P Dobie, A.M. Kilminster and D.Boulter. (1979). Biochemical basis of insect resistence in Vigna unguiculata. J. Sci. Food. Agric., 34 : 345.

Gatehouse, A.M.R. and D. Boulter. (1983). Assessment of the Antimetablic effects of trypsin inhibitors from cowpea (Vigna unguiculata), and other legumes on development of the Bruchid Beetle (Callosobruchus maculatus). J. Sci. Food. Agr., 34 : 345.

Gemmell, R.P. (1977). "Colonization of industrial waste land". London : Arnold., PP1-67.

Grill, E., E.L. Winnacker and M.H. Zenk. (1985). Phytochelatins: the Principle heavy-metal complexing peptides of higher plants. Science., 230 : 674-676.

Hain, R., P.Stabel, A.P.Czernilofsky, H.H.Steinbiss, M.Herrera Estrella and J.Schell. (1985). Uptake, integration, expression and genetic transmission of a selectable Chimeric gene by plant protoplasts. Mol. Gen. Genet., 199 : 161-168.

Hammer, R.E., C.G. Pursel, C.E.Rexroad, R.J. Wall. D.J Bolt, J.M. Ebert, R.D. Paimiter and R.L. Brinster (1985). Production of transgenic rabits, Sheep and pigs by micro injection, Nature, 315 : 680-683.

Herskowitz, I. (1987). Functional inactivation of genes by dominant negative mutations. Nature., 329 : 219.

Hilder, V.A., A.M.R. Gatehouse, S.E. Sheerman, R.D. Barker and D. Boulter (1987). A novel mechanism of insect resistance engineered into tobacco. Nature. 330 : 260.

Horsch, R.B., R.T. Fraley, S.G. Rogers, P.R. Sanders, A. Lloyd and N. Hoffmann. (1984). Inheritance of functional Foreign genes inplants. Science., 223 : 496–498.

Horsch, R.B.J. Fry, N.L. Hoffmann, M. Wall roth, D. Eichhollz, S.G. Rogers and R.T. Fraley (1985). A Sample and general method for transferring genes into plants. Science., 227 : 1229-1231.

Khush, G.S. (1991). Approaches and Strategies in Breeding for pest resistance. Golden Jubilee Symposium on Genetic Research and Education: Current trends and the Next fifty years, Abstracts : 1 : 101-102.

Lamb, C.J. and R.N. Beachy (1989). Plant Gene transfer : UCLA Symposia on Molecular and Cellular Biology. Vol. 129. John Wiley and Sons, inc. publication, New York.

Lawrence, W.A. and D.R. Davies, (1985). A. method for the micro-injection and culture of protoplasts at very low densities. Plant Cell Reports: 4 : 33-95.

Lloyd, A.M., A.R. Baramaspm. S.G. Rogers, M.C. Byrne, R.T. Fraley and R.B. Horsch. (1986). Transformation of Arabidospsis thaliana with Agrobacterium tumeifaciens using a gene conferring hygromycin resistance. Science.

Lorz, H. S. Banker and J. Schell (1985). Gene transfer to Cereal Cells mediated by protoplast transformation. Mol. Gen., 199 : 178-182.

Mariotti, D., M.R. Davey, J. Draper, J.P. Freeman and E.C. Cocking (1984). Crown gall. Tlumorigenesis in the forage legume Medicago sativa. L. Plant. Cell. Physiol., 25 : 473-482.

Meyer, P., S. Kartzte, I. Niedenhof, I. Heidmann, K.Bussmann and H. Saeedles (1988). A genomic DNA segment from Petunia hybrida leads to increased transformation frequencies and simple integration patterns, Proc. Natt. Acad. Sci. USA., 85 : 8566-8572.

Misra, S. and L. Gedmau. (1989). Heavy metal tolerant transgenic plants of N. tabacum and B. napus. Theor. Appl. Genet.

Murai, N., D.W. Shutton and M.G. Murai, (1983). Phaseolin gene from bean is expressed after transfer to sunflower via. tumour-inducing plasmid vectors. Science., 222 : 476-482.

Nelson, R.S., P.A. Powell and R.N. Beachy (1981). Lesions and Virus accumulation in inoculated transgenic tobacco plants expressing the coat protein gene of tobacco mosaic virus. Virology., 158 : 126.

Nelson, R.S., S.M. Mc Cormick, X. Delannay, P. Dube, J. Layton, E.J. Anderson, M. Kaniwska, R.K. Proksch, R.B. Horsch, S.G. Rogers, R.T. Fraley and R.N. Beachy (1988). Virus tolerance, plant growth, and field performance of transgenic tomato plants expressing coat protein from tobacco mosaic virus. Bio. technology., 6 : 403.

Nester, E., M. Cordone, R. Amasino and M. Yanofasky. (1984). Ann. Rev. Plant. Physiol., 35 : 387-413.

Ogata, C., M. Hatada, G.Tomlinson, W.C. Shin and S.H. Kim. (1987). Crystal structure of the intensely wheat protein monellin. Nature (Lond)., 328 : 739-742.

Ohta, Y. (1986). High efficiency genetic transformation of maize by a mixture of pollen and exogenous DNA. Proc. Natl. Acad. Sci. USA., 83 : 715-719.

Petolino, J.G. and G.B. Collins. (1984). Cellular approaches to environmental stress resistance. In : Collins, G.B. and J.G. Petolino (Eds) : "Applications of Genetic Engineering to Crop improvement :. Junk publishers, pp. 341-499.

Pons, F. and G. Bruening. (1986). Mechanisms of resistance to plant Viruses. Ann. Rev. Phytopathol 24 : 355.

Potrykus, I., J. Petruska, J. Paszkowski, and R.D. Shillito. (1985). Direct gene transfer to cereals to cells of graminaceous monocot. Mol. Gen. Genet., 199 : 183-188.

Potrykus, I. and R.D. Shillito. (1985). Protoplasts : Isolation culture and plant regeneration. In : Weissbach, A. and H. Weissbach (eds.) Methods in Enzymology, 118. Academic Press. USA.

Powell, P.A., P. Sanders, R.T. Fraley and R.N. Beachy, (1989). Protection against tobacco mosaic virus in transgenic Plant requires accumulation of capsid protein rather than capsid protein RNA sequences. Proc. Natl. Acad. Sci. USA.

Powell, P.A., R.S. Nelson, B.De., N.H. Offmann, S.G. Rogers, R.T. Fraley and R.N. Beachy (1986). Delay of disease development in transgenic plants that express the tobacco mosaic virus coat protein gene. Science., 232 : 738.

Primrose, S.B. (1987). Modern Biotechnology, Black well publication. Oxford.

Puertas, M.J., A. Dela Pena, B.N. Estades and F.Merino. (1984). Early sensitivity to colchicine in developing anthers of rye. Chromosoma. 89 : 121-126.

Reddy, A.R., L. Britsch, F. Salamini, H. Saedler and W. Rohde. (1987). The A1 (Anthocyanin-1) locus in Zea mays encodes Dihydroquercertin Reductase. Plant. Sci., 52 : 7-12.

Register III, J.C. and R.N. Beachy (1988). Resistance to TMV in transgenic plants results from interference with an early event in infection. Virology., 166 : 524.

Register III, J.C., P.A. Powell, R.S. Nelson and R.N. Beachy (1988). Genetically engineered cross protection against TMV. Interferrers, with initial infection and long distance spread of the virus. In. Stagkawicz, B., P.A. Hyquistand, O.Y. Oder (eds) : "Molecular Biology of plant–pathogen interactions" Alan R. Liss, Inc. New York., P.269.

Ryan, C.A. (1973). Proteslytic enzymes and their inhibitors. Ann. Rev. Plant. Physiol., 24 : 173.

Scherer, S. and R.W. Davis. (1979). Replacement of Chromosome Segments with altered DNA sequences constructed in vitro-proc. Natl. Acad. Sci. USA., 76 : 4951.

Schroder, G., S. Waffenschmidt, E. Weiler and K. Schroder (1984). The T-region of Ti plasmids codes for an enzyme synthesizing indole-3-acetic acid. Eur. J. Biochem., 138 : 387-391.

Schwarz-Sommer, Z., Shepherd, E. Tacke, A. Gierl, W. Rohde, L. Lectereq, M. Mattes, R. Berndtgem, A. Peterson and H. Saedler (1987). Influence of transposable elements on the structure and function of the A1 gene of Zea mays. EMBO. J., 2 : 287-294.

Sherlock, J.C. (1984). Cadmium in foods and the diet. Experientia., 40 : 152-156.

Shillito, R.D., M. W. Saul, J. Paszkowski, M. Muller and I. Potrykus (1985). High efficiency direct gene transfer to plants. Biotechnology. 3 : 1099-1103.

Smirnov, S.P., E.C. Teveroyskaya, V.A.P. Uhalsky, L.V. Krashe ninnikova, G.V. Rassadina and L.M. Khromova. (1991). Introduction into the Tobacco and Potato Plants of recombinant Human Alfa Interferoncrene, Golden Jubilee symposium on Genetics; Research and Education; Current and the next fifty years. Abraotes ; 1 : 843-844.

Stark, D.M. and R.N. Beachy (1989). Protection against potyvirus infection in transgenic plants. Evidence for broad Spectrum resistance-protoplasma.

Stinbiss, H.H. and P.Stabel (1983). Protoplast derived tobacco Cells can survive capillary micro injection of the fluorescent dye lucifer yellow. Protoplasma. 166 : 223-227.

Vack, M., A. Reynaerts, H. Hofte, S. Jansens, M.D. De Beukleer, C. Dean, M. Zabecm, M.V. Montagu and J. Leemans (1987). Transgenic plants protected from insect attack. Nature., 328 : 33.

Varshney, U. and L. Gedamu. (1984). Human metallothionein MT-1 and MT-11. Processed genes. Gene., 31 : 135-145.

Wilson, T.M.A. (1984). Cotranslational disassembly of tobacco mosaic virus in Vitro. Virology., 137 : 255.

Yamada, Y., Y. Zhi-Qi and T. Ding-Tai (1986). Plant regeneration from protoplast derived callus of rice. (Oryza sativa L), Plant. Cell. Rep. 5 : 85-88.

Yang, S.F. and N.E. Hoffman. (1984). Ethylene biosynthesis and its regulation in higher plants. Ann. Rev. Plant. Physiol., 35 : 155.

Zambryski, P., H. Joos, C. Genetello, J. Lecmans, M. Van Montagu and J. Schell (1983). Ti plasmid vector for the introduction of DNA into plants cells without alteration of their normal regeneration capacity. EMBO. J., 2 : 2143.

CHAPTER 16

Protoplast Fusion and Somatic Hybridization

Introduction

The application of somatic cell hybridization to crop improvement became a possibility with the discovery by Cocking (1960) that large numbers of protoplasts could be released through use of cellulose from the fungus *Myrothecium verrucaria*, the demonstration by Take, be et al. (1971) that isolated tobacco leaf protoplasts could be regenerated into complete plants, and the development of the polyethelene glycol (PEG) fusion method by Kao and Michayluk (1974) and Wallin et al. (1974). These pioneering efforts and many others have advanced experimentation on somatic hybridization towards transferring genes for crop improvement. Any two protoplasts, no matter how distantly related can be fused. First somatic hybrid plants were recovered by Carlsons et al. (1972) between *Nicotiana tabacum* X *N.longsdorfii*.

In early protoplast fusion studies, workers concentrated on making very wide hybridization. In these wide combinations, as was observed in mammalian cell hybrids, selective elimination of one or the other parental chromosome occurred. In addition, regeneration capacity was either reduced or absent in such hybrids. Those hybrid cell lines from which plants were regenerated produced mostly infertile plants, thus precluding their integration into crop breeding programmes. These problems associated with somatic hybridization subsequently led researchers to concentrate on producing hybrids between more closely related species. Currently, the greatest promise for crop improvement via protoplast fusion is thought to be through production of cytoplasmic hybrids (Cybrids) for transfer of agriculturally useful traits that are cytoplasmically encoded or through directed transfer of a few nuclear genes from one species to another (Evans, 1983). Both approaches circumvent the gross genetic incompatibilities often associated with fusing together the entire genomes of two parental cells. These approaches combined with the ever-expanding list of crops that can be regenerated from protoplasts promise to produce novel germplasm that could be of value of plant breeders for integration into crop improvement programmes.

Sources of Protoplasts

Protoplasts are prepared from tissues of various plant organs – leaves, roots, tubers, petals, fruits, leguminous plant root nodules – and from tumors of different origin and tissues and cells cultivated in vitro. Protoplasts can be isolated from microspores at the tetrad stage or from unicellular microspores.

Leaf mesophyll is commonly used to obtain large amounts of homogeneous and homoploidal protoplasts. From one gram of potato leaf mesophyll 5 – 8 + 10^6 protoplasts can be isolated, of which 70-80% are viable. The age of the plant, the age of the leaf, and the growth conditions affect the quantity of viable protoplasts and their behaviour in cultivation. Leaf mesophyll protoplasts can be readily separated and purified from tissue and cell debris, but sometimes they may fail to divide.

Protoplasts isolated from callus or suspension cultures are better adapted to division and growth on nutritive media. But their disadvantage are morphological heterogenity and heteroploidy. The separation of protoplasts and cells in suspensions is sometimes easy, but in other cases it is very difficult. Cell suspensions growing at a high rate maintained by frequent subculturing (every 3 days) into fresh media are preferable.

Isolation of Protoplasts

Protoplasts can be produced through mechanical means, but the yields are low and the resulting protoplasts are highly vacuolated and have reduced viability. Digestion of the cell wall with enzymes that degrade the cellulose, hemicellulose, and pectin components has become the preferred method for rapid isolation of viable protoplasts in high yield. (Fig. 16).

The conditions for isolating protoplasts, that is, the type and concentration of enzymes and osmotic stabilizer, pH, temperature, fine and the use of protectors, determine the quality and quantity of the resulting protoplasts. The goal is to degrade the cell wall without impairing the living protoplasts. This is achieved by creating a water stress resulting in plasmolysis. The plasmolysis depends on the type of plasmolytic agent used and its concentration ; 0.45 – 0.8 M mannitol or sorbitol is routinely used, as is a mixture of 0.3 M sorbitol and 0.3 M mannitol (Kao et al., 1974). The application of metabolising osmotic agent, sucrose or glucose can be more advantageous, as a decrease in its concentration during the cultivation of protoplasts can be beneficial for division and growth of the cells (Kao et al., 1974).

On the contrary, when the aim is to decrease the influx of an osmotic stabilizer in the cell, a high molecular-weight substance, for example, polyethlene glycol (PEG) (MW, 3000) is used. Mineral salts are not frequently employed for this purpose, because in the presence of certain salts some of the enzymes used for degradation of the cell decrease their activity, besides, salts readily penetrate into the cell. Protoplasts were successfully isolated under comparatively low osmotic pressure provided by 0.3 M sucrose with 2% polyvinyl pyrolidone that served as a stabilizer (Shepard and Toiten, 1977).

Enzyme Preparations

The primary cell wall can be degraded by hydrolyticenzymes, that is, cellulose, hemi cellulase and pectinase. The composition of the enzyme mixture, pH, and duration of treatment of a tissue are derived empirically with due regard to the species, age and physiological state of the plant. The age of the plant or tissue determines the composition of the primary cell wall, the older the plants or tissue, the higher the cellulose content and the lower the hemic cellulose content. When the aromatic polymer appears in the cell wall the isolation of plant protoplasts by enzymatic techniques becomes impossible.

Consecutive and simultaneous enzymes treatments of the tissue are also employed. With the former method, (Takebe et al., 1971) a tissue is first treated with pectinase, to induce maceration into individual cells, and then placed on a solution containing cellulase to obtain isolated protoplasts. With the latter method (Cocking, 1960) the tissue is treated with pectinase and cellulase at the same time, and hemi cellulase is sometimes added. In this case, maceration of the tissue into the cells and isolation of the protoplast occur simultaneously.

The advantage of consecutive treatment is the reduced possibility of spontaneous fusion of protoplasts and a more homogeneous population from leaf tissue, since the cells of spongy parenchyma are destroyed. The disadvantage of the method is that the protoplasts inside the cell may undergo-degradation during maceration of the tissue. The disadvantage of the simultaneous treatment is greater spontaneous fusion of protoplasts, results in a large number of multinucleate cells.

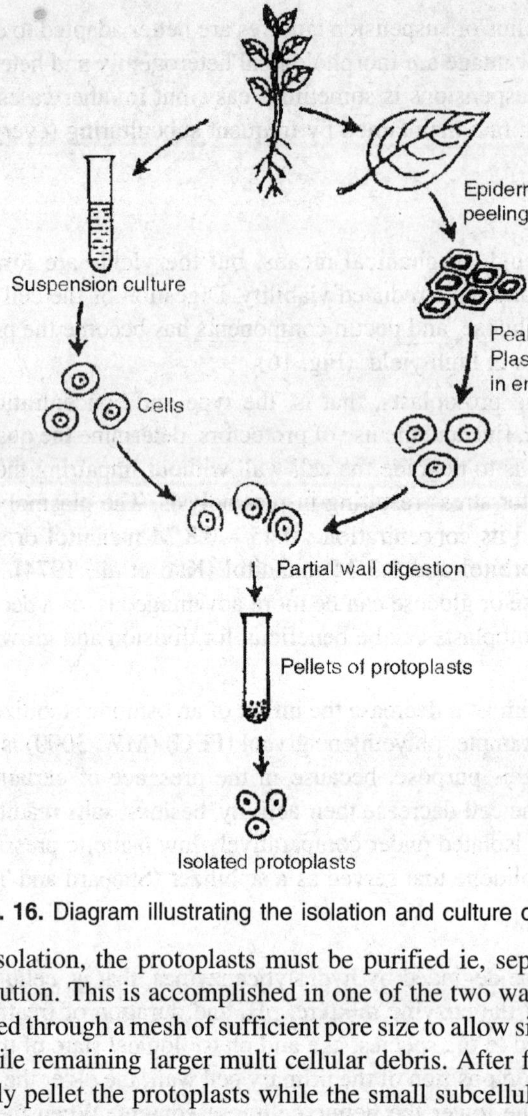

Fig. 16. Diagram illustrating the isolation and culture of plant protoplasts.

Following isolation, the protoplasts must be purified ie, separated from the undigested tissue and the enzyme solution. This is accomplished in one of the two ways. First, the unpurified protoplast mixture can be passed through a mesh of sufficient pore size to allow single protoplasts and cell fragments to pass through while remaining larger multi cellular debris. After filteration, the collected mixture is centrifuged to gently pellet the protoplasts while the small subcellular debris remain suspended.

Through repeated centrifugation, the protoplasts are eventually purified. However, filtration, particularly using stainless steel filters, may damage fragile protoplasts. The second commonly used method, floatation may be preferable for fragile protoplasts. The second commonly used method, floatation maybe preferable for fragile protoplasts. In this method, the crude protoplast mixture is centrifuged to sediment protoplasts and debris. Then the mixture is centrifuged again in a solution containing a high concentration of sucrose (Gamborg et al., 1981). Viable, intact protoplasts float after this second centrifugation and can be removed using a pipette and subsequently diluted with culture medium.

Isolated protoplasts are thoroughly washed from the enzyme solution with a medium containing as osmotic stabilizer and calcium chloride. The optimal conditions for isolation, washing and purification (centrifugation in the washing medium) are determined by calculating the number of viable protoplasts.

Viable protoplasts are detected with the use of Evans blue, methylene blue, or fluorescein disacetate (Larkin, 1976).

Methods for Fusion of Plant Protoplasts

(*a*) **Spontaneous Fusion.** Once the protoplasts obtained from leaf mesophyll or cell culture are purified, they can be fused. Indeed, lacking a cell wall, the protoplasts often fuse spontaneously, apparently through expansion of the plasmodermata of adjacent cells (Boss et al., 1983) in enzyme solutions or of meiotic protoplasts through simple contact and without the help of any inducing agents. Such homokaryotic prototplasts can be stimulated tc divide and to form plants. The frequency of spontaneous fusion differs and seem to be dependent on the plant species. (Figure 17).

(*b*) **Induced Fusion.** Chemical fusion methods have been used often for somatic hybridization studies. Early success was reported by Carlson et al. (1972) utilizing sodium nitrate as fusogen. However, since sodium nitrate was found to be toxic to protoplasts at fusogenic concentrations and caused a fusion frequency only slightly higher than that of spontaneous fusion, researchers subsequently turned to other methods. With the method of Keller and Melchers (1973), fusion rates greater than 25% were obtained by combining a pH of 10.5 and high Ca^2 + (50 m M) concentration in buffer solution.

The most effective cell fusion method to date is an extension of the high pH./Ca^{2+} method. Kao and Michaluk (1974) first reported that polyethylene glycol (PEG) agglutinates protoplasts and facilitates fusion. Generally, the protoplasts to be fused are mixed and aggregated with PEG. During elution in the presence of high pH and Ca^{2+} the cell membrane is altered so that protoplasts fuse (Keller and Melchers, 1973).

Critical factors for success in the fusion process include the molecular weight of the PEG (the optimum is between 1000 and 6000) and the PEG concentration (ideally 25-50%). Temperature, which affects the fluidity of the cell membrane, should be carefully controlled (Klebe and Mancuso, 1981). Also the protoplasts should be preincubated (allowed to settle) before the PEG solution is applied. The duration of the PEG treatment required is dependent upon the sensitivity of protoplasts being treated (Glimelius et al., 1981). Fusion frequencies of 100% can be achieved using the PEG method ; however, to avoid multiple fusions, most researchers alter conditions so not to exceed a 10-20% fusion rates.

The frequency of polyethlene glycol (PEG) - induced fusion can be enhanced by addition of concanavalin A (Con A) to the PEG solution. The con A reportedly strengthens physical attachment of the protoplasts to surface, allowing for greater recovery of fusion products during the washing procedure. Addition of dimetheyl sulfoxide (DMSO) to the PEG solutions also reportedly increases the fusion frequencies by making the protoplasts more susceptible to the PEG. Good fusion frequencies have also been reported following treatment with a positively charged phospholipid, 1, 2-0 dipentadecyl-methylidene glycerol-3-hosphoryl (N-ethylamino) - ethianolanine, PEG - induced fusion are non specific and are therefore useful for intra and interspecific fusion, for fusion of animal cells, and for inter kingdom fusion of mammals and plant protoplasts (Vasil et al., 1975). The actual mechanism of fusion is not clearly understood although some theories have been proposed (Kao et at., 1974).

Other chemicals have also been observed to promote cell fusion. 15% solution of poly vinyl alcohol (PVA) in combination with 0.05 m and $CaCl_2$ and 0.3 M Mannitol may be used to fuse plant cells. PVA is a nonionic surfactant and is apparently not toxic as cells can grow in upto 2% PVA. PVA is also available with different molecular weights and it has been shown that for fusion, PVA with (molecular weight) MW-500 is preferable to either 1400 or 2000. Dextran sulfate has also been used to induce protoplast fusion, which has limited use due to its toxicity to plant cells. Lectins are also known to agglutinate protoplasts, but it has been not tested as fusogenic agent.

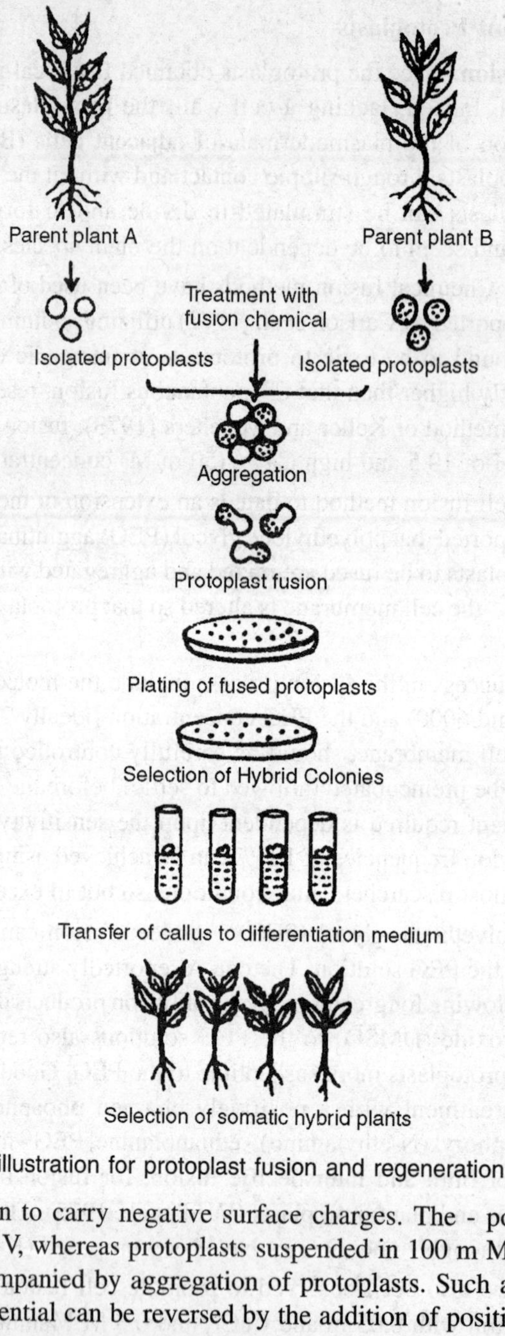

Fig. 17. Diagramatic illustration for protoplast fusion and regeneration of somatic hybrids.

Protoplasts are known to carry negative surface charges. The - potential of normal tobacco protoplasts is - 25 to - 30 m V, whereas protoplasts suspended in 100 m M $CaCl_2$ have a potential of 0 mV. This decrease is accompanied by aggregation of protoplasts. Such aggregation is essential for fusion to take place. The-potential can be reversed by the addition of positively charged polymer.

Plant Protoplasts Fusion by Electro Manipulation

Since the first reports of Zimmerman and Scheurich (1981) electro, fusion, that is cell fusion in an electric field, has become of great interest as powerful technique for the fusion of plant, animal and microbial cells, as it is simpler, more rapid and less harmful to the cells, than the conventional PEG method, and is available for mass production of fusion products. It requires alternative current (AC) for cell alignment.

This method has certain definite advantage over chemical fusion in that it is.

(1) non toxic,

(2) encourages higher fusion frequency, and

(3) is reproducible.

Electric fusion is based on the reversible electrical breakdown of cell membrane (Zimmerman and Scheurich, 1981). This breakdown results in a tremendous increase of the membrane permeability and conductivity. As a consequence, fusion is initiated where electric field pulse of 50 us duration and of sufficiently high intensity induced reversible electrical breakdown of the membrane. Various modifications in technology have been employed (Zachrisson and Bornman, 1986) and currently electro fusion is being utilized routinely in a number of plant species, notable among which are tobacco (Bates, 1985) and potato (De Veries and Templeaar, 1987).

Although chains of protoplasts are formed in electro fusion and in many cases this fusion product involves many protoplast, which is not desirable, yet it is possible to control the dielectrophoretic adhesion of cells in such a way so as to produce aggregate consisting of two desired protoplasts. Recent work of Spangenberg and Schweigher (1986) has demonstrated that it is possible to select two protoplasts from two desired parents to fuse them. The application of this method to fuse individually paired Brassica protoplasts in microdrop cultures has been found convenient. Pairing and fusion of individual protoplasts provided a means of controlling the number and type of cells being fused, which also eliminates the need for selection systems.

Electrofused protoplasts have the same level of viability, division potential, and differentiation potency as those fused with polyethlene glycol (PEG). Furthermore, it has been reported that a supply of a direct current pulse with almost similar duration and field strength required for optimum fusion stimulated the division and differentiation of plant protoplasts. Such a phenomenon is called *electro stimulation.* Perhaps the electro fused protoplasts can divide and regenerate more rapidly and efficiently than those fused with PEG. One of the remaining problems in connection with electrofusion technology is to establish a simple physical method for the efficient sorting and collection of diheterokaryons after fusion to replace conventional screening methods. A symmetric cell hybridization has become popular and seems most effective for breeding viable hybrid plants through cell fusion.

Following fusion, the heterokaryons can be cultured in various ways. They are generally cultured in a defined culture medium such as Murashige and Skoog or B5, modifications thereof.

Selection hybrid cells

Since the PEG-high pH/Ca^{2+} fusion method is nonspecific, a mixture of unfused, protoplast, homokaryons, heterkaryons and multiple cell fusion is obtained. Thus, a selection system must be employed to recover the desired hterokaryons. Selection systems are most efficient when used at the cellular level, thus permitting early identification of hybrids. Later, when plants are regenerated from selected fusion products, they can be subjected to several tests to verify hybridity. (Figure 18).

The most widely utilized selection system is based on genetic complementation of non allelic mutants. Melcher and Labib (1974) were the first to regenerate hybrid plants using this selection method by combining two recessive albino mutations of *Nicotiana tabacum* virecent and chlorophyll deficient (vc). Through this method, somatic hybrid plants were selected as dark green regenerating shoots.

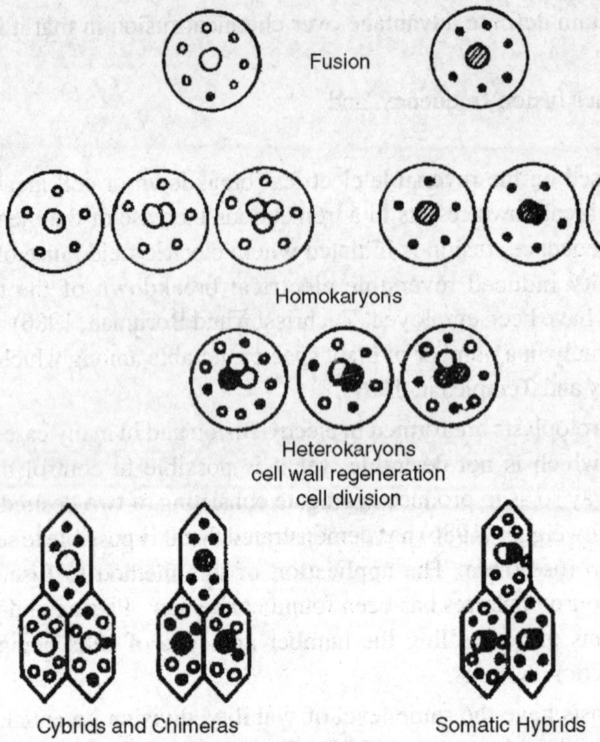

Fig. 18. The range of fusion products possible following the fusion of protoplasts from two cell lines. Cybrids, Chimeras and true hybrids may result as a result of heterokaryotic fusions.

Albino aB + albino Ab green Aa Bb

Auxotroph aB + auxotroph Ab autotroph Aa Bb

Evans et al. (1980) selected somatic hybrids in *Nicotiana* based on regeneration of light green shoots following fusion of protoplasts of a semidominant albino (Su/Su) tobacco mutant with normal green (su/su) protoplasts. An additional observation, dark green or albino spots, the result of mitotic crossing over, was detected on light green hybrid shoots while still in tissue culture, thus permitting early selection of hybrid plants. Non allelic auxotrophic mutants have also served as donor material for somatic hybridization studies. For example, two nitrate reducatse-deficient mutants of N.tabacum were fused and fusion products were selected based on their ability to form shoots on medium containing nitrate as the sole nitrogen source, (Glimelius et al. 1978). Although fusion of auxotrophic mutants is a very efficient method of selection, it has not been employed extensively due to the lack of well characterized auxotrops in higher plants.

Genetic complementation is often combined with other methods to select by hybrid cells. For example, albino mutant Petunia hybrida protoplasts, which could regenerate to form plants, were fused with P. *parodii*-wild-type protoplasts, which could not regenerate callus in the culture medium used (Cocking et al., 1977) putative somatic hybrids were selected as green callus.

While no genetic studies of regeneration capability have been reported, regeneration appears to behave as a dominant trait in fusion products. Several workers have used intermediate morphology to verify hybridity in plants regenerated from fused protoplasts (Binding and Nehls, 1978). In these cases, essentially no selection systems was applied.

Stable resistance mutants and variants, also difficult to isolate and establish in higher plants, have been used to select for fusion products. Horn et al. (1983) fused a 5 - methyl tryptophan (5 MT) resistant mutant of *N.tabacum* with a normal *N.glutinosa* and selected hybrid cells partially based on their ability to grow in the presence of 5 MT. Natural hybrid plants were recovered from this fusion. Harms et al. (1982) used 5 MT and aminoethyl cystine resistance to select for somatic hybrid plants in Daucus carota.

Selective media can also be used to recover somatic hybrid cells. Carlson et al. (1972), in the first report of somatic hybrid plant production, fused *N.glauca* and *N.longsdorifii protoplasts* that both required hormone supplemented media to divide. Somatic hybrid cells were selected based on their ability to grow on hormone-free medium.

Microisolation, the physical separation and culture of protoplast fusion products, is the most tedious but perhaps the most efficient universal selection method available. Kao (1977) first described this method. In most cases, microisolation must be performed before the contents of the two protoplasts mix completely and the two parent cellular components become indistinguishable within fused protoplasts. In the case petal and cell culture derived protoplasts, the petal pigment in vacular and diffuses quickly. However, fusion products are easily identified as having a colour that can be distinguished from the unfused petal protoplasts. Once a somatic hybrid cell is microisolated, it must be either cultured in a very small volume of rich culture medium or placed into a nurse culture to support cell division and continued grown.

Also, the two parental protoplast types can be treated with fluorescent dyes. Once fused, the hybrid cells with a unique fluorescence can be separated using microisolation or a cell sorter. Once somatic hybrid cells are placed under selection pressure and putative hybrid plants have been regenerated, these plants must then be verified as true somatic hybrids. One frequently used method of verification is determination of chromosome number and chromosome morphology of reiterated plants. Somatic hybrids contain the summation number of chromosomes of both parents.

The expression and genetic transformation of morphological characters can also be examined. The morphological characters include leaf shape, flower shape, colour, pollen viability, etc. However, it should be interpreted continuously since there is a great range of variation in plants regenerated even from unfused protoplasts (Shepard, 1982). Among the most reliable methods of verifying the hybridity of a somatic hybrid plant is the examination of isoenzyme patterns. Isoenzymes are enzymes that perform a similar function but which exhibit different molecular forms. Because of these differences, isoenzymes exhibit differing electrophoretic mobilities following electrophoresis in standard mobilities following electrophoresis is standard polyacrylamide of starch gels (Tranksley and Orton, 1983). A somatic hybrid may have the sum of parential isoenzyme bands or unique hybrid bands together with the parental bands. Specific isoenzymes studied include esterases, peroxidases, amylases, phosphates, alcohol dehydrogenases, aspartate aminotransferases, and alanine aminopeptidases. Expression of isoenzymes often differs depending on plant tissue used and stage of growth of the plant. Therefore, workers must take care to collect samples for isoenzyme analysis from identical plant tissues at the same stage of development.

Barriers of Protoplast Fusion

The major barrier is the lack of plant regeneration from protoplasts of major crops. Protoplast isolation has been reported in many cereal crops such as maize, rice, oat, barely, wheat, and leguminous crops such as bean, peas, and soybean. In most cases, the extent of development in these protoplasts culture was limited to a few cell divisions. For these crops, no plant regeneration has yet been achieved.

However, some progress in important crop species has been achieved. Plant regeneration has been reported from protoplasts of rape seed, cabbage, alfalfa, lettuce, cauliflower, tomato and flax. Important progress has also been made in Napier grass and in forage legumes.

A great amount of genetic variability and incompatibility occurs during protoplast fusion, especially in most distant hybridization. Two nuclear (diploid) and two cytoplasmic genomes are combined in the same cell during somatic hybridization, whereas in sexual hybridization, two haploid nuclear genomes and generally only the maternal cytoplasm are combined. Sexual hybrids are uniform. However, a great deal of variability has been observed in population of somatic hybrid plants. Generally, this variability has been ascribed to nuclear genetic instability, mitotic recombination, somaclonal variation, or cytoplasmic segregation.

Nuclear genetic instability was observed in early somatic hybridization studies between distantly related species. Directional loss of *Nicotiana* glauca chromosomes was documented in *soybean N.glauca* hybrids (Kao, 1977). This type of chromosome elimination, observed in many somatic hybrid combinations, lead to production of aneuploid plants that are sterile and, therefore, not useful for integration into crop improvement programmes. Such aneuploids also interfere with regenerated. For example, in addition to morphological abnormalities, aneuploid plants are often difficult to root.

Even in somatic hybridization between more closely related species, nuclear genetic instabilities have been observed. For example, in the combination *Daucus carota* + *D.capillifolius,* plants were recovered with chromosome numbers in the range of 34 - 54. Cultured cells accumulate aneuploid and polyploid cells. Since such cell cultures are often used as donar material for protoplast fusion experiments, another reason for the recovery of aneueploid plants following protoplast fusion may be due to the use of such cell culture lines. Some workers have, however, been able to avoid aneuploidy in somatic hybrids. Several *Nicotiana* somatic hybrid exhibited the amphidiploid chromosome complement (Evans et al., 1983).

Evidence suggests that although two types of cytoplasm are initialy mixed during protoplast fusion, cytoplasmic segregation usually occurs.

Cytology of Somatic Hybrids

In general, diploid protoplasts have been used for the production of somatic hybrids in higher plants. Therefore, predominantly tetraploid somatic hybrids should be expected. Indeed, the double-diploid on amphidiploid chromosome set in intra specific as well as interspecific somatic hybrid plants could be found in *Nicotian, Datura, Petunia* and *Daucus* together with somatic hybrid lines that showed either ploidy levels or aneuploid chromosome numbers.

Somatic hybrids with ploidy levels higher than tetraploid may be the result of fusion events in which more than two protoplasts have fused or endomitosis during postfusion development. Hexaploid somatic hybrids, which are often selected must originate predominantely by triple fusion events. The aneuploid somatic hybrids indicate irregular mitosis during culture or they may be attributed to the use of cell cultures for protoplast fusion, because in cell cultures instabilities of chromosome numbers are quite common.

Somatic hybrids of distantly related species encounter different situation. The somatic hybrid cell lines of *N.glauca* and *G.max* produced by Kao (1977), in which the chromosomes of both parents could be easily distinguished by their different sizes, were found to be unstable. In these somatic hybrids, chromosomes elimination was found to be unstable. In these somatic hyrbrids, chromosomes elimination was random, and chromosomes of one or the other partner were shown to have been lost in three investigated hybrid calluses.

In animal somatic cell hybrids, chromosome elimination occurs frequently. This phenomenon has been used for chromosome mapping of mammalian genes. Similar studies in somatic hybrid cell lines of distantly related plant species should also be possible.

Fusion of Haploid Protoplasts

The fusion of haploid protoplasts is some what similar and analogues to the fusion of egg and sperm. Moreover, it would ensure the involvement of the male cytoplasm in the creation of more cytogenetic variability and the production of "cybrids" and somatic hybrids. The protoplasts from the maturing pollen when centrifuged and left undisturbed in the enzyme mixture tend to fuse. In some cases multinucleate tuber like giant protoplasts are formed.

Cytoplasmic Hybrids (Cybrids) : (Fig. 19)

Sexual hybridization results only in the mixture of nuclear genes and not the plasmagenes. They undergo strict material inheritance. On the other hand, during protoplast fusion, all of the genetic information of the two donor cells is mixed together resulting in heterokaryons or hybrids cells with two plastomes (Units of Chloroplast DNA) and two chondriosomes (Unit of mitochondrial DNA).

However, since cytoplasmic replication is not synchronized with mitosis, daughter cells produced following protoplast fusion can be cloned for the nuclear genes but may contain an uneven distribution of the parental chloroplast and mitochondrial DNA. After several cell divisions, hybrid cells that originally contained mixed organelles begin to sort out to yield cell lines with one or other parental organelle type, and that sorting out occur independently for chloroplast and mitochondria (Galun, 1982). Also their is evidence of co segregation when selective pressure is exerted on fusion products. Hence it is possible to recover plants that contain chloroplasts from one parent and mitochondira from the second parent. In addition, mitochondrial DNA recombination may occur and cell lines with the hybrid mitochondria can be established.

The production of cybrids following protoplast fusion is one of the most existing application of protoplast fusion research. In cybridization, the nuclear genome of one parent is combined with the *orgenells* of a second parent.

Cytoplasmic male sterility (CMS) has been used to follow transfer of mitochondria in *Nicotiana.* Aviv and Galun (1980) restored fertility to a CMS lines of *N.sylvestris* following protoplast fusion of irradiated fertile *N.tabacum* with CMS *N.sylvestris.* Resistance to triazine herbicides has been transferred via somatic hybridization and expressed in *S.tuberosum* + *S.nigrum* somatic hybrid.

Potential for Crop Improvement

Protoplast fusion has already been used to produce new species combination from interspecific or inter generic combination. Most research has been directed towards the production of novel hybrid, but not breeding line development. Hence, much more research must be done in this area before new cultivars are produced using protoplast fusion.

Emphasis must be placed on agriculturally important traits to be transferred using protoplast fusion. Also emphasis should be placed on achieving gene combinations that can only be achieved by protoplast fusion. Thirdly, somatic hybrids must be integrated into conventional breeding programmes. Fourthly, protoplast regeneration must be extended to a wider range of crop species.

Most of the somatic hybrids of early days were sterile intergeneric plants of no use in crop improvement. But now-a-days, emphasis is given on agriculturally important traits. (Table 30)

Table 30. Agriculturally important genetic traits transferred via protoplast fusion.

Trait	*Somatic hybrid*	*Reference*
Tobacco Mosaic virus resistance	Nicotiana tabacum + N.nesophila	Evans et al. (1981)
potato virus resistance	Solanum tuberosum + S.chacoense	Butenko et al. (1982)
Tobacco horn worm resistance	Nicotiana tabacum + N.nesophila	Evans, unpublished
Cytoplasmic male Sterility	N.tabacurr. + N.sylvestris	Aviv and Galun, (1980)
Atrazing (Herbicide) resistance	Solanum tuberosum + S.nigrum	Binding et al. (1982)

Each of these lines could result in new breeding materials. Agriculturally useful traits present in divergent wild species should be indentified and procedures outlined to permit transfer by protoplast fusion into a cultivated speices. In most cases traits controlled by a small number of nuclear genes or plasmogenes are transferred by protoplats fusion. Other traits that have been suggested as prospect for transfer by protoplast fusion include disease resistance in *Arachis*, alkaloid content in *Solanum*, drought tolerance in *Linum*, Bremia resistance in *Lactuca* and altered linolenic acid content in *Brassica* (Cocking, 1983).

It has been suggested that somatic hybridization is more useful for over coming pre zygotic incompatibilities than post zygotic incompatibilities. Also, protoplast fusion can be used to achieve many unique nuelcar - cytoplasmic gene combinations that cannot be achieved using conventional breeding. Such is the case with the recently reported *Brassica-Raphanus* cybrides (Pelletier et al., 1983) that contain the nucleus of *Brassica napus.* Chloroplasts of atrazine - resistance *B.campestris,* and mitorchondria that confer male sterility from *Raphanus sativus.*

To ensure new cultivar development, somatic hybrids must be integrated into a conventional breeding programme. Wide sexual hybrids have been used to transfer useful genes to develop new crop cultivars in toabcco, tomato and other crops. As with sexual hybrids, certain barriers must be overcome prior to development of valuable germplasm using protoplast fusion. First, the somatic hybrid must be at least partially fetile to permit backcrossing to the cultivated crop. If the hybrid is completely sterile, it represents a dead end for breeding. In some species, somatic hybrids that are tertaploid may have an advantage over sexual hybrids. Tetraploid interspecific hybrids are more fertile than diploid hybrids (Evans et al. 1983). Second in a back cross programme. Effort msut be directed towards transfer of a small number of genes. If one attempts to transfer multi-genic traits, the back cross programme is likely to be lengthy and difficult. This is unfortunate, because yield, flavour and many other valuable economic characters have been shown to be controlled by multiple genes. Third, the back cross programme can be facilitated if recombinant lines are developed. Hence partial pairing and recombination of homologous chromosomes would be useful for the development of new breeding lines with reconstructed chromosomes.

The potential for crop improvement using protoplast fusion is widespread and is just now beginning to be realized. To date, no new crop cultivar has been produced from breeding lines developed using products of protoplast fusion. The prospects for creation of new nuclear-cytoplasmic gene mixture is particualry exciting. Interesting possibilities exist for transfer of herbicide resistance in *Brassica* and *Solanum*, and disease resistance and CMS transfer in several *Nicotiana* combinations.

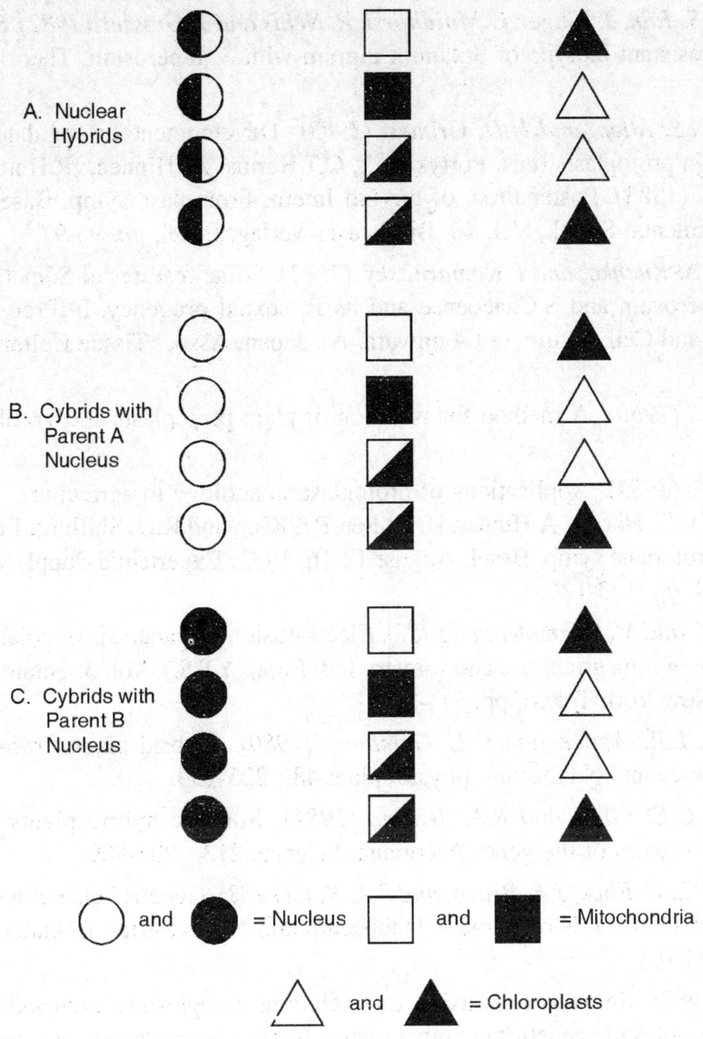

Fig. 19 : Gentic outcomes known to be possible via Somatic Hybridization that are not possible via sexual hybridization.

REFERENCES

Aviv, D. and E. Galun. (1980). Restoration of fertility in cytoplasmic male sterile (cms) Nicotiana sylevestris by fusion with X-irradiated N.tabacum protoplasts. Theor. Appl. Genet. 58 : 121-127.

Bates, G.W. (1985). Electric fusion for optimum formation of protoplast heterokaryons in Nicotiana. Planta. 165 : 217-224.

Binding, H. and R. Nehls. (1978). Somatic cell hybridization of Vicia faba + Petunia hybrida. Molec. Gen. Genet. 164 : 137-143.

Binding. H., S. Jain, J. Finger, G.Mordhorst, R. Nehls and J. Gressel. (1982). Somatic restoration of atrazing resistant biotype of Solanum nigrum with S. tuberosum. Theor. Appl. Genet. 63 : 272-277.

Boss, W.F., N.S. Allen, and H.D. Grimes. (1983). Developmentally regulated fusion of carrot protoplasts. In protoplasts (eds. Portykus, I., C.T.Harms, A. Hinnen, R.Huttner, P.J. King, and R.D. Stillito. (1983). Poster Proc. of the 6th Intern. Protoplast Symp. Basel, August 12 - 16, 1983. Experimental Suppl., Vol. 46, Birkhauser Verlag, Basel, pp. 96-97.

Butenko, R., A. Kuchko, and I. Komarnitsky. (1982). Some features of Somatic hybrids between Solanum tuberosum and S.Chacoense and its F_1 sexual progency. In Proc. 5th Intern. Cong. plant Tissue and Cell culture. (ed. Fujiwara, A). Japan. Assoc. Tissue Culture, Tokyo, pp. 643-644.

Cocking. E.C. (1960). A method for isolation of plant protoplasts and vaculoes. Nature. 187 : 962-963.

Cocking, E.C. (1983). Applications of protoplast technology to agriculture. In protoplasts (ed. Postrykus, I) C.T. Harms, A.Hinner, R. Hulter, P.J. King and R.D. Shillito), Lecture Proc., of the 6th Intern. Protoplast symp, Basel, August 12-16, 1983. Experientia Suppl. Vol. 46. Birkhauser Verlag, Basel. pp. 123-126.

De Vries, S.K. and M.J. Templeaar. (1987). Electrofusion and analysis of potato somatic hybrids. In Biotechnology in agriculture and forestry (ed. Bajaj, Y.P.S.), Vol. 3. Potato, Springer, Berline Heidelberg New York Tokyo, pp. 211-222.

Evans, D.A., L.R. Wetter, and O.L. Gamborg. (1980). Hybrid plants from Somatic cells of Nicotiana glauca and N.tabacum. physiol plant 48 : 225-230.

Evans D.A., C.E. Flick, and R.A. Jensen. (1981). Somatic hybrid plants between sexually incompatible speices of the genus Nicotiana. Science. 213 : 907-909.

Evans. D.A., C., E. Flick, J.E. Bravo, and S.A. Kut. (1983). Genetic behavoiur of somatic hybrids in the genus Nicotiana, N.otophora + N.tabacum and N.sylvestries + N.tabacum. Theor. Appl. Genet. 65 : 93-101.

Galun, E. (1982). Somatic cell fusion for including cytoplasmic exchange: a new bological system for cytoplasmic genetics in higher plants. In plant improvement and somatic cell genetics (ed. Vasil, I.K., W.R. Scowcroft and K.J. Frey) Academic Press, New York. pp. 205-220.

Gamborg, O.L. J.P. Shyluk, and E.A. Shahir, (1981). Isolation, fusion and culture of plant protoplasts. In plant tissue culture : methods and application in agriculture (ed. Thorpe, T.A.) Academic Press. New York. pp.115-154.

Glimelius, K., A., Walling, and T. Eriksson, (1978). Concanavalin A improves the polyethylene glycol method for fusing plant protoplast. Physiol. Plant. 44 : 92-96.

Glimelius. K., K.Chen and H.T. Bonnett. (1981). Somatic hybridization in Nicotiana segregation of organelle traits among hybrid and cybrid plants. Planta. 153 : 504-510.

Kao, K.N. and M.R. Michayluk. (1974). A method for high frequency Intergeneric fusion of plant protoplasts. Planta. 115 : 355-367.

Kao, K.N.F. Constabel, M.R. Michayluk, and O.L. Gamborg. (1974). Plant protoplast fusion and growth of intergeneric hybrid cells. Planta 120 : 215-227.

Keller, W.A. and G.Melchers. (1973). The effect of high pH and Calcium on tobacco leaf protoplast fusion. Z. Naturforschung 28C : 737 - 741.

Klebe, R.J. and M.G. Mancuso. (1981). Chemicals which promote cell hybridization, Som. Cell Genet. 7 : 473-488.

Larkin, P.J. (1976). Purification and viability determinations of plant protoplasts, Planta. 128 : 213 - 216.

Melchers, G. and G. Labib. (1974). Somatic hybridization of plants by fusion of protoplasts. I. Selection of light resistance hybrids of "hyploid" light sensitive varieties of tobacco. Molec. Gen. Genet. 135 - 277 - 294.

Pelletier, G., C.Primard, F. Vedel, and P. Chetrit. (1983). Integeneric Cytoplasmic hybridization in Cruciferea by protoplast fusion. Molec. Gen. Genet. 191 : 244 - 250.

Shepard, J.F. (1982). Cultivar dependent cultural refinements in potato protoplast regeneration. Plant Sci. Lett. 26 : 127 - 132.

Shepard. J.F., and R.F. Totten, (1977). Mesophyll cell protoplasts of potato : isolation, proliferaton and plant regeneration, Plant physiol. 60 : 313 - 316.

Spangenberg, G. and H.G. Schweiger. (1986). Controlled electrofusion of different types of protoplats and subprotoplasts including Cell reconstitutio in Brassica napus L.Eur. J. Cell Biol. 41 : 51 - 56.

Takebe, I., G. Labib, and G. Melchers, (1971). Regeneration of whole plants from isolated mesophyll protoplasts of tobacco. Naturwissenschaften. 58 : 318 - 320.

Transksley, S.D. and T.J. Orton. (1983). Isozymes plant genetics and breeding. Part A and B. Elsevier, Amsterdam.

Vasil, I.K., V. Vasil, W.D. Sutton, and K.L. Giles. (1975). Protoplasts as tools for genetic modification of plants. In Proc. Intern. Symp, On yeast and other protoplasts, Univ. of Nottingham. pp. 82.

Zaccharisson, A. and C.H. Bornman. (1986). Electromanipulation of plant protoplasts. Physiol. Plant. 67 : 507 - 516.

Zimmermann, U. and P.Scheurich. (1981). High frequency fusion of plant protoplasts by electric fields. Planta. 151 : 26 - 32.

CHAPTER 17

Nucleic Acid Hybridization

Introduction

In 1960, Julius Marmur and Paul Doty discovered that the two strands of a DNA double helix can be dissociated from each other and then put back together in proper alignment under appropriate experimental condition. This pioneering discovery has formed the basis for the development of a diverse family of techniques, collectively referred to as nucleic acid hybridization, in which the base sequences of two nucleic acid chains are compared by testing their ability to form hybrids with each other. Such hybrids can form only when the two chains have complementary sequences that are capable of hydrogen bonding to one another. This approach which has been utilized to study DNA-DNA, DNA-RNA and even RNA-RNA interactions, has had a profound impact on our understanding of gene organization and expression.

Physical Changes of DNA

When DNA is exposed to high temperature or alkaline pH, the hydrogen bonds holding the two strands together are disrupted and the molecule dissociates into two separate poly nucleotide chains. DNA molecules that have been dissociated in this way are said to be denatured. Denaturation is accompanied by several changes in the physical properties of the DNA solution, including a decrease in viscosity, an increase in density, and an increase in the ability to absorb ultraviolet in the 260 nm region of the electromagnetic spectrum.

Of the three physical changes mentioned above, the increase in absorption at 260 nm is most easily used to monitor the progress of denaturation. If the absorbance of a DNA solution is monitored as the sample is gradually heated, a temperature is eventually reached at which the absorbance begins to increase. The absorbance will then continue to increase until denaturation is complete.

A graph showing the increase in the absorption of ultraviolet light that accompanies the heat induced conversion of DNA from the double stranded to single stranded state. (Fig. 20)

The total increase in absorbance, referred to as the hyperchromic effect, varies among DNA samples, but generally ranges from 20 to 60% of the initial value. For DNA samples that are properly base paired, the temperature range over which denaturation occurs is relatively small. Since denaturation occurs over a narrow temperature range, just like the melting of a simple crystal, thermal denaturation of DNA is often referred to as 'melting'. The melting temperature or Tm is defined as the temperature at which the transition from double-stranded to single stranded DNA has proceeded halfway to completion.

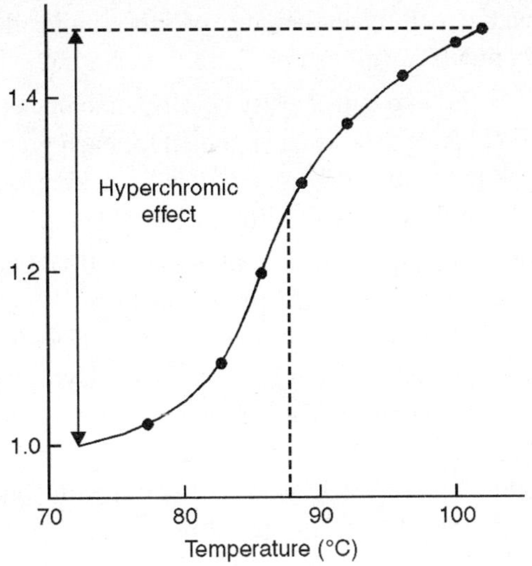

Fig. 20. A graph showing the increase in the absorption of ultraviolet light that accompanies the heat induced conversion of DNA from the double stranded to single stranded state. Source : Principles of Cell Biology

Effect of G-C Content on 'Melting'

One prominent factor influencing the denaturation of double stranded DNA molecules is relative content of A-T and G-C base pairs. The G-C base pair, held together by three hydrcgen bonds is tighter and more compact than the A-T base pair, which is held together by only two hydrogen bonds. Since more energy is therefore required to disrupt G-C based pairs, DNA samples with high G-C contents will melt at higher temperatures than samples with lower G-C contents. Fig. 21 and 22.

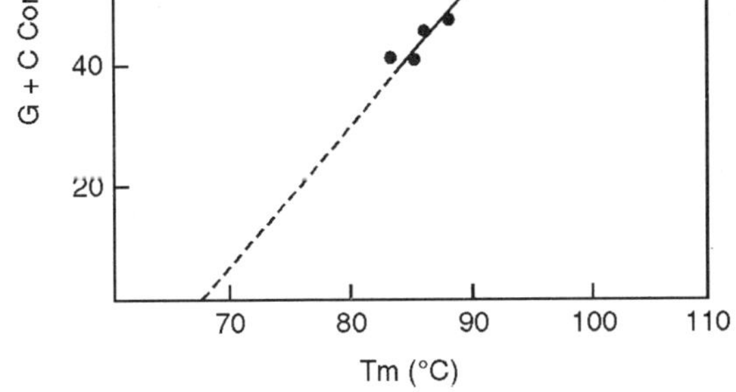

Fig. 21. Effect of GC content on the melting temperature of DNA molecules. Data obtained from various types of DNA differing GC contents have been pooled for each graph.

Effect of GC content on the melting temperature of DNA molecules. Data obtained from various types of DNA differing GC contents have been pooled for each graph.

Effect of GC content on the buoyant density of DNA molecules. Data obtained from various types of DNA with differing GC contents have been pooled for each graph. The greater compactness of the G-C base pair also increase the buoyant density of DNA, making it possible to employ isodensity centrifugation to separate DNA molecules with differeing G-C contents.

In nucleic acid hybridization experiments double-stranded DNA must first be dissociated into its individual strands before the ability of these strands to from complementary hybrids with other DNA or RNA molecules is carried out by mixing the two nucleic acids together and incubating them under appropriate conditions of temperature and ionic strength. Hybridization is generally carried out either in solution or with one of the nucleic acids is immobilized on an inert support such as a synthetic memberane surface. It has been empirically determined that a complementary region as short as a dozen bases is sufficient to cause two nucleic acids to hybridize to one another. Since nucleic acid molecules are generally much longer than this, a small region of complementarity could conceivably serve to hold together two molecules whose sequences are largely noncomplementary. For this reason nucleic acid hybridization is usually estimated after hybridized sample has been treated with an enzyme, such as s1 nuclease, that selectively degrades single stranded nucleic acids. Only base sequences that have successfully hybridized to one another will survive treatment with such nucleases (Fig. 23).

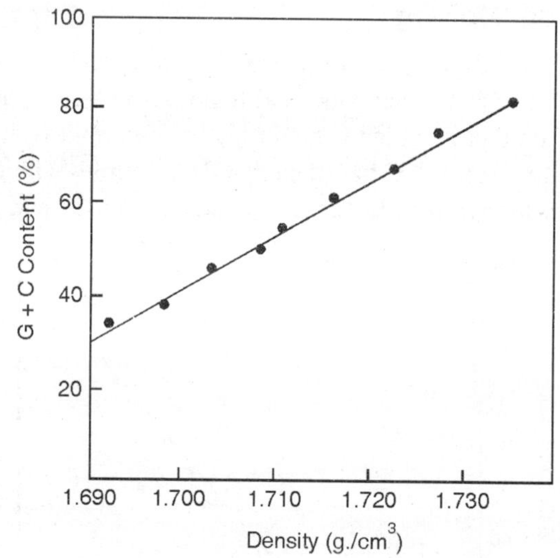

Fig. 22. Effects of GC content on the buoyant density of DNA molecules. Data obtained from various types of DNA with differing have been pooled for each graph.

An outline of one of the many ways in which nucleic acid hybridization experiment can be carried out. In this example, denatured DNA is immobilized on a membrane filter to prevent its individual strands from hybridizing back to one another. After RNA has been hybridized to the immobilized DNA non-hybridized nucleic acid is destroyed with a nuclease that degrades single stranded nucleic acid. Similar procedures can be employed for DNA-DNA hybridization.

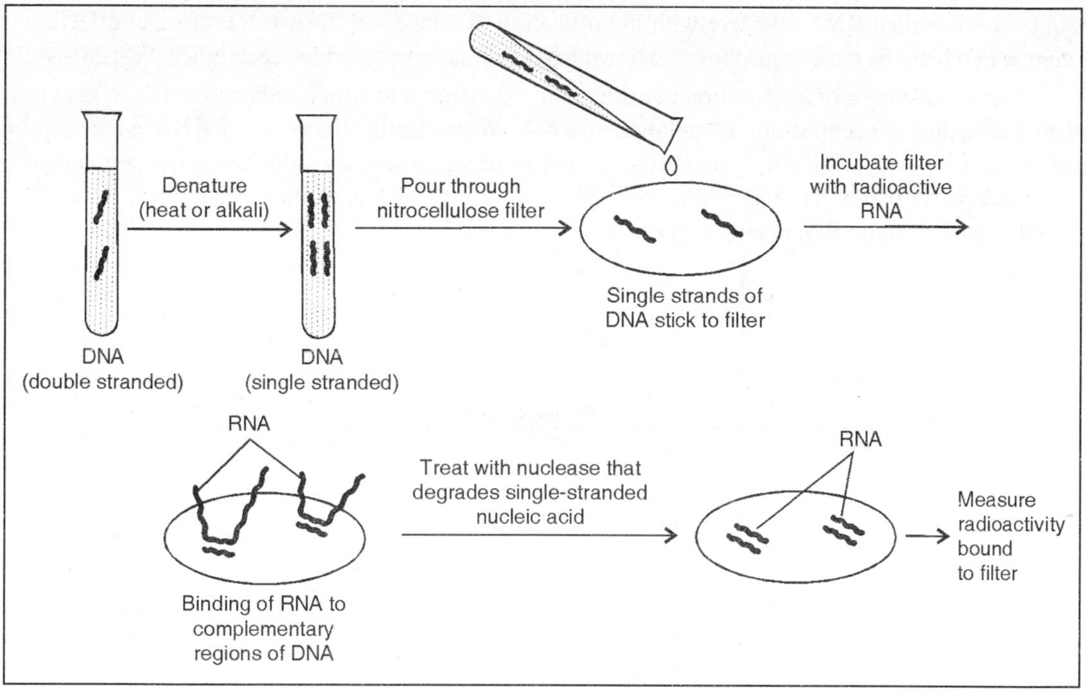

Fig. 23. Outline of nucleic acid hybridization.

The principles of nucleic acid hybridization have been applied to biological problems in a wide variety of ways. These various approaches can be subdivided, into major categories summarized below.

Saturation Hybridization

One common application of nucleic acid hybridization is in determining the relative abundance of a particular type of sequence in a DNA sample. To cite an example, this approach has been used to determine the percentage of an organism nuclear DNA that is complementary to ribosomal RNA. In this case, denatured DNA is hybridized to increasing amounts of radioactive ribosomal RNA. When the RNA concentration reaches a level at which no further increase in hybridization occurs, the DNA is said to be saturated. At this stage all the DNA sequences complementary to ribosomal RNA have become hybridized. By measuring the initial amount of DNA and quantity of radioactive RNA that has become hybridized, the percentage of the total DNA that is complementary to ribosomal RNA can be calculated. Since the length of the ribosomal RNA molecule is known, such data can be used to estimate that total number of ribosomal RNA genes present in DNA.

Competition Hybridization

In contrast to saturation hybridization, competition hybridization is employed to determine the extent to which two nucleic acid samples resemble each other in sequence. This approach is most commonly used to compare two different sample of RNA one of which has been radioactively labelled. If the two RNA samples resemble each other in sequence, they would be expect to hybridize to the same regions of a denatured DNA molecule. Because the two RNAs are competing for hybridization to the same DNA sites, increasing the amount of unlabelled RNA in the hybridization mixture causes a decrease in the amount radioactie RNA that can hybridize. RNA sample that differ in sequence on the other hand, hybridize to different DNA regions and so do not compete with each other. The extent to which increasing amounts of unlabelled RNA interfere with hybridization of a labelled RNA is therefore a reflection of the degree to which the base sequences of the two RNA populations resemble each other (Fig. 24 and 25)

(Fig. 25) Kinetics of a saturation hybridization experiment in which radioactive DNA, is hybridized to increasing concentrations of unlabelled RNA. When further increase in RNA concentration cause no additional hybridization, saturation is said to be achieved. By calculating the percentage of the total radioactivity that has become hybridized at this point, one can estimate how much of the DNA is complementary to the RNA sample in question.

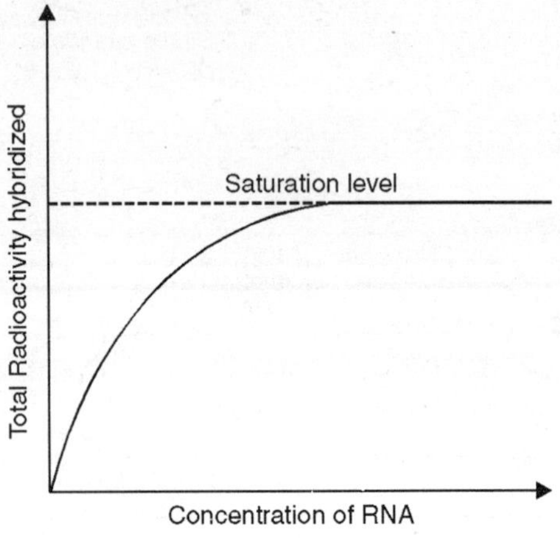

Fig. 24. Kinetics of a Saturation Hybridization experiment.

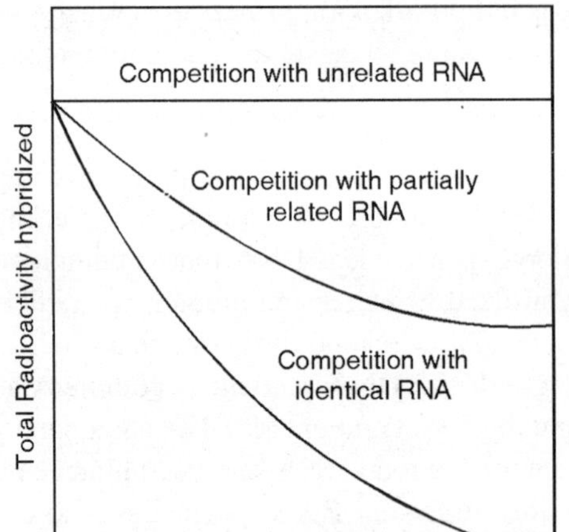

Fig. 25. Kinetics of competition of RNA.

(Fig. 26) Kinetics of a competition hybridization experiment in which a sample of radioactive RNA is hybridized to DNA in the presence of varying concentrations of another type of unlabelled RNA. The degree to which hybridization of the radioactive RNA is decreased by the presence of the unlabelled RNA indicates how closely related the two RNA sample are to each other.

Southern and Northern Blotting

Hybridization is used to localize specific nucleic acid sequences that have been separated by electrophoresis. In the southern blotting procedure, DNA that have been fractioned by electrophoresis is transferred to a membrane filter by "blotting" the gel with the membrane. This process yields a replica of the gel that can then be hybridized with a radioactive nucleic acid. Autoradiography of the hybridized membrane allows one to localize DNA fragments that are complementary to the radioactive nucleic acid employed as probe. An analogous procedure, termed Northern blotting, utilize a similar approach to analyze electrophoretically fractionated RNA samples for the presence of RNA molecules that are complementary to a radioactive nucleic acid probe.

In situ and Colony Hybridization

In addition to being used to characterize isolated nucleic acid molecules, hybridization techniques can also be applied to situations in which nucleic acids reside in their original location (*in situ*) within cells. This approach of *in situ* hybridization has been used, for example, to determine the subcellular localization of the DNA sequences coding for ribosomal RNA. To accomplish such an objective, cells are first fixed and sectioned for microscopic examination. This tissue section is then incubated at high pH to denature the DNA, and radioactive ribosomal RNA is added. After incubation under suitable hybridization conditions, unhybridized radioactivity is washed away and the section is examined by autoradiography.

The principle of nucleic acid hybridization has also been utilized to determine whether or not a given population of cells contains a particular nucleic acid sequences. In one widely used version of this approach, termed colony hybridization, an agar plate containing bacterial colonies is blotted with a piece of nitrocellulose membrane or filter paper to make a replica of the pattern of colonies. The replica is then hybridized with a specific radioactive nucleic acid and analysed by autoradiography.

Isolating Individual Genes from the Genome

One of the principal uses of cloning technology is to isolate specific genes directly from the genome. Any particular gene represents a very small part of a eukaryotic genome, In a typical mammal, the size of the genome is 10^9 bp, so that a single gene of (say) 5600 bp represents only 0.00005% of that total nuclear DNA.

To identify such a tiny proportion, we need a very specific probe that reacts, only with particular sequence in which we are interested, to pick it out from the vast excess of other sequences. The usual technique is to use a highly labelled radioactive probe of RNA or DNA, whose hybridization with the gene is assayed by autoradiography.

Applications for Labelled Hybridization Probes

Southern blots	Detection of gel-fractioned DNA molecules following transfer to a membrane.
Northern blots	Detection of gel-fractioned RNA molecules following transfer to a membrane.
Dot blots	Detection of unfractioned DNA or RNA molecules immobilized on to a membrane.
Colony/Plaque blots	Detection of DNA reduced from lysed bacteria or phage and immobilized on a membrane.
S1/RNase mapping	Positional mapping of termini of target molecules.
Hybridization	Cytological preparations.
In-situ	Detection of DNA or RNA molecules in

It is to be emphasised that labelling of nucleic acids has to be done before hybridization for without a label the detection of hybrid is not easy. There are five methods for labelling nucleic acids.

1. Nick Translation
2. Primer extension;
3. Methods based on RNA polymerse
4. End labelling methods and
5. Direct labelling methods.

Among the above methods, nick translation is the commonest and it has the advantage to produce large quantities of probe for use in multiple hybridization reactions.

Prime extension method is ideal for situations such as single copy gene detection on a Southern Blot.

RNA probes have been reported to show higher sensitivity than equivalent nick translated probes in both Northern blots and in-situ-hybridization. Most probably this is due to abvoidence of probe reannealing during hybridization and because RNA-RNA hybrids are more stable than DNA-DNA hybrids.

The choice of label may be radioactive labels which are most widely used or non radioactive labels like Biotin. Radioactive labelled probes are detected by authoradiography and non radioactive labelled probes by a variety of signal-generating systems.

Formats of Hybridization Reactions (Fig. 26)

There are two basic formats for hybridization reactions. Either one of the two nucleic acid molecule in the hybridization reaction is immobilized to a solid support, as in filter hybridization or the hybridization is carried out with both nucleic acids in solution. In solution hybridization both the target and the probe nucleic acid are free to move, thus maximising the chance that complementary sequences will align and bind. Consequently solution hybridizations go to completion five to 10 fold faster than those on solid supports.

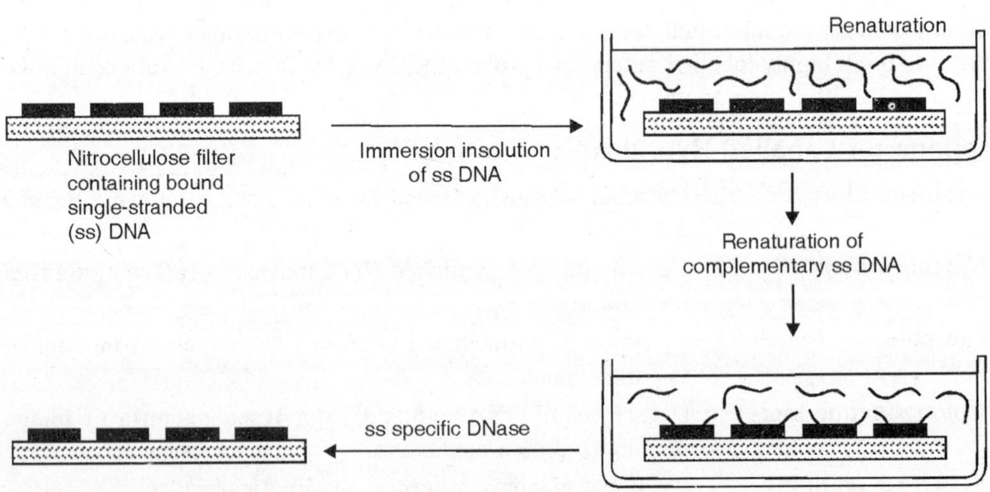

Fig. 26. Formats for hybridization reactions.

Fig. 26 Very thin filters (membrane filters) made of nitrocellulose are commercially available. These filter bind single-stranded DNA very tightly but fail to bind either double stranded DNA or RNA. A sample of denatured DNA is filtered. The single strands bind tightly to the filter along the sugar phosphate backbone but bases remain free. The filter is then placed in a vial with a solution containing a reagent that prevents additional binding of single standard DNA to filter and a small amount of radioactive denatured DNA. After a period of renaturation, the filter is washed. Radioactivity is found on the filter only if renaturation has occured.

Filter binding can be used to determine whether two organisms have common base sequences. For example, if excess *E. coli* DNA is on the filter and a very small amount of denatured 14c – labelled DNA isolated from Salmonella bacteria is applied the fraction of the applied 14c that is retained on the filter is the fraction of Salmonella DNA that has sequences with sufficient complementarity with that of *E.coli* to hybridize under the annealing conditions used.

A variation of the filter hybridization is to attach the probe to the bottom of the microtitre plate well or to a tube. This facilitates the washing step, reduces the total volume of hybridization and facilitates the automated reading of the result if a calorimetric detection system is used.

If solution hybridization has been used, removal of unbound probe is not easy. One method is to digest the single-stranded nucleic acid that remains after hybridization with an appropriate nuclease. An alternative method is to separate duplex from single stranded by means differential binding to hydroxyapatite or antibodies specific for double stranded nucleic acids.

In another approach by Gingeras et al, (1987) the target DNA is hybridized in solution to a labelled probe and to an unlabelled caputre probe which is immobilized by covalent attachment at its 5' end to a solid support.

Urdea et al (1987) has devised a method employing solution phase hybridization of unmodified synthetic oligonucelotides with the target nucleic acid followed by solid-phase capture, labelling and signal amplification steps. Probes A and B carry sequences complementary to different parts of the targets genome as well as unique sequences not found in target DNA. After hybridization of probes A and B, a third probe C is added which is linked to a solid support and which is complementary to the unique sequences of A. Thus problem C serves to capture the hybridization complex and to immobilize it. Next probe D is added. Probe D serves as a amplification multimer and is a chemically cross-linked oligonucleotide complex. Finally probe E is added which hybridizes to probe D and carries the detection label.

The ultimate choice of format will depend on the specific and the speed and sensitivity required. Filter hybridization is very convenient but the rate of hybridization is slow by comparison with reaction in solution. The sensitivity of hybridization is affected by the choice of label, regardless of the formate used, but in the case of filter hybridization the efficiency of retention of target DNA by the filter is also important. These problems can be minimized by use of the polymerase reaction to increase the target DNA concentration. As the target DNA concentration increases so does the rate of hybridization and the need for the sensitive detection system decreases.

Conclusion

Hybridization is a diagonestic tool used in many fields of biology. It has the greatest potentiality in clinical microibiology connected with the identification of causative organizm of a disease in a very shorter time compared to other usual analysis. Apart form this, there are many applications in plant pathology, particuarly in the identification of viruses in annual crops, prevention of infection in planning stock in monitoring disease control methods and in diagonising disease in plants held in quarantine. Hybridization is also being used as a tool in microbial ecology.

Its role in detection of genetic disorders and DNA profiling or finger printing are excellent.

DNA Profiling (Finger Printing)

DNA probes have been developed that at low stringency hybridize simultaneously to a number of tandom repeats in the genome to produce individual specific fingerprints. This technique has been shown to have considerable potential in forensic applications.

REFERENCES

Freifelder, D.(1987). Micodibal Genetics. Jones and Bartlett Publishers, Inc. U.S.A. Page: 37-39.

Gingeras, T.R., D.Y.Kwoth and G.R. Davis, (1987). Hybridization properties of immobilized nucleic acids. Nucleic Acid Res. 15 : 5373 - 90.

Kleinsmith, L.J., and V.M. Kish. (1988). Principles of Cell Biology. Harper and Row Publishers New York. P.P. 371.

Old R.W. and S.E. Primrose, (1989). Principles of Gene Manipulation. An introduction to Genetic Engineering, Fourth Edition. Black well Scientific Publications. 319 - 343.

Prescott. D.M. (1988). Cells, Jones and bartlett Publishers, Iuc. U.S.A. 330 - 331.

Urdea, M.S.J.A.Ruming, T. Horn, J. Clyne, L. Ku, and B.D. Warner (1987). A novel method for rapid detection of specific nucleotide sequences in crude biological samples without blotting or radioactively application to the analysis of hepatitis B virus in human serum. Gene 61, 253 - 264.

Weaver, R.F. and P.W. Hedrick, (1989). Genetics. W.M.C. Brown Publishers, Dubuque, IOWA. 152-427.

CHAPTER 18

RFLP in Plant Breeding

Introduction

In recent years genic cloning technology has led to fundamental advances in many areas of biology. One of the most exciting of these in the use of cloned fragments of chromosomal DNA as genetic markers, usually termed "RFLP mapping", with RFLP denoting Restriction Fragment Length Polymorphism. In this technique, which depends on natural variation in DNA base sequence, DNA is digested with a restriction enzyme. Homologous restriction fragments of DNA which differ in size, or "length", can be used as genetic markers to follow chromosome segments through genetic crosses. This new technology promises to revolutionize some areas of plant genetics and plant breeding.

Restriction Fragment Length Polymorphism (RFLP)

The genetic information which makes up the genes of higher plants is stored in the DNA sequences of the nuclear chromosomes and the organelle genomes. Plants are able to replicate their DNA with high accuracy and rapidity, but many mechanisms are operative which cause changes in the DNA. Simple base pair changes may occur, or larger scale changes as a result of inversions, translocations, deletions or transpositions. There is such an enormous amount of DNA in a higher plant cell, that no two organisms are likely to be identical in DNA base sequence. Thus there is a tremendous amount of DNA variation present in natural populations of plants, but until recently there has been to direct way to utilize this variation in plant genetics.

Natural variation in DNA sequence can be detected in several ways. One way, of course, is to directly sequence the DNA and make detailed comparisons. Unfortunately, this method is still very cumbersome and time consuming. Another way to access this variation is by using a special class of enzymes called restriction enzymes. These are nucleases produced by a variety of microorganisms, and they have the ability to recognize target sites (called restriction sites) made up of specific base seqeunces in the DNA. If the required base sequence is present in the target DNA, the restriction enzyme will cleave the DNA at the target site. A large piece of DNA will thus be reduced to a series of smaller fragments of defined size by digestion with a restriction enzyme. The number of fragments produced and the sizes of each fragments will reflect the distribution of restriction sites in the DNA. The fragments produced will thus be specific for each target DNA or restriction enzyme combination and can be used as a "finger print" specific for a given target DNA (or for the organism containing that DNA).

Relatively small DNAs, such as chloroplast DNA, will usually produce about 40 discrete restriction, fragments when digested with a typical restriction enzyme, such as EcoRI. The restriction fragments produced by digestion of purified chloroplast DNA can be separated according to size by subjecting the DNA to agarose gel electrophoresis after digestion with a restriction enzyme. After the gel is stained with ethidium bromide, the pattern of restriction fragments can be directly observed or photographed in ultraviolet light. Chloroplast DNAs which differ from one another in base sequence, or have been rearranged by insertions, deletions, or inversions will produce restriction fragments of different sizes. Such differences in fragment size, arising from restriction enzyme digestion of nuclear, organellar, or total DNA, are called restriction fragment length polymorphisms (RFLP) and can be used as a direct measure of genetic variability. Many studies have utilized RFLPs of chloroplast DNA to study phylogeny and systematics in various plant groups. These studies have shown that the analysis of RFLP variation in chloroplast DNA is very useful in unravelling systematic relationships in plants. Unfortunately, however, the utility of chloroplast DNA in plant breeding is extremely limited. Most genes of agronomic importance are located on nuclear chromosomes; few are found in chloroplast DNA.

RFLP analysis can be applied to chromosomal DNA also, but is more complex because of the much greater complexity of nuclear DNA. Digestion of the nuclear DNA from a higher plant with a typical restriction enzyme produces millions of discrete DNA fragments in a continuous range of sizes. If digested DNA is subjected to gel electrophoresis and stained with ethidium bromide no distinct fragments can be visualized. So many fragments are produced that the DNA appears to run as a continuous smear. However, individual restriction fragments are still well resolved in the gel, and RFLPs are still present between DNAs from different organisms. More complex techniques have to be used to resolve them. These techniques involve the use of cloned DNA probes and DNA hybridization.

Libraries of Cloned Probes

Since RFLPs of nuclear DNA cannot be directly visualized, the usual procedure is to use small pieces of chromosomal DNA as probes to detect individual restriction fragments. Using the high specificity of DNA-DNA hybridization, such probes can detect individual restriction fragments in the complex mixture of fragments of nuclear DNA present in a restriction digest. To use this techniques a set of chromosomal DNA fragments is prepared to use as probes. Such a set of probes is called a library. DNA isolated from the species of interest is digested with a restriction enzyme, and relatively small fragments (usually 2-5 kb) are used as DNA hybridization probes. Individual restriction fragments can be used as a probe, but a supply of the individual fragments in pure form is needed. It would be very difficult to directly isolate individual restriction fragments, but fortunately it is not necessary to try to do this. Instead, one utilizes gene cloning technology and the ability of a bacterium such as *E.coli* to very accurately replicate DNA. In this technique, individual restriction fragments are ligated into a bacterial plasmid, and the plasmid is then transformed into a bacterial cell. The bacterium replicates the plasmid as it grows and divides. By growing a culture of these transformed bacteria and then isolating the plasmid, one obtains a large supply of a single plant DNA restriction fragment which is suitable for use as a hybridization probe. The bacterial strain carrying a fragment of interest can be maintained for a long period of time, and can thus be repeatedly used for distributed to other researchers for use in their experiments. Using cloned probes to detect RFLPs in nuclear DNA preparations is done in the following manner. DNA from the plants to be compared for RFLP difference is isolated, digested with a restriction enzyme, and then fractionated on an agarose gel. As explained above, the DNA wll now be in the form of millions of restriction fragments fractionated in the gel by molecular weight. In order to use DNA-DNA hybridization to detect specific fragments, the probe DNA and the DNA in the gel must be single-stranded (denatured). The gel is soaked in a base such as NaOH to denature the DNA and to facilitate hybridization, the DNA is transferred out of the gel onto a membrane filter by a procedure called a "Southern" transfer (or blot). A filter cut to the same size as the gel is placed directly against the gel, and the DNA is eluted out of the gel onto the filter. Since the filter is in direct contact with the gel, the pattern or restriction fragments in the gel is maintained on the filter. The denatured

DNA binds very tightly to such filters, and the filter can be repeatedly used for hybridization experiments.

After a filter is prepared by Southern transfer, it can be probed with one of the cloned probes from the library. *Basicaly*, this procedure consists of denaturing the cloned probe and allowing it to hybridize with the filter containing the nuclear DNA. Under the proper conditions of temperature and salt concentration, the denatured probe will specifically hybridize to any nuclear restriction fragments on the filter which are homologous to it. However, in order to see where the probe hybridized, it is necessary to label it in some way. The most common way to do this has been to radioactively label the probe with phosphorus-32 (^{32}P). The labeled probe is allowed to hybridize to the DNA filter, excess unhybridized probe is washed off the filter, and the filter is subjected to autoradiography to detect the restriction fragments to which the probe has hybridized. In this way, individual restriction fragments can be detected in complex preparations of higher plant DNA. Currently, probe labeling techniques which do not require the use of radioisotopes are being developed.

Construction of RFLP Genetic Maps

In order to discuss the utility of RFLP markers in genetic mapping, it is useful to think of genetic mapping in a slightly different fashion. Consider a genetic cross in a diploid organism in which two parents are crossed to produce an F_1 and the F_1 is selfed to produce an F_2 population. To simplify the discussion we will assume the parent plants have been selfed through several generations and are homozygous at all loci. It is clear that the F_1 hybrid from two such parents will have one set of chromosomes from each of the parents and that the two members of a homologous chromosome pair will be different from one another to the degree which the parents differ in DNA base sequence.

When the F_1 plant undergoes meiosis to produce gametes, its chromosomes will undergo recombination by crossing over. This recombination process will form gamete chromosomes which are mosaics with segments from each of the two parental chromosomes, and no two chromosomes will have an identical array of segments. In F_2 plants produced by selfing the F_1, homologous pairs will contain different mosaics of parental chromosome segments.

This recombination process is the basis of conventional genetic mapping, which depends on two observations. (1) Chromosome segments which are not on the same chromosome will undergo random recombination. (2) Chromosome segments on the same (homologous) chromosome will undergo recombination according to a function which depends on their physical distance apart. Chromosome segments which are close together (closely linked) will undergo recombination less commonly than will segments which are further apart. Thus genetic distance or map distance is defined as a function of the recombination which occurs during gametic formation.

Previously, the only way to follow chromosome segments and observe their recombination during gamete formation was to observe the phenotype caused by the action of genes which happened to be on the segments. By observing a phenotype such as flower colour, plant height, insect resistance, or characteristics of the endosperm, one could infer the behaviour of the respective chromosome segments during meiosis and recombination. Following chromosome segments in this indirect manner has been utilized to construct fairly detailed genetic maps of several plant species. However, the method is cumbersome and extremely time consuming.

Fortunately, RFLP markers provide a way to directly follow chromosome segments during recombination and greatly simplify the construction of this type of genetic map. Instead of looking at the phenotype caused by the presence of a gene on a chromosome segment, one looks directly at an RFLP marker on the segment itself. Thus one is looking directly at the genotype of the plant, rather than indirectly, through the phenotype produced by gene action. This has a number of advantages, which are discussed below:

When one considers the inheritance pattern of RFLP markers, it wll be seen that they segregate exactly in the same way as do conventional gene markers and follow strict Mendelian rules. This means that segregation patterns of RFLP markers can be analyzed by conventional Mendelian methods, and that maps of RFLP markers can be constructed in the same fashion as can maps of conventional markers.

Comparison of Conventional Genetic Markers and RFLP Markers

When conventional genes and RFLP markers are compared, it is apparent that RFLP markers have many advantages for the construction of genetic maps. Some of the most obvious of these are :

1. Natural Variation

To construct a genetic map by following chromosome segments through segregation, one must be able to distinguish segments from each of the two parental chromosomes. If conventional genes are used to follow the chromosome segments, then each parent must have different alleles of the gene in question (sometimes referred to as wild type and mutant). Often, one cannot depend on natural variation of conventional markers, and variants will have to be produced by mutagenesis. RFLP mapping, on the other hand, utilizes abundant natural variation in DNA sequence, and no mutagenesis necessary. Unlimited numbers of RFLP markers are available in virtually any plant species, whether or not any formal genetics has been done with it.

2. Developmental Stage of Plant Organ

To infer the segregation of a chromosomal segment by analyzing the segregation of conventional genetic markers, one has to depend on gene expression. This can lead to many complication. For example, genes may be expressed only in certain tissues (flower colour genes can normally be scored only when one has flowers present), or at certain developmental stages, or under certain environmental condition. In some cases, scoring for a conventional gene may be expensive or time consuming. For example to score for an insect resistance gene, it is often necessary to build cages around the plants and introduce insects (which must be the right biotype in a suitable developmental stage). Since RFLP markers directly detect RFLP variation in the chromosomal DNA, any developmental stage and virtually any plant part from which a small amount of DNA can be isolated is suitable for analysis.

3. Phenotypic Effect

RFLP markers are reflections of the natural variation present in the DNA sequence of plants. In the vast majority of cases, they will have no phenotypic effect at all. An effectively unlimited number of RFLP markers are naturally present in any plant. Contrast this with phenotypic markers. In most cases only a few phenotypic markers can be maintained in any one plant, either because the mutant phenotypes are harmful to the plant or because the mutations have been induced in different plants, and it would be very time-consuming to try to assemble these by crosses into any one plant line. These facts have important consequences for the construction of genetic maps. Since individual plants contain a great deal of RFLP variation, one can often construct a detailed RFLP map utilizing only one cross, one mapping population, and one modest-sized research group to gather and analyze the data. Conventional genetic maps, however, always require the cooperation of many research groups and represent years of effort and hundreds of crosses.

4. Environmental Effect

Conventional genes depend on gene expression for scoring. The DNA sequence comprising the gene must be transcribed into mRNA, the mRNA must be translated to protein, and the protein must assume the proper configuration and localization to exert its effect. The effect exerted by the protein is often due to its enzymatic activity or ability to bind to some cell component. All these stages can, and often are, affected by the environment. Since scoring of RFLP markers does not depend on gene expression, it is not affected by the environment. RFLP markers simply represent the presence or absence of characteristic base sequence.

5. Interaction with Other Genes

Since conventional gene markers depend on gene expression to be scored, their phenotypic effect is often modulated by other genes. One of the simplest cases is dominance; many mutations of common genes are recessive. These can only be scored in plants homozygous for the mutant allele, and heterozygotic plants cannot be distinguished from homozygotic wild-type plants. RFLP markers are codominant; the three possidle genotypes for one marker can always be distinguished. In addition, conventional marker will often display different phenotypes in different genetic backgrounds due to gene interactions. Since RFLP markers are a direct reflection of the genotype, they are independent of one another, and an RFLP marker will maintain its identity in any genetic background.

Construction of Genetic Maps Using RFLP Markers

To construct an RFLP map a fairly well- defined series of steps is followed. These are outlined below. For several crop species, such as rice, tomato, and maize, RFLP maps are already available, and one can simply use these without having to construct a new one.

1. Select the Parent Plants

As parent plants in a cross for RFLP map construction, one wants to select plants which are genetically divergent enough to exhibit sufficient RFLP, but not so far distant as to cause sterility of progeny. It would be advantageous also to select a cross in which some desirable agronomic traits are segregating, but this has not always been possible. In most plants thus far mapped, it has been possible to find sufficient variation between varieties of a single species. For intance, in the Rockfeller rice RFLP mapping project, a cross between indica and javanica types of rice was used. In tomato, however, the genetic base of the cultivated is too narrow, and to obtain a usable level of polymorphism, a cross had to be made between a cultivar and a wild species

Since one does not know a priori how much RFLP variation will be present in a crop species, it is necessary to make a survey using a random selection of cloned probes. Such a survey should include cultivars or plant introductions thought to represent the spectrum of variability present in the taxon. Related wild species which can be crossed to the cultivar should also be included. One should utilize any available information to select the plants to be surveyed. Existing systematic or evolutionary studies should be consulted; isozyme surveys be available.

After selecting a range of plants to be surveyed, DNA is isolated from individual plants of each of the accessions, digested with restriction enzymes, and screened for polymorphisms by the usual method. Each accession can then be scored for RFLP alleles present, and two accessions showing a usable amount of polymorphism can be selected for crossing.

2. Produce a Mapping Population

The selected parent plants are crossed to produce an F_1 plant or plants. In the case of a selfing plant such as rice, all f_1 plants should be alike and heterozygous for all RFLPs exhibited between the parent plants. A mapping population can be derived by selfing the F_1 to produce an F_2, which is then scored for segregation of the RFLPs, or by backcrossing the F_1 to one of the parents and observing segregation in the first backcross generation. It is better to use an F_2 population if this is possible, since more information can be gained than from a back-cross population of comparable size. A mapping population of about 50 F_2 or backcross plants is sufficient for a fairly detailed map. With plants in which plant breeding studies are actively being carried out, suitable mapping population might already be availabe as remnant seed from previous crosses. If this is the case it may be possible to make a map without having to make any crosses. This can greatly speed up the process.

3. Score RFLPs in the Mapping Population

Once the mapping population is obtaied, DNA is isolated from each individual plant in the population. It is important to realize that the chromosomes of each plant in the mapping population contain a unique array of parental chromosome segments. To derive an RFLP map it is necesary to determine which parent's chromosome segments are present in each plant in the mapping population. Since the unique makeup of chromosomal segments would be destroyed by recombination if the F_2 or backcross plants were to be allowed to reproduce sexually, every effort should be made to keep the individual plants in the mapping population alive so that repeated DNA extractions can be made, and a large amount of DNA can be accumulated. Mapping proceeds by sequentially scoring RFLPs in the individual plants of the mapping population by the following series of procedures.

A. Screening for Polymorphism

Probes are selected sequentially from the library and tested against the parents in an effort to determine which restriction enzymes will detect a polymorphism between the parents. The number of enzymes required will depend on the amount of variation present in the parent plants. Very polymorphic plants, such as inbred lines of maize, will usually require only one or two enzymes to detect polymorphism with most probes. In rice, 11 enzymes were used and about 75% of the probes were polymorphic with these.

B. Scoring

When a polymorphic probe or enzyme combination is detected, it can then be scored in the mapping population. To accomplish this a series of agarose gels must be prepared and filters prepared from them for DNA hybridization. These filters are generally called F_2 survey filters or mapping filters. There must be one filter set for each restriction enzyme to be used in the mapping project, and each filter set must contain digested DNA from each individual plant in the mapping population. Thus one set of mapping filters would contain DNA from each of the F_2 plants digested with EcoRI, and another set would contain DNA from each of the F_2 plants digested with PstI. If a polymorphism between the parent plants has been detected with a given probe or enzyme combination, then that probe will be scored on the corresponding mapping filter set. For example, if probe # 36 is found to be polymorphic between the parents using PstI, then it would be scored on the PstI mapping filter set. In an F_2 mapping population scored for a single RFLP, only three types of plants can be present. There are two homozygotes (one similar to each of the parents) and the heterozygotes. Thus each plant in the mapping population can be scored as a heterozygote or as one of the two possible types of homozygotes when tested with each library probe.

C. Linkage Analysis

Data accumulated from scoring the mapping population sequentially with probes from the library is used to construct the linkage map. Linkage analysis is based on the degree to which probes tend to co-segregate. The first probe scored in a mapping study will, of course, provide no information on linkage, but beginning with the second probe one can determine whether linkage is indicated. If the second probe is linked to the first one, they will tend to co-segregate (F_2 plants which are heterozygous for the first probe will tend to also be heterozygous for the second probe ; plants homozygous for the allele found in one of the parents for the first probe will also tend to be homozygous for the second probe). If no linkage is indicated, distribution of homozygotes and heterozygotes for the first two probes will tend to be random. Simple statistical tests, such as a chi square analysis, will test for randomness of segregation and hence linkage. The first few probes, since they are randomly selected, are unlikely to be linked and hence to show co-segregation. As one sequentially adds probes, however, linkages will be detected. At first there will be many more linkage groups than there are chromosomes, but the two members will tend to converge as more markers are added. As each probe is screened, it is tested for possible linkage to all the other markers which have been mapped before. Thus when one is testing probe #100, one can test it against the 99 other probes that have been scored before to see if it is linked to any of them. This is essentially like making a 100 point cross. It is evident that a great amount of data is rapidly accumulated in an RFLP mapping project, and a computer is required for efficient data storage and analysis. When a new probe is scored, if linkage with one or more markers is indicated by the data, the map distance is determined by computer analysis using an algorithm such as maximum likelyhood. Recombination data can be converted to map distances using any of the commonly used mapping functions.

RFLP Maps and Conventional Maps

One of the goals of the RFLP mapping project is the production of a "saturated" map, which is a map with RFLP markers spaced every 10-20 centimorgans (cm) over all the chromosomes. On such a map, any conventional gene will be within a few cm of an RFLP marker. Since conventional plant genetic maps are about 1500 cm, about 150 well-spaced markers would be sufficient for a saturated map. Since probes are randomly selected (and will represent random chromosome locations), however, it will be necessary to map several hundred to achieve this degree of saturation.

1. Rationalization of RFLP Maps with Conventional Maps

After construstion of a saturated or near saturated RFLP map, the normal next step is to associate the linkage groups of RFLP markers with the conventional genetic map, if one is available.This can be done in several ways depending on the particular situation. If aneuploid lines of the plant are available, these can be very useful for rationalizing RFLP and conventional genetic maps. This technique depends on the fact that one can usually determine the number of RFLP loci present from the intensity of the hybridization signal. Thus if an RFLP marker is on chromosome 1, one would expect to see about half the signal intensity in at plant monosomic for chromosome 1 when compared with a normal plant having two copies of chromosome 1. If a complete set of monosomic is available, each of the RFLP linkage groups can be assigned to a chromosome. Trisomic plants can be used in a parallel fashion.

2. Placing Conventional Genes on the RFLP Map

Once an RFLP map is constructed, conventional genes can be placed on the map. In order to do this it is necessary to have a mapping population in which both the conventional gene and the RFLP markers are segregating. By determining which RFLP marlers show co-segregation with the conventional marker, one can place the conventional marker on the map.

Uses of RFLP in Plant Breeding

Construction of an RFLP map can be very interesting, and during map construction one can gather data which is useful in systematic or evolutionary studies. However, an RFLP map is not useful by itself in plant breeding. It is only sueful when it is used in conjunction with analysis of conventional markers. One way in which RFLP maps can used to supplement reguler plant breeding protocols is by utilizing indirect selection. This procedure is useful when one wants to select for a conventional gene, but derectly selecting for that, gene would be expensive, difficult, or time-consuming. With indirect selection, one does not directly select for the gene of interest, but rather for one or more closely linked RFLP markers. If the RFLP markers are indeed closely linked, they will remain associated with the gene of interest during segregation. This allows one to select for the RFLP marker with confidence that the conventional gene will also be present, since only relatively rare recombination events would separate the two. For example, if a conventional gene is flanked by two conventuional markers each 10 cm away, 99% of the time one or another of the markers would remain associated with the conventional gene of interest.

Indirect selection for an RFLP marker, rather than selecting direvtly for the gene, can have advantages in several plant breeding scenarios. One such stituation might involve selection for a recessive gene. Suppose one were trying to introgress a recessive gene into a cultivar by using a backcross breeding programme. Backcross protocols involve alternate backcross and selection phases. It is necessary to select progeny bearing the desired gene at several points in the backcross cycle. If a gene is dominant it can be directly selected, but recessive genes will not be expressed in any of the backcross palnts, and it would be necessary to carry out progeny testing (probably by selfing the backcross plants to test for the presence of the recessive gene). A recessive allele could be indirectly selected, however, by selecting for a linked RFLP marker. No progeny testing would be necessary, and the process could be geratly speeded up. In some cases indirect selection for dominant genes could also be profitable in a backcross programme. Plants can be scored for a linked RFLP marker while in the juvenile stage, since only a little DNA needs to be isolated, and the plant does not have to be destroyed. Plants not bearing the desired allele could be weeded out early, saving space and expense. Another situation in which indirect selection could be of great use would be in selecting plants in a breeding protocol which have two or more independent genes which give a similer phenotype (sometimes called pyrmiding). For example, it might be advantageous in breeding insect resistant crops to include more than one resistance gene in a newly developed cultivar.However, selection for more than one gene conferring resistance to a given insect is difficult since the phenotype is the same whether one, two or more resistance genes is present. Again, some sort of progeny testing would be necessary, but where more than one gene is involved progeny testing quickly becomes impractical. However, once linkage of the individual insect resistance genes to RFLP markers is accomplished, one can easily follow any number of them through a cross. Individual plants can be scored for the presence of one, two, three or more resistance genes with no progeny testing needed.

Quantitative Traits

Unfortunately, most of the traits of agronomic significance in plants are not independent single genes, but are controlled by a number of genes scattered over the chromosome complement. Each of the individual genes of such a polygenic system contributes a small positive or negative effect to the trait of interest. Clear dominance is not exhibiited and the phenotype has a large environmental component. All these characteristics conspire to make quantitative traits very difficult to analyze. Conventional Mendelin methods of analysis which are suitable for single gene traits cannot be applied to analysis of quantitative traits, and one is forced to use biometrical methods and extensive testing over different years and in different environments in an effort to advance towards the desired state. Fundamantal advance in this area of plant breeding seemed unlikely before the advent of RFLP mapping techniques. Since one can use RFLP markers to simultaneously follow the segregation of all chromosome segments during a cross, the basic idea is to look for correlations between the quantitative traits of interest and specific chromosome segments marked by RFLP. If correlations exist, then the chromosome segent must be involved in the quantitative trait (one or more genes determining the trait must be on that chromosome segment). The difficult part in the procedure is establishing the correlations between the trait and specific chromosome segments. The RFLP markers are easily scored, but the quantitative trait must be scored in the conventional fashion. Once this time-consuming process is completed, however, and specific chromosome segments are implicated in the train, specific chromosome segments with a positive effect on a quantitative trait can be selected from a population of plants and incorporated into single plants with high efficiency. This is possible because of the ability to score for several RFLP markers simultaneously in a single plant in a manner which is free from environmental influence or gene interactions. Thus the analysis of quantitative traits can be carried out by conventional Mendelian analysis, and much of the uncertainty can be removed.

Conclusion

It is evident that RFLP markers can be directly used as probes for the presence or adsence of certain chromosome segments. The ability to directly follow chromosome segments has many fundamental applications in plant breeding and moleculer biology. During the next few years RFLP analysis will increasingly take its place as another tool which plant breeders will routenly use in their efforts to produce superior plants.

DNA Terminology

DNA is a giant molecule, far bigger than any other moleule in the cell. DNA amounts are usually expressed as the absolute amount per cell (in picograms) or the number of base pairs (often expressed as thousands of base pairs, kb, or millions of base pairs mt). Since DNA molecules are all quite similar (they all consist of alternating A-T or G-C nucleotide pairs), it is easy to convert from one form of measurement to another using the following relationship;

One picogram = 0.965×10^9 bp = 6.1×10^{11} dalton = 29 cm

It can be seen from this relationship that human cells, which contain about 3×10^9 base pairs per haploid genome, each contain 1.8 meters of DNA. Plants show a great deal of variation in DNA content; the table below lists DNA content for several organisms and organelles.

DNA content of some representative organisms and Organelles

Organism	**DNA** content (per genome)	
	picograms	kilobase pairs
Escherichia coli	0.0046	4.2×10^3
Chloroplast(*Zea mays*)	0.0002	1.6×10^2
Mitochondrion(*Zea mays*)	0.0007	5.7×10^2
Arabidopsis thaliona	0.070	7.0×10^4
Oryza sativa	0.600	5.8×10^5
Lycopersicon esculentum	0.7000	7.1×10^5
Zea mays	7.500	7.2×10^6
Homo sapiens	3.2000	3.9×10^6

Restriction Enzymes

Restriction enzymes are a class of DNA endonucleaes which are found in various micro organisms. Type II restriction enzymes, which are able to cleave DNA at sites where a specific base sequence (restriction site) occurs, are the type used most often in molecular biology. Restriction enzymes are thought to have evolved as a defence mechanisms against foreign DNA, such as that introduced into a bacterial cell by a bacteriophage. Such exogenous DNA is rapdidly cleaved into smaller fragments by restriction enzymes and the bacterium is thus protected from viral infection. Bacteria protect their own DNA from being degraded by attachig methy groups to nucleotides in potentially sensitive restriction sites.

The first three letters in the name of a restricion enzyme generally denote the genus and species of the organism in which the enzyme was first characterized. Subsequent letters and numbers denote the specific strain of the organism or the serial order in which restriction enzymes were characterized. For example :

PstI—the first restriction enzyme characterized from *Providencia stuart*

EcoRII—the second restriction enzyme from *Escherichia coli* strain R 245

Restriction sites for various restriction enzymes vary from 4 to 8 base pairs length, and are usually palindromic. The restriction site for PstI is thus :

$$\begin{array}{l} \qquad\qquad\;\downarrow \\ \text{C T G C A G} \\ \text{G A C G T C} \\ \;\uparrow \end{array}$$

(The arrows denote the site where the phosphodiester backbone of the DNA is cleaved.)

Plant Genome Organization

Plant chromosomes contain a large amount of DNA, and its organization is very complicaed. One such complication involves the copy number of individual sequences. Early in the study of eukaryotic genomes it was discoverded that DNA sequences may be present in virtually any number from one copy to millions of copies per haploid genom. Such a repeated DNA sequence may be present in tandem copies at one chromosomal locus, or may be dispersed throughout the genome. Repeated DNA sequences are often observed to be concentrated in centromeres, telomeres, or other heterochromatic parts of chromosomes, and may thus have some structural role.

Most RFLP research projects use single copy sequences as hybridization probes, because markers specific to single chromosome segments are desired. It is therefore necessary to screen probe libraries to remove repeated DNA sequences (and organelle sequences).

Agarose Electrophoresis

Since DNA has a large number of phosphate groups which are negatively charged at neutral pH, it will migrate towards the anode in an electric field. The basic phosphodiester backbone of all DNAs is the same so they have a uniform charge density. Therefore, if the DNA is placed in a porous medium such as agarose or polyacrylamide, it will migrate towards the anode at a rate which is proportional to its molecular weight. This is the technique which is used to separate DNA fragements of different sizes produced by digestion of DNA with restriction enzymes.

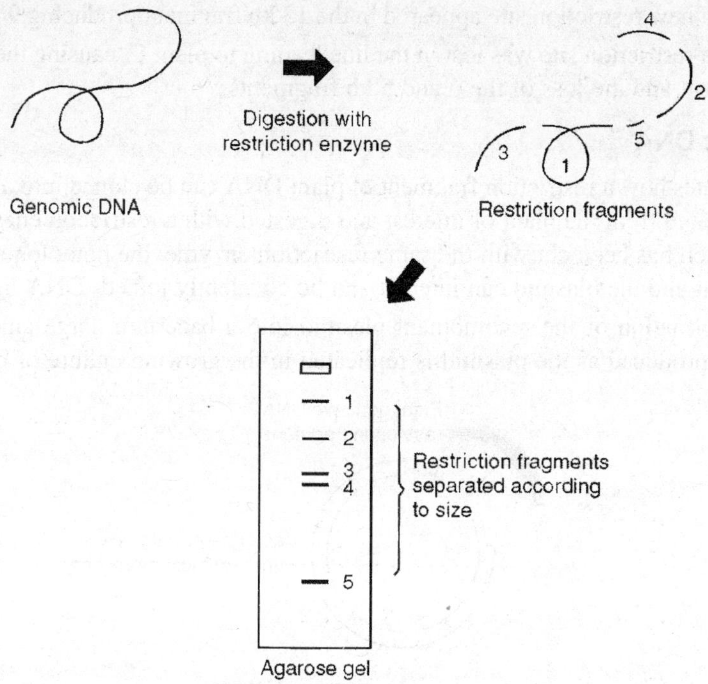

PHYLOGENETIC ANALYSIS USING RFLP

In the simplified example illustrated here, RFLP analysis is applied to a phylogenetic problem. The restriction map for a small region of the chloroplast genome of plant A is shown, along with RFLP patterns for plant A and two other plant accessions (B and C). If plant A is assumed to be least changed from the progenitor species, two mutations can be deduced;

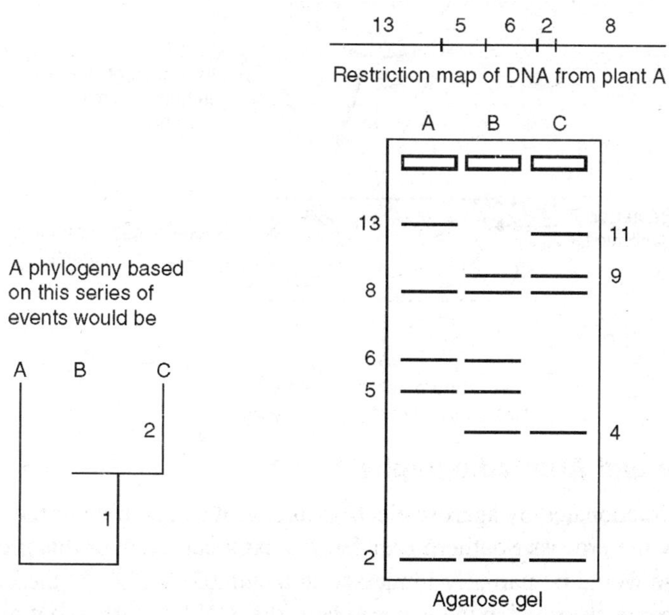

Mutation 1, a new restriction site appeared in the 13 kb fragment producing 9 and 4 kb fragment Mutation 2, a restriction site was lost in the line leading to plant C, causing the appearance of an 11 kb fragment and the loss of the 6 and 5 kb fragments.

Cloning of Plant DNA

The diagram illustrates how a restriction fragment of plant DNA can be cloned into a becteriel plasmid. DNA would be isolated from the plant of interest and digested with a restriction enzyme. When mixed with a plasmid, which has been cut with the same restriction enzyme, the homologous single-stranded ends of the fragment and the plasmid can interact and be covalently joined DNA ligase.

After transformation of the recombinant plasmid into a bacerium, large amounts of the plant DNA fragment are produced as the plasmid is replicated in the growing culture of bacteria.

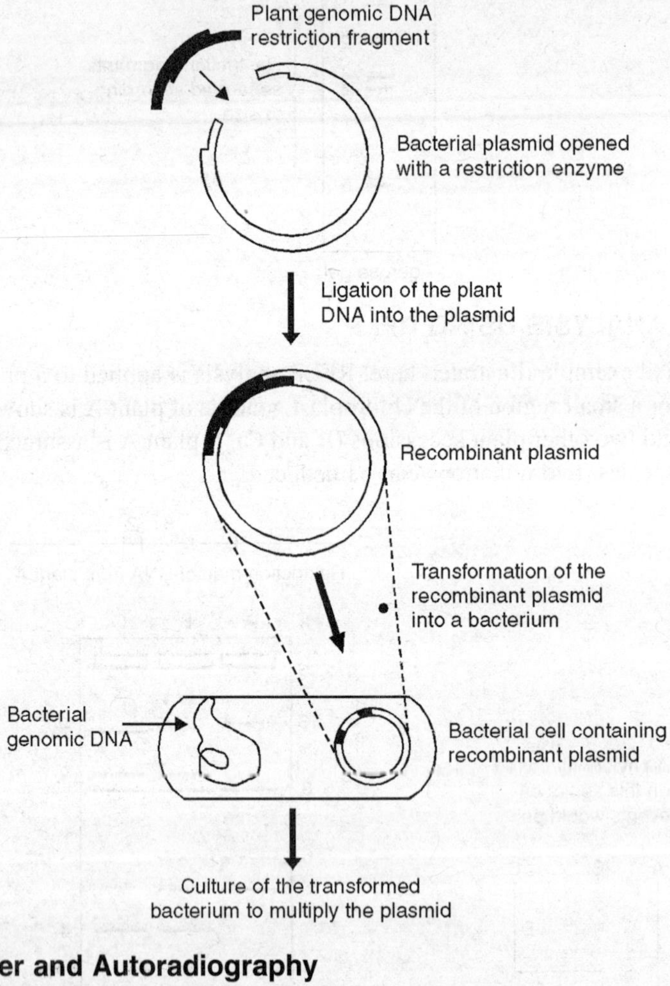

Southern Transfer and Autoradiography

After DNA has been fractionated by agarose electrophoresis, it can be transferred out the gel and on to a memberane filter by the process Southern transfer. An exploded view of this process is diagrammed next page. The sponge would be partially immersed in a suitable buffer. As the buffer passes through the gel on its way to being absorbed in the paper towels, the DNA is carried out of the gel and binds to the filter.

After the filters are prepared, radioactive RFLP probes are hybridized to them. Location of restriction, fragments homologous to the probes is determined by autoradiography, and RFLP data can be scored from the developed film.

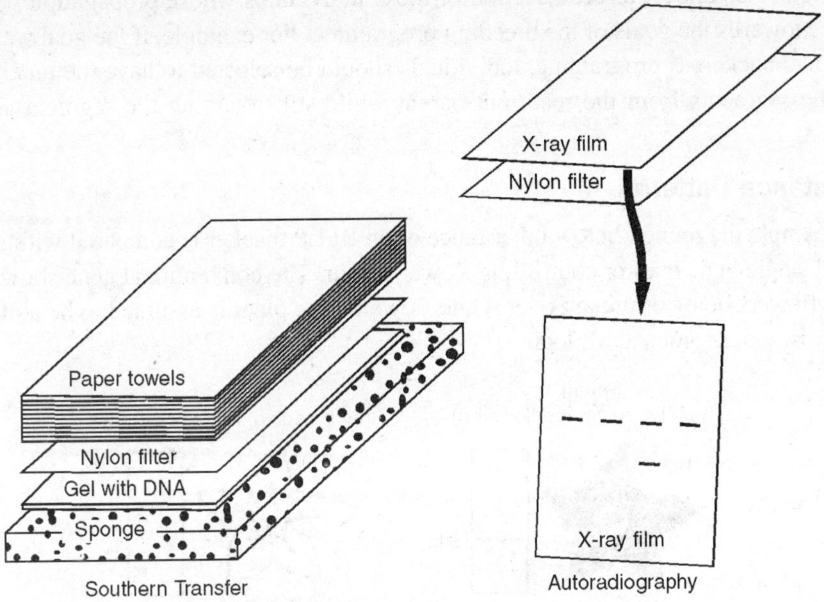

Inheritance of Chromosome Segments

This diagram illustrates the way chromosome segments are inherited in a genetic cross. Two inbred, completely homozygous parents are crossed to poroduce an F_1,which is heterozygous at all loci (only a single chromosome pair is shown). During selfing of the F_1, recombination produces gamete chromosomes which contain varying parental segments. Four F_2 plants illustrate the distribution of chromosome segments possible in such a segregating population.

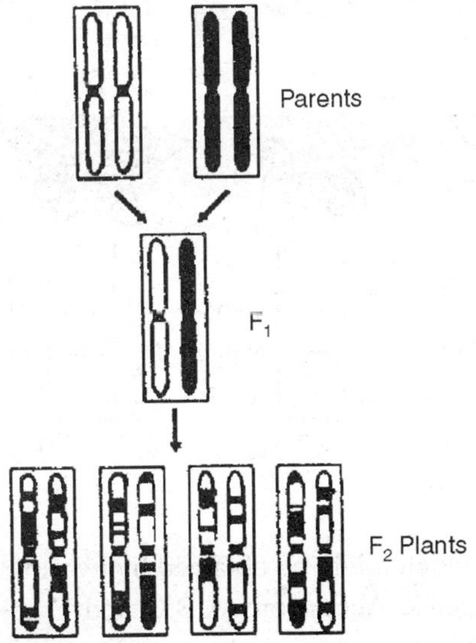

Since large numbers of RFLP markers can be analyzed in a single mapping population, it is possibile to construct chromosome segment maps of a segregating population. These maps show the distribution of chromosome segments from each of the parents for the entire genome of each individual in the population. This allows direct selection of those individuals whose propagation will lead to the fastest advance towards the goals of the breeding programme. For example, if the goal was to introduce a single gene in a backcross programme, individuals should be selected to have the maximum number of chromosome segments from the recurrent parent while still retaining the segment containing the gene of interest.

RFLP Inheritance Patterns

In the exmple diagramed below inheritance of an RFLP maeker is compared with inheritance of a conventional single gene marker controlling flower colour. The conventional gene shows dominance, with coloured flowers being dominant over white flowers. The plant is assumed to be a diploid, selfing species, which is homozygous at all loci.

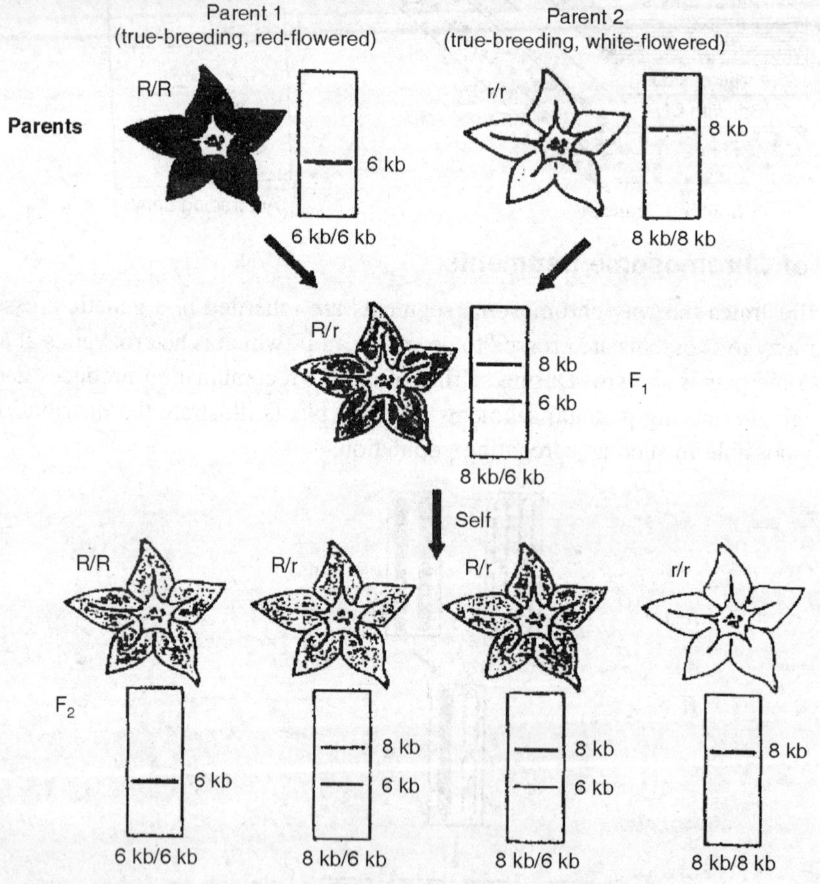

Polymorphism Survey

This figure shows results which might be obtained by hybridizing a cloned probe to DNA from several accessions of a plant species. Accessions A,B,C and E are homozygous for a 6 kb allete, accessions D and F are homozygous for a 4 kb allele, accession G is heterozygous for the two alleles.

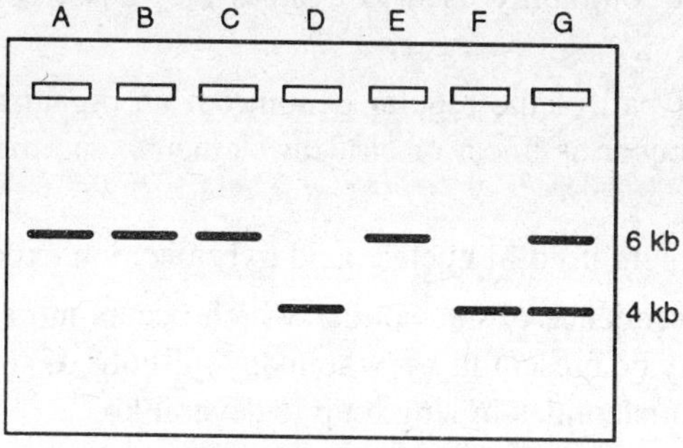

Quantitative Traits

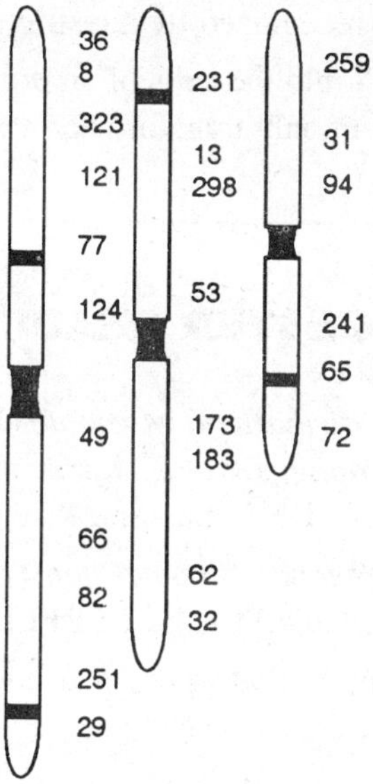

This figure illustrates how RFLP mape can be used to analyze quantitative traits in plants. Three chromosomes of a hypothetical plant are shown at the left. Numbers to the right of the chromosomes indicate the position of mapped RFLP markers. The position of some genes affected a quantitative trait are shown as shaded areas on the chromosomes.

Since all areas of the chromosomes are "tagged" with RFLP markers, it is possible to follow them through a genetic cross and to analyze their effect on the trait of interest. In this example, RFLP markers numeber 77, 29, 231, and 65 will show strong association with the trait.

Glossary of RFLP

cDNA-complementary DNA, a fragment of DNA which has been produced from an RNA sequence by reverse transcription. Messenger RNA is commonly used to synthesize cDNA.

kb - kilobase, 1000 bases of base pairs. DNA sizes are often expressed in kilobases.

library - a collection of cloned DNA fragments. The library may consist of cDNAs or genomic cloned (fragments cloned directly from cellular DNA).

ligation - joining of DNA fragments to produce a single DNA molecule. Ligases are enzymes which perform this reaction. fragment of plant DNA are ligated into baterial plasmids during the cloning of probes to use for RFLP analysis.

pcogram - 10^{-12} grams, a unit commonly used to express the DNA content per cell or per nuclear genome in a plant.

plasmid - DNA which is not part of the regular genome of an organism, and self-replicates during growth. Plasmids commonly occur as linear or cirulcar elements, and may be present in one to many copies per cell.

probe - a DNA or RNA fragment used in nucleic acid hybridization experiments to detect RFLPs.

repeated DNA sequence - a sequence of nucleotides which occurs more than once in a genome. Repeated seqeated sequences may be present in a few to many millions of copies. The individual repeated sequence may be only a few nucleotides in length up to several kb.

restriction enzyme - an endonuclease with the ability to cleave DNA at the point where a certain base sequence occurs.

restriction site - a DNA base sequence recognized by a restriction enzyme.

transformation- the introduction of DNA into the cells of an organism by a method other than conventional sexual crossing. Plasmids are commonly transformed into bacterial cells by direct DNA uptake through the cell wall and membrane.

SUGGESTED READINGS

Antonio, B.A., Y. Harushima, T. Kirihara, Y. Myamoto, Y, Nagamura, T. Shimizu, A. Shomura, N.Sue, Y, Tamure, T. Toyama, Z.Wang, J.Wu, K. Yammamoto, M. Yano and N. Kurata.1992. An RFLP likage map with informative DNA markers. Rise Genome. 1(2):4-7

Chareyazie, B., N. Huang, G. Second, J.Bennet and G.S. Khush. 1995. Classfication of rice germplasm I. Analysis using ALP and PCR based RFLP. Theor.Appl. Genet 91 : 218-227

Hittalmani, S.Foolad, M.R. Mew, T. Rodriguez, R.L, and Huang, N. 1995. Development of a PCR based marker to identify rice blast resistant gene, PI-2(t) in a segregating population. Theor. Appl. Genet. 91: 9 -14

Inoue, T., L.s. Zhong, A. Miyaro, I. Ashikawa, L.Monna, S. Fukuoka, N. Myadera, Y. Nagamuran. Kurata, T. sasaki and Y. Minobe. 1994. sequence-tagged-sites(STSs)as standerd landmarkers in rice genome. Theor.Appl.Genet. 89 : 734.

Karl, S.A., B.W. Brain and J.C Avise. 1992. Global population genetic structure and male mediated gene flow in the green turtle (Chelonia mydas): RFLP analysis of anonymous loci. Genetics.131:163-173.

Monna, L., A. Miyao, T. Inoue, S. Fukuoika, M. Yamazaki, H.S. Zhong, T. Sasaki and Y. Minobe. 1994. Determination of RAPD markers in rice and their conversion into sequence tagged sites (STSs) and STS specific primers. DNA research 1:139-148.

Tragoonrung, S., V. Kanazn, P.M. Hayes and T.K. Blake. 1992. Sequence-tagged-site facilitated PCR for barley genome mapping. Theor. Appl.Genet. 84 : 1002 - 1008

Weining, S. and P. Langridge. 1991. Identification and mapping of polymorphisms in cereals based on the polymerase chain reaction. Theor appl.Genet. 82 : 209 - 216.

Williams, M.N.V., N. Pande, S. Nair, M. Mohan and J. Bennet. 1991. Restriction fragment length polymorphism analysis of polymerase chain reaction products amplified from mapped loci of rice (Oryza sativa L.) genomic DNA. Theor. Appl.Genet. 82 : 489 - 498.

Glossary

2n. Somatic chromosome number of a species. The chromosome number present in somatic cells of a species.

ADAPTATION. The process by which individuals (or parts of individuals),population, or species change in form or function in such a way to better survive under given environmental conditions. Also the result of this process.

ADDITION LINE. An addition line has one pair of chromosomes from another variety or species in addition to the normal somatic chromosome complement (2n) of the species.

ALIEN-ADDITION LINE. It has one pair of chromosomes from a related wild species in addition to the normal somatic chromosome complement (2n) of the species.

ALLELE OR ALLELOMORPH. One of a pair or series of forms of a gene which is alternative in inheritance because they are situated at the same locus in homologous chromosomes.

ALLOGAMY. In allogamy, pollen grains from flowers of one plant pollinate the flowers of other plants (Syn., cross-pollination).

ALLOPOLYPOID. A polyploid containing genetically different sets of chromosomes, for example, sets from two or more different species.

AMPHIDIPLOID. A polyploid whose chromosome complement is made up of the entire somatic complements of two species.

ANEUPLOID. An organism whose somatic chromosome number is not an even multiple of the haploid number.

ANTHER CULTURE. Culture of anthers (or pollen grains) on a suitable medium for the production of callus and/or haploid plants.

ANTHESIS. The first opening of a flower.

ANTIBIOSIS. The Adverse effect of feeding on a resistant host plant on the development and reproduction of an insect pest.

ANTIMETABOLITE. A chemical which interferes with the metabolic activities of an organism, e.g., antibiotics.

APOGAMY. Development of embryo from synergids or antipodal cells without fertilization; a form of apomixis.

APOMIXIS. Reproduction in which sexual organs or related structures take part but fertilization does not occur, so that the resulting seed is vegetatively produced.

APOSPORY. A form of apomixis in which the embryo sac develops from a vegetative cell of the ovule.

ARTIFICIAL SELECTION. Selection by man.

ASEXUAL REPRODUCTION. Reproduction which does not involve the union of gametes.

ASYMMETRICAL HYBRIDS. Cells/plants produced through somatic hybridization, and containing the full somatic complement of one fusion partner but only part of the somatic complement of the other.

ASYNAPSIS. Failure of pairing of homologous chromosomes during meiosis.

AUTOGAMY. Pollen grains of a flower pollinate the same flower (Syn., self-pollination).

AUTOGAMY. Self-fertilization.

AUTOPOLYPLOID. A polyploid arising through multiplication of the complete haploid set of a species.

AVOIDANCE. Escaping of a susceptible host plant from damage due to an insect pest, usually, as a result of the host plants being at a much less susceptible developmental stage when the pest population is at its peak.

B_1 AND B_2. Backcrosses of F_1 to P_1 (first parent) and P_2 (second parent), respectively, of the hybrid.

BACKCROSS BREEDING. A system of breeding whereby recurrent backcrosses are made to one of the parents of a hybrid, accompanied by selection for a specific character or characters.

BALANCE. The condition in which genetic components are adjusted in proportions that give satisfactory development. Balance applies to individuals and populations.

BASIC NUMBER. The number of chromosomes in ancestral diploid ancestors of polyploids, represented by x.

BASIC SEED. Original seed evolved by a plant breeder by crossing selected genotypes.

BC_1, BC_2, BC_3 etc. Progeny from F_1 x recurrent parent, BC: x recurrent parent, BC_2 x recurrent parent etc., respectively.

BIAS. A consistent and false departure of a statistic from its proper value.

BIOCONTROL. Use of biological agents, e.g., viruses, bacteria, fungi etc., for the control of pathogens, insect pests, weeds etc.

BIOLOGICAL PESTICIDES. Biological entities or their products used to control insect pests, e.g., *Bacillus thuringiensis* spores or a preparation of the crystal protein produced by it.

BIOMETRY. The branch of science which deals with statistical procedures in biology.

BIOTECHNOLOGY. Utilization of biological agents or their components for generating products or services for the welfare of mankind.

BIOTRANSFORMATION. Conversion of a less useful compound into a more valuable and useful one through the action of an organism or (more precisely) by an enzyme produced by an organism.

BIOTYPE. A group of individuals with the same genotype. Biotypes may be homozygous or heterozygous.

BIPARENTAL CROSS ANALYSIS. A design for genetic analysis based on progenies derived from the matings involving a random sample of F_2 or later generation plants from a cross between two pure lines; the four designs based on such progenies are NCD I, NCD II, NCD III and TTC.

BIVALENT. A pair of homologous chromosomes united in the first meiotic division.

BREEDER SEED. Seed produced by the agency sponsoring a variety and used to produce foundation seed.

BREEDING. The art and Science of changing plants or animals genetically.

BRIDGING SPECIES. A species used in gene transfer from one species to another sexually incompatible species; bridging species is compatible with both the donor and recipient species; it may be natural or synthetic species.

BUD MUTATION. Mutation in somatic tissues, usually affecting an axillary bud.

BUD SELECTION. A form of clonel selection in which mutant buds are selected.

BULK BREEDING. The growing of genetically diverse populations of self-pollinated crops in a bulk plot with or without mass selection, followed by single-plant selection.

CENTRES OF DIVERSITY. Areas where cultivated plant species and/or their wild relatives show much greater variation than anywhere in the rest of the world.

CENTRES OF ORIGIN. Areas where cultivated plant species are supposed to have originated. Based on centres of diversity.

CENTROMERE. The localized region of a chromosome where spindle fibers attach, and which is responsible for chromosome movement during cell division.

CERTIFIED SEED. Seed used for commercial crop production produced from foundation, registered, or certified seed under the regulation of a legally constituted agency.

CHARACTER. A morphological, anatomical or physiological feature of an organism; usually a product of the actions of both genotype and environment.

CHARACTER. An attribute of an organism resulting from the interaction of a gene or genes with environment.

CHIASMA. An exchange of partners between paired chromatids in the first division of meiosis.

CHIMERA. An individual that is composed of cells of two or more genotypes.

CHROMATID. one of two threadlike structures formed by the longitudinal division of a chromosome during meiotic prophase and known as a daughter chromosome during anaphase.

CHROMOSOME MANIPULATION TECHNIQUES. Techniques used for promoting gene transfers from chromosomes of a related species into those of a cultivated species.

CHROMOSOMES SUBSTITUTION. The procedure or the process of replacing one pair of chromosomes of a variety with those of another variety or a related species. The latter is alien substitution.

CHROMOSOMES. Structural units of the nucleus which carry the genes in linear order. Chromosomes undergo a typical cycle in which their morphology changes drastically in various phases of the life cycles of organisms.

CLEISTOGAMY. Flowers do not open at all, i.e., there is no anthesis.

CLONAL MULTIPLICATION. Multiplication through asexual reproduction.

CLONAL SELECTION. A method of selection based on clones. It does not involve sexual reproduction.

CLONE. A group of organisms descended by mitosis from a common ancestor.

COEFFICIENT OF VARIATION. The ratio of standard deviation of a sample to the mean of that sample; usually expressed in per cent [(S/X) × 100].

COMBINATION BREEDING. It involves the transfer of one or a few characters, usually governed by oligogenes, from one variety into another.

COMBINING ABILITY. General, average performance of a strain in a series of crosses. Specific, deviation from performance predicted on the basis of general combining ability.

COMPETITIVE ABILITY. The ability of a genotype to survive in a mixture of several genotypes.

COMPOSITE CROSS. A hybrid population generated from a large number of pure lines either by compositing equal seeds from the F_1 or F_2 generations from all possible crosses among them, by ordering them into a complex cross, or allowing them to cross with a male sterile line and harvesting the seeds from only the male sterile plants in each generation.

COMPOSITE VARIETIES. Varieties produced by open-pollination among a number of outstanding strains usually not tested for combining ability with each other.

CONTIGUOUS CONTROL. A control or check variety planted in an adjacent plot; design for the evaluation of a large number of lines/progenies in which the control variety is planted in every 5th or 7th row/plot (augmented design), and the comparisons of the line/progeny means are made with the control.

CONTROL. In plant breeding; a known variety used for comparing the performance of newly developed strains during their evaluation.

CONVERGENT CROSS. A cross involving more than two parents (Syn., complex cross).

COORDIANTED VARIETAL TRAIL (CVT). Multilocation trials (usually for two year) conducted under the respective All India Coordinated Crop Improvement Projects for the evaluation of the performance of newly developed strains; in many crops, one year of IET precedes CVT. (Syn., URT, Uniform regional trial). (IET = Initial Evaluation Trial).

CORRELATED RESPONSE TO SELECTION. Change in one or two more quantitative characters due to selection for another character.

COUPLING. Linked recessive alleles occur in one homologous chromosome and their dominant alternatives occur in the other chromosome. Opposed to repulsion in which one dominant and one recessive occur in each member of the pair of homologous chromosomes.

COVARIANCE. The mean of the product of the deviation of two variants from their individual means. A statistical measure of the interrelation between variables.

CROSS. Mating of two or more individuals or strains having different genotypes. Also the product of such a mating.

CROSSING OVER. Exchange of homologous segments between nonsister chromatids of homologous chromosomes; takes place during pachytene.

CROSSING OVER. The exchange of corresponding segments between chromatids of homologous chromosomes during meiotic prophase. Its genetic consequence is the recombination of linked genes.

CYBRIDS. Somatic hybrids obtained by fusion of the complete protoplast of one species with such protoplasts of another species which either have no nucleus or have an inactive nucleus, e.g., protoplast of species A + cytoplast of species B = cybrid.

CYTOGENETICS. Study of chromosomes in relation to genetics.

CYTOPLASMIC INHERITANCE. Transmission of hereditary characters through the cytoplasm as distinct from transmission by genes carried by chromosome. Detected by differing contribution of male and female parents in reciprocal crosses.

CYTOPLASMIC MALE STERLITY. Male sterility showing cytoplasmic inheritance.

CYTOPLASMIC-GENETIC MALE STERILITY. Cytoplasmic male sterility for which a restorer gene is available/known.

CYTOPLAST. A protoplast containing either no nucleus or an inactive nucleus.

DEFICIENCY. The absence or deletion of a segment of chromosome.

DEGREES OF FREEDOM, NUMBER OF. The number of independent comparisons that can be made in a set of data.

DETASSEL. Removal the tassel (male inflorescence) as in maize.

DETASSELLING. Removal of the tassel (the male inflorescence) before it sheds pollen; in maize. An easy method of emasculation.

DEVIATION. Departure of an observation from its expected value.

DIALLEL CROSS, COMPLETE. The crossing in all possible combinations of a series of genotypes.

DIALLEL SELECTIVE MATING SCHEME. A breeding scheme in which several selected pure lines are mated according to the diallel scheme; the promising F_2 populations are advanced by selfing for the isolation of pure lines to be used as varieties; simultaneously the selected plants in the F_2 generations are intermated to generate populations which will again be handled as above. This process continues.

DICHOGAMY. Maturation of male and female reproductive organs of a hermaphrodite flower at different times.

DIHAPLOID. Haploid derived from tetraploid, i.e., monoploid for two distinct genomes.

DIHYBRID. Heterozygous with respect to two genes.

DIOECIOUS. Plants in which staminate an pistillate flowers occur on different individuals.

DIPLOID. An organisms having two copies of a single genome, i.e., with chromosome number of 2x.

DIPLOID. An organism with two chromosomes of each kind.

DIPLOSPRAY. A form of apomixis in which the embryo sac develops from the megaspore.

DIPLOTENE. The stage of meiosis which follows pachytene and during which the four chromatids of each bivalent move apart in two pairs but remain attached in the region of chiasmata.

DIRECT SELECTION. The selection agent or condition permits the survival and proliferation of only the variant cells, e.g., selection for resistance to antibiotics.

DISARMING OF TI PLASMID. Removal of that DNA segment of the T-DNA of Ti plasmid which is responsible for the production of tumors in plants by producing auxins and/or kinins.

DISEASE. An abnormal condition produced by an organism.

DISJUNCTION. The separation of chromosomes at anaphase.

DISOMIC. An individual with two homologues for each chromosome of the genome or genomes.

DISTANT CROSS. See, Wide Cross.

DISTANT HYBRIDIZATION. Hybridization between individuals belonging to two different species of the same genus or of different genera.

DOCKAGE. Per cent impurity in a seed sample.

DOMESTICATION. The process of bringing a wild species under human management.

DOMINANCE. Intra-allelic interaction such that one allele manifests itself more or less, when heterozygous, than its alternatives allele.

DONOR PARENT. The parent from which one or a few genes are transferred to the recurrent parent in backcross breeding.

DOUBLE CROSS. A cross between two F_1 hybrids.

DOUBLED HAPLOID TECHNIQUE. A technique in which homozygous lines are obtained by doubling the chromosome numbers of haploid plants extracted from heterozygous (usually F_1) plants.

DOUBLED HAPLOID. A plant or line obtained by doubling the chromosome number of a haploid plant/individual.

DRIFT. Changes in gene and genotypic frequencies in small populations due to random processes.

DUPLEX. (See nulliplex.)

DUPLICATION. Occurrence of a segment of a chromosome twice in the haploid set.

EARLY TESTING. In self-pollinated crops, evaluation of the Performance of early segregating generations (usually, F_2 or F_3 bulks) or F_3/F_4 individual plant progenies in order to assess the worth of pure lines that would be isolated from them; in cross-pollinated crops, the estimation of GCA or SCA of plants/progenies early in the process of inbreeding.

ELECTROFUSION. Induction of fusion between two protoplasts by exposing them to a very short (few milliseconds) duration high voltage current.

ELECTROPORATION. Induction of fusion between two protoplasts by exposing them to a very short duration high voltage electric current; this induces transient pores in the plasmalemma.

EMASCULATION. Removal of immature anthers (or androecium) from a hermaphrodite flower.

EMASULATION. Removal of the anthers from a flower.

EMBRYO. CULTURE. Removal of developing young embryos from seed and their cultivation in vitro.

EMBRYO. SAC. CELL DERIVED FROM MEGASPORE. It contains egg cell, synergids, antipodals and polar nuclei (Syn. megagametophyte).

ENDONUCLEASES. Enzymes which produce single/double-stranded cuts within DNA molecules (that is, they do not digest DNA molecules from their ends).

ENTRY. A newly developed line/strain included for evaluation in the multilocation trails of the various All India Coordinated Crop Improvement Projects.

ENVIRONMENT. The sum total of the external conditions which effect growth and development of an organism.

ENZYME ENGINEERING. Enhancing the activity or changing the substrate specificity of enzymes by changing their amino acid sequences, e.g., through site-directed mutagenesis.

ENZYME TECHNOLOGY. Immobilization of enzymes in different matrices with a view to enhance their efficiency, reusability or hygiene.

EPIDEMIC. Development of a disease from a low infection level to a high intensity of infection. Artificial epidemics are created by man by providing the inoculum and the suitable environment for the disease development and used in disease resistance tests.

EPIGENETIC CHANGES. Changes that are suitable through mitotic divisions, but not to sexual progeny.

EPIPHYTOTIC. See, Epidemic.

EPIPHYTOTIC. An unarrested spread of a plant disease.

EPISTASIS: Dominance of one gene over a nonallelic gene. The gene suppressed is said to be hypostatic. More generally, the term epistasis is used to describe all types of interallelic interaction whereby manifestation at any locus is affected by genetic phase at any or all other loci.

ERROR VARIANCE. Variance arising from unrecognized or uncontrolled factors in an experiment with which the variance of recognized factors is compared in tests of significance.

ESCAPE. More commonly disease escape. Phenomenon of susceptible plants avoiding disease attack. Also plants showing escape.

EUPLOID. Individuals whose chromosome number is an exact multiple of the basic number.

EVALUATION. The process of assessing the performance (yield, quality, disease and insect resistance, etc.) of newly developed strains of a crop through appropriate multilocation trails.

EXCITATION. Movement of an electron to an outer orbit of a higher energy level.

EXOTIC. A variety or species introduced from a foreign country.

EXPLORATION. A trip for collection of germplasm of cultivated and related wild species (Syn., expedition).

EXPRESSIVITY. The degree of manifestation of a genetic character.

EXTREME HOMOZYGOTE. A homozygote isolated from a hybrid, usually produced by crossing two or more parents, that has either all the positive or negative alleles for a trait for which the hybrid was heterozygous.

F_1. The first generation of a cross.

F_2. The second filial generation obtained by self-fertilization or crossing inter se of F_1 individuals.

F_3. Progeny obtained by self-fertilizing F_2 individuals.

FACTOR. Same as gene.

FAMILY. A group of individuals directly related by descent from a common ancestor.

FERTILITY. Ability to produce viable offspring.

FERTILIZATION. Fusion of the nuclei of male and female gametes.

FOUNDATION SEED. Seed stock produced from breeder seed by or under the direct control of an agricultural experiment station. Foundation seed is the source of certified seed, either directly or through registered seed.

FREE RADICAL. Electrically neutral molecules with an unpaired electron in the outer orbit. Free radicals are highly reactive.

FUSOGEN. An agent that induces fusion among protoplasts, e.g., PEG (polyethylene glycol)

GAMETE. Cell of meiotic origin specialized for fertilization.

GAMETIC CHROMOSOMES NUMBER. Chromosomes number present in gametes of a species.

GAMETOGENESIS: Production of gametes from spores.

GENE BANK. Large collection of germplasm representing materials from various parts of the world (Syn., world collection)

GENE FREQUENCY. The proportion in which alternative alleles of a gene occur in a population.

GENE INTERACTION. Modification of gene action by a nonallelic gene or genes

GENE POOL. Sum total of genes present in a Mendelian/panmictic (random mating) population.

GENE SECONDARY. An area within the centre of diversity protected from human interference.

GENE. The unit of inheritance. Genes arc located at fixed loci in chromosomes and can exist in a series of alternative forms called alleles.

GENERAL COMBINING ABILITY. Average performance of a strain in a series of cross combinations. Estimated from the performance of F_1 s from the crosses.

GENETIC ADVANCE (UNDER SELECTION). Improvement in the performance of selected lines over the original population.

GENETIC EMASCULATION. Use of genetic factors to make the male gamete nonfunctional in self-and or cross-pollination.

GENETIC ENGINEERING. Isolation of the desired gene from a organisms, its integration into a suitable vector and then its introduction into another organisms (called host) with a view to either obtain multiple copies of the desired gene or the protein produced by this gene. The gene may remain in the vector or may get integrated into the chromosome of the host; the latter produces transgenic organisms.

GENETIC EQUILIBRIUM. The condition in which successive generations of a population contain the same genotypes in the same proportions with respect to particular genes or combinations of genes.

GENETIC EROSION. Gradual disappearance of various forms of a cultivated species and of its wild relatives.

GENETIC LOAD. The sum total of deleterious (harmful) alleles present in a random mating or Mendelian population.

GENETIC PURITY OF SEED. Freedom of seed from seeds for weeds, other crops and other varieties of the same crop.

GENETIC VARAINCE. Variance due to genotypes of different plants or strains. It has additive, dominance and epistatic components.

GENETIC VULNERABILITY. The susceptibility of most of the cultivated varieties of a crop species to a disease, insect pest, or some other stress due to similarities in their genotypes; usually due to the presence of one (generally more) common parent(s) in their ancestry.

GENETICS. A study of the mechanism of transmission of characters from parents to their offspring, origin of variation and gene action.

GENERATION MEAN ANALYSIS. The estimation of additive, dominance and epistatic gene effects from the means of certain specific families, usually, P_1, P_2, F_1, F_2, B_1 and B_2; F_3 may be used in the place of B_1 and B_2, but the j effect cannot be estimated in such a case.

GENOME. A set of chromosomes corresponding to the haploid set of a species.

GENOTYPE. The entire genetic constitution of an organism.

GERMPALSM. The sum total of the hereditary materials in a species.

GERMPLASM COMPLEX. Open-pollinated progeny from a mixture of a number of strains of diverse origin.

HAPLOID. A cell or organism with the gametic chromosome number (n).

HERITABILITY. The proportion of observed variability which is due to heredity, the remainder being due to environmental causes. More strictly, the proportion of observed variability due to the additive effects of genes.

HERMAPHRODITISM. Reproductive organs of both sexes present in the same individual or in the same flower in higher plants.

HETEROSIS. Hybrid vigor such that an F_1 hybrid falls outside the range of the parents with the respect to some character or characters. Usually applied to size, rate of growth, or general thriftiness.

HETEROSTYLY. Occurrence of styles and stamens of different lengths in flowers from different plants of a single species.

HETEROTHALLY. Haploid incompatibility in fungi (opposite of homothally).

HETEROZYGOUS. Having unlike alleles at one or more corresponding loci (opposite of homozygous).

HETERPLOID. An individual with a chromosome number other than the normal diploid (2x) number.

HETERROCARYOSIS: The presence of two or more genetically different nuclei within single cell of a mycelium.

HOMEOLOGOUS CHROMOSOMES. Partially homologous chromosomes; generally show reduced or lack of pairing.

HOMOGENEOUS. Consisting of individuals of the same genotype or phenotype.

HOMOLOGY OF CHROMOSOMES. Applied to whole chromosome or parts of chromosome which synapse or pair in meiotic prophase.

HOMOZYGOUS. Having like alleles at corresponding loci on homologous chromosomes. An organism can be homozygous at one, several, or all loci.

HORIZONTAL RESISTANCE. Resistance governed by polygenes and is pathotype nonspecific.

HOST. The organism attacked by a pathogen.

HYBRID SUBSTANCE. New isozyme present in hybrid which is different from those present in its presents.

HYBRID VARIETIES. A hybrid variety is the F_1 generation from a cross between two different strains.

HYBRID VIGOUR. See, Heterosis.

HYBRID. The product of a cross between genetically unlike parents.

HYBRIDIZATION. Mating of two different strains.

I_1, I_2, I_3,..... Symbols used to designate first, second etc. inbred generations (See S_1, S_2....)

IDENTIFICATION OF A VARIETY. Recommendation by the annual workshop of the concerned All India Crop Improvement Project for the release of a strain or an entry as a new variety.

IMMUNE REACTION. Complete absence of symptoms of a disease even after the host is exposed to the pathogen.

IMPROVED SEED. Seed of an improved variety having a very high genetic and physical purity and germination.

IN VITRO TECHNIQUES. The technique of culturing the plants cells and organs on artificial media in vitro.

INBRED-VARIETY CROSS. The F_1 cross of an inbred line with a variety.

INBREED LINE: A line produced by continued inbreeding. In plant breeding a nearly homozygous line usually originating by continued self-fertilization, accompanied by selection.

INBREEDING COEFFICIENT. a quantitative measure of the intensity of inbreeding.

INBREEDING DEPRESSION. Loss in vigor due to inbreeding.

INBREEDING. The mating of individuals more closely related than individuals mating at random.

INDEPENDENCE. The relationship between variables when the variation of each is uninflucnced by that of others, that is, correlation is zero.

INERT MATTER. In a seed sample; all the nonliving matter, disease and insect infested seeds and broken (damaged) seeds.

INFESTATION. The attack by an insect pest on the host plant.

INITIAL EVALUATION TRIAL. The preliminary evaluation of a relatively large number of newly developed strains in the multilocation trails of the concerned All India Coordinated Crop Improvement Project; the purpose is to reduce the number of strains to a manageable level for a relatively more extensive evaluation in URT or CVT; omitted in crops where the number of entries is not large.

INSECT PEST. An insect species which causes damage to a crop species; they may be tissue feeders or sucking pests.

INSECT RESISTANCE. The property of a host variety to be attached by an insect pest to a noticeably smaller degree than some other varieties of the same host species.

INTERFERENCE. The effect of one crossover influencing the probability that an other will occur in the immediate vicinity.

INTERGENOTYPIC COMPETITION. Competition between or among plants of different genotypes; important in the segregating generations, varietal mixtures, etc.

INTERPLANT COMPETITION. Competition among plant of the same genotype; important in the stands of pure lines, single cross hybrids, clones etc.

INTERSPECIFIC HYBRIDIZATION. Hybridization between plants belonging to two different species of the same genus.

INTRODUCTION, PLANT. Taking of a few genes from one species into the full diploid chromosome complement of another species.

INVERSION. A rearrangement of a chromosome segment so that its genes are in reversed linear order.

ION. An atom that has lost or gained an electron in comparison to the number of electrons normally present.

IONIZATION. Loss or grain of an electron by an atom.

IRRIDIATION. Exposure of plants or plant parts to X-rays or other radiations to increase mutation rates.

ISOGENIC LINES. Two or more lines differing from each other genetically at one locus only. Distinguished from clones, homozygous lines, identical twins, etc., which are identical at all loci.

ISOLATION. The separation of one group from another so that mating between or among groups is prevented.

ISOALLELES. Alleles indistinguishable except by special tests.

KINETOCHORE. See, Centromere.

LD_{50}. The dose of a mutagen (or any other agent) which will kill 50 per cent of the treated individuals.

LETHAL GENE. Usually recessive; lethal genes kill each and every individual that carries them in the appropriate genotype.

LINE A group of individuals having common parents or ancestors.

LINE BREEDING. A system of breeding in which a number of genotypes which have been progeny tested in respect to some character or group of characters are composited to form a variety.

LINE X TESTER ANALYSIS. The estimation of GCA and SCA effects and variances from progenies developed by mating each of the several lines to each of the same selected testers; additive and dominance variances are estimated from GCA and SCA variances.

LINKAGE MAP. Map of position of genes in chromosomes determined by recombination relationship.

LINKAGE VALUE. The recombination fraction expressing the proportion of cross overs versus parental types in a progeny. The recombination fraction can vary from zero to one half.

LINKAGE. Association of characters in inheritance due to location of genes in proximity on the same chromosome.

LOCUS. The position occupied by a gene in a chromosome

M_1, M_2, M_3,..... Symbols use to designate first, second third, ... generations after treatment with a mutagenic agent.

MAINTAINER LINE. Line used for maintaining a cytoplasmic male sterile line. It has the same nuclear genotype as the male sterile line.

MALE STERILITY. Absence of nonfunction of pollen in plants.

MASS SELECTION. A form of selection in which individual plants are selected and the next generation propagated from the aggregate of their seeds.

MASS-PEDIGREE METHOD. A system of breeding in which a population is propagated in mass until conditions favourable for selection occur, after which pedigree selection is practiced.

MATING SYSTEM. Any of a number of schemes by which individuals are assorted in pairs leading to sexual reproduction. Random, assortment of pairs is by chance. Genetic assortative mating, mating together of individuals more closely related than individuals mating at random. Genetic disassortative mating, mating together of individuals less closely related than individuals mating at random. Phenotypic assortative mating, mating individuals more alike in appearance than the average. Phenotypic disassortative mating, mating of individuals less alike in appearance than individuals mating at random.

MEAN. The arithmetic average of a series of observations.

MEDIAN. The value of the variate on each side of which there is an equal number of larger and smaller variates.

MEGAGAMETOGENESIS. Production of the female gametophyte embryosac from megaspore (through meiosis)

MEIOSIS. A double mitosis occurring in sexual reproduction which results in production of gametes with haploid (n) chromosome number.

MERICLONING. Vegetative multiplication through meristem culture; not in common use.

MERISTEM CULTURE. Cultivation of apical meristems, particularly of shoot apical meristem, for the production of shoots and plantlets.

METAPHASE. The stage of meiosis or mitosis at which the chromosomes lie on the spindle.

METAXENIA. Influence of pollen on maternal tissues of the fruit. (See xenia)

MICROCENTRE. Area within a centre of origin that shows a greater diversity than the remaining centre of origin.

MICROGAMETOGENESIS. Production of the male gametophyte from microspore (through mitosis)

MICROSPOROGENESIS. Production of pollen grains (microspores) from microspore mother cells (through meiosis).

MIGRATION. Movement of individuals from one population into another.

MITOSIS. The process by which the nucleus is divided into two daughter nuclei with equivalent chromosome complements, usually accompanied by division of the cell containing the nucleus.

MODE. The value of variate in the class of greatest frequency in a frequency distribution.

MODIFYING GENES. Genes that affect that expression of a nonallelic gene or genes

MONOECIOUS. Staminate and pistillate flowers borne separately on the same plant.

MONOHYBRID. Heterozygous with respect to one gene.

MONOPLOID. An organisms with the basic (x) chromosome number. (See haploid)

MONOSOME. An organism lacking one chromosome of the diploid complement, hence, having 2n - 1 chromosomes.

MULTILINE VARIETIES. Mixtures of several similar pure lines having different genes for disease resistance.

MULTILOCATION TRAILS. Trials conducted at several locations under the various All India coordinated crop improvement projects.

MULTIPLE ALLELE. A number of a series of more than two alternative forms of a gene.

MULTIVALENTS. (See univalent)

MUTAGEN. A chemical or physical agent that induces mutation.

MUTAGENESIS. Induction of mutations with the aid of mutagens.

MUTATION BREEDING. Breeding method utilizing variation created through mutagenesis.

MUTATION. A sudden heritable variation in a gene or in chromosome structure.

n. Gametic chromosome number of a species.

NATIONAL EVALUATION TRIAL (NET). In oilseeds; the plot size is three times the size of that in CVT; the entries are identified for prerelease multiplication only after NET.

NATIONAL TRAIL. In some crops, e.g., wheat; the topmost entry in CVT or URT of each zone is evaluated in an IET throughout the country to assess if an entry developed in one zone may perform well enough for release as a variety in another zone of the country.

NATURAL SELECTION. Selection due to natural forces, i.e., the environment.

NEGATIVE SELECTION. The selection condition does not allow the variant cells to divide while normal cells go on dividing. A counter selection agent is employed which gets incorporated into the dividing wild type cells. Later when the selection condition and the counter selection agent are removed, the variant cells proliferate, while the wild cells are killed by the counter selection agent present in them.

NOBLISATION OF SUGARCANE. Introgression of genes and chromosomes from noble canes and wild Saccharum species into India canes (S. barberi).

NONPREFERENCE. A host variety being unsuitable or unattractive to an insect pest for colonization, oviposition or both.

NONRECURRENT PARENT. In backcross method; the parent from which one or few genes are transferred to the recurrent parent (Syn., donor parent).

NORTH CAROLINA DESIGN I (NCD I). A mating design used for genetic analysis; four random plants are mated as females and one random plant used as male to generate one male group; four male group make a set and several sets may be generated in one experiment; applicable to the segregating generations (self-pollinated species) or random mating populations.

NORTH CAROLINA DESIGN II (NCD II). Equal number of male and female plants selected randomly and mated in a diallel fashion to yield m x f number of crosses (m = number of males, f= number of females); applicable to the segregating generations (self-pollinated species) or random mating populations.

NORTH CAROLINA DESIGN III (NCD III). Several random plants selected from an F_2 are backcrossed to the two parents to the concerned cross.

NOTIFICATION OF A VARIETY. A notification from the Director, High yielding Varieties, Ministry of Agriculture and Irrigation, Govt. of India, to the concerned authorities for the seed multiplication of a variety just released for cultivation by the Central of a State Variety Release Committee.

NUCLEUS SEEDS. See, Basic Seed.

NULL HYPOTHESIS. Hypothesis that there is no discrepancy between observations and expectation based on some set of postulates.

NULLIPLEX. The condition in which a polyploid is recessive in all chromosomes in respect to a particular gene. Simplex denotes recessiveness at all loci except one, duplex two, triplex three, quandriplex four, etc.

NULLISOME. An otherwise 2n plant that lacks both members of one specific pair of chromosome, hence with 2n-2 chromosomes.

OFF-SEASON NURSERY. Growing the breeding materials during the off-season(a season in which the crop is not growth in the concerned locality) in a location suitable for the purpose; generally to advance the generation.

OLIGOGENES. Genes having large individual effects producing distinct phenotypes.

OPEN-POLLINATION. In cross-pollinated species, pollination occurring naturally without restriction.

OUTCROSS. A cross, usually natural, to a plant of different genotype.

P1, P2, P3... . First, second etc. Generations from a parent. Also used to designate different parents used in making a hybrid or series of hybrids.

PACHYTENE. The double-thread or 4 strand stage of meiosis.

PANMIXIS. Random mating without restriction.

PARAMETER. A numerical quantity which describes some characteristic of a population.

PARAMETER. A numerical quantity which specifies a population in respect to some characteristic.

PARTHENOGENESIS. Development of an organism from a sex cell but without fertilization.

PARTIAL DIALLEL ANALYSIS. A diallel analysis based on a specified sample of crosses from among the all possible crosses (diallel scheme) among a number of parental lines.

PATH ANALYSIS. A standardized partial regression coefficient analysis which splits the various correlation coefficients into the measures of direct and indirect effects of a set of independent variables on a dependent variable (usually, yield).

PATHOGEN. An organism producing a disease.

PATHOTYPE. Strain of a pathogen virulent toward a specific resistance gene of the host.

PEDIGREE BREEDING. A system of breeding in which individual plants are selected in the segregating generations from a cross on the basis of their desirability judged individually and on the basis of a pedigree record.

PEDIGREE. A record of the ancestry of an individual, family or strain.

PENETRANCE. The frequency with which a gene produces a recognizable effect in individuals which carry it.

PHENOTYPE. Appearance of an individual as contrasted with its genetic make-up or genotype. Also used to designate a group of individuals with similar appearance but not necessarily identical genotypes.

PHYSICAL PURITY. Freedom of seed from inert matter, including broken seeds.

PHYSIOLOGICAL RACES. Pathogens of the same species with similar or identical morphology but differing pathogenic capabilities.

PLANT BREEDING. The branch of biology concerned with changing the genotype of plants so that they become more useful to man.

PLEIOTROPY. Phenomenon of a single gene affecting two or more different characters.

POLLEN CULTURE. Cultivation of pollen grains in vitro for producing haploid plants (generally referred to as anther culture).

POLLINATION. It consists of pollen grains reaching the stigma of a flower.

POLYCROSS. Open pollination of a group of genotypes (generally selected) in isolation from other compatible genotypes in such a way as to promote random mating inter se.

POLYGENES. Genes whose effects are too slight to be identified individually but which, through similar and supplementary effects, can have important effects on total variability.

POLYGENIC INHERITANCE. Inheritance of quantitative characters due to the action of polygenes.

POLYMORPHISM. The occurrence together in the same population of two or more distinct forms at frequencies too great to be explained by recurrent mutation.

POLYPLOID: An organism with other than two basic sets of chromosomes that is monoploid. triploid, tetraploid and various aneuploids.

POPULATION CROSS. A cross between two open-pollinated varieties (Syn., varietal cross).

POPULATION IMPROVEMENT. Improvement of random mating population through a scheme of selection with or without progeny test. It is essential to keep inbreeding to a low level.

POPULATION. In genetics, a community of individuals which share a common gene pool. In statistics, a hypothetical and infinitely large series of potential observations among which observations actually made constitute a sample.

POSITIVE SELECTION. The selection condition allows only the variant cells to survive and proliferate.

PROPAGULE. Plant part used for propagation.

PREPOTENCY. The capacity of a parent to impress characteristics on its offspring so they are more likely than usual.

PROBABILITY. The proportion of times in which an event occurs in an infinitely large and hypothetical series of cases, each capable of producing the event.

PROGENY TEST. A test of the value of genotype based on the performance of its offspring produced in some definite system of mating.

PROTANDRY. Maturation of anthers before pistils.

PROTOGYNY. Reverse of protandry.

PURE LINE. A strain homozygous at all loci, ordinarily obtained by successive self-fertilization in plant breeding.

PURELINE SELECTION. Isolation of pure lines from a mixture of pure lines (Syn., individual plant selection)

QUADRIALLEL ANALYSIS. Analysis based on double crosses obtained by crossing n homozygous lines in a diallel fashion, and then consisting the F_1's so generated according to a diallel scheme with the restriction that in any double cross a homozygous line must nor occur more than once as a parent.

QUADRIPLEX. (See nulliplex)

QUADRIVALENT. Structure produced by pairing among four homologous chromosome during meiosis.

QUADRIVALENT. (See univalent)

QUALITATIVE CHARACTER. A character in which variation is discontinuous.

QUANTITATIVE CHARACTER. A character in which variation is continuous so that classification into discrete categories is not possible.

QUARANTINE. Isolation of an organism for observation on weeds, diseases and pests and for preventing their spread.

RANDOM DRIFT. See, Drift.

RANDOM MATING. A system of mating in which an individual has equal chance of mating with every other individual of the same population.

RANDOM. Arrived at by chance without discrimination.

RANDOMIZATION. Process of making assignments at random.

RANGE. The differences between the highest and the lowest values of observations in a sample.

REAL VALUE OF SEED. The percentage (by weight) of the seed sample which would produce seedlings of the variety under certification.

RECESSIVE. The member of an allelic pair which is not expressed when the other (dominant) member occupies the homologous chromosome

RECIPIENT PARENT. In backcross breeding, the parent to which one or few genes from the donor parent are transferred (Syn., recurrent parent).

RECIPROCAL CROSSES. Crosses in which the sources of male and female gametes are reversed.

RECOMBINATION. Formation of new combinations of genes as a result of segregation in crosses between genetically different parent. Also the rearrangement of linked genes due to crossing over.

RECURRENT PARENT. The parent to which successive backcrosses are made in backcross breeding.

RECURRENT SELECTION. A method of breeding designed to concentrate favourable genes scattered among a number of individuals by selecting in each generation among the progeny produced by mating inter se of the selected individuals (or their selfed progeny) of the previous generation.

REGISTERED SEED. The progeny of foundation seed normally grown to produce certified seed.

REGRESSION, COEFFICIENT OF. A numerical measure of the rate of change of the dependent on the independent variables.

RELEASE OF A VARIETY. The recommendation by the Central or a State Variety Release Committee of an improved strain of a crop species for commercial cultivation in the whole or a part of the country.

REPORTER GENE. A gene which produces a product or phenotype which can be easily and unequivocally detected through simple tests.

REPRODUCTION. Production of a new generation of individuals (progeny) by sexual or asexual means.

REPULSION PHASE. Linkage between the dominant allele of one gene with the recessive allele of another gene.

RESISTANCE BREAKING BIOTYPE. The biotype of an insect pest which has developed the ability or adaptation to infest a resistant variety of the host.

RESISTANCE. Ability of a host strain to restrict or even prevent the production of disease symptoms by a pathogen, due to gene for resistance.

RESTORER GENE. A gene usually dominant, that effectively overcomes the effect of male sterile cytoplasm on male fertility, i.e., produces functional male gametes even in the presence of the male sterile cytoplasm.

RESTRICTION ENZYMES. An endonuclease which cuts a DNA molecule within or near a site having a specific nucleotide sequence; used in restriction mapping, RFLP analysis and recombinant DNA technology/genetic engineering.

ROGUE. A variation from the standard type of a variety or strain. Roguing, removal of undesirable individuals to purify the stock.

S_1, S_2, S_3, ... Symbols for designating first, second and third, etc., selfed generations from an ancestral plant (So).

SAMPLE: A finite series of observations taken from a population.

SAMPLING ERROR. Deviation of a sample value from the true value owing to the limited size of sample.

SATELLITE RNA. A small RNA molecule present in many RNA viruses in addition to their normal genome; not necessary for their function; modifies, usually reduces, their virulence; utilized in generating resistance to the viruses in which they occur naturally.

SEED CERTIFICATION. Seed Certification consists of field inspections and seed tests to ensure genetic and physical purity and good germination of seed lots, and issuance of a certificate to that effect.

Seed produced through mass selection (with progeny test) in a pure line variety or clone: source of breeder seed.

SEED TEST. A series of tests on purity, moisture content and germination of a seed sample to determine its quality.

SEED. Part of a plant which appears after the flowers and then can produce a new plant.

SEGREGATING GENERATIONS. The F_2 and the subsequent generations obtained through continued selfing of a hybrid between two or more strains.

SEGREGATION. Separation of paternal from maternal chromosomes at meiosis and consequent separation of genes leading to the possibility of recombination in the offspring.

SELECTION. In genetics, discrimination among individuals in the number of offspring contributed to the next generation. In statistics, discrimination in sampling leading to bias. Opposed to randomness.

SELF FERTILITY. Capability of producing seed upon self-fertilization.

SELF FERTILIZATION. Fusion of male and female gametes from the same individual.

SELF-INCOMPATIBILITY. Genetically controlled physiological hindrance to self fruitfulness.

SHORT-TERM INBRED. A line derived by one or few generations of inbreeding. Such a line is not homozygous or even newly homozygous.

SIB MATING. The mating of sibs.

SIBS. Progeny of the same parents derived from different gametes. Half sibs, progeny with one parent in common.

SIGLE SEED DESCENT METHOD. A method of advancing the generations; one seed from each F_2 plant of a cross and the subsequent generations is bulked to raise the next generation.

SIGNIFICANCE, TEST OF. Statistical test designed to distinguish differences due to sampling error from differences due to discrepancy between observation and hypothesis.

SIMPLE CORRELATION COEFFICEINT. A measure of the relationship between two variables.

SINGLE CROSS. A cross between the two genotypes, usually two inbred lines, in plant breeding.

SNY.1. The population derived by intermating among the Syn.O lines in all possible combinations. Also, by open-pollination among the Syn.O lines (parental lines).

SOMATIC BACKCROSSING. Production of a somatic hybrid cells to proliferate, retain the normal somatic complements of both the species involved in fusion, or to regenerate normal plants.

SOMOTOPLASMIC STERILITY. Collapse of zygotes during embryonic stages due to disturbances in embryo-endosperm relationships.

SPECIES: The unit of taxonomic classification into which genera are subdivided. A group of similar individuals different from other similar arrays of individuals. In sexually reproducing organisms, the maximum interbreeding group isolated from other species by barriers of sterility or reproductive incapacity.

SPECIFIC COMBINING ABILITY. Deviation in performance of a cross combination from that predicted on the basis of the general combining abilities of the parents involved in the cross.

SPOROGENESIS. Production of mega and micro-spores from mega-and micro-spore mother cells, respectively (through meiosis)

STABILITY ANALYSIS. An analysis to estimate the adaptability of a genotype; adaptability is the ability of a genotype to produce a relatively narrow range of phenotypes in the different environments.

STANDARD DEVIAITON. A measure of variability. Mathematically, the distance along the abscissa from the mean to the point of inflection of a normal curve.

STANDARD ERROR. A statistic which is the estimated value of the standard deviation (a parameter).

STATION TRIAL. The trial conducted by a breeder to assess the performance of strains evolved by him as compared to a standard check; usually at one location. (Syn., preliminary yield trail).

STATISTIC. Estimate of a parameter made from a sample. Statistic is to sample what parameter is to population.

STOCK SEED. The seed produced by a breeder of a strain which has been identified but has not yet been released it will be known as breeder seed once the strain is released and notified.

STRAIN BUILDING. Improvement of cross-fertilizing plants by any one of a number of methods of selection.

STRAIN: A group of similar individuals within a variety.

SUBSTANDARD SEED. Seed having the quality parameters, particularly, germination, below the standard prescribed for the certified seed of the concerned species.

SUBSTITUTION LINE. A line is which a pair of chromosomes has been replaced by a pair from another variety of the same species. Used in aneuploid analysis.

SYMMETRICAL HYBRIDS. Somatic hybrid cells/plants containing the complete somatic complements of both the species involved in fusion.

SYN2, SYN3 ETC. Generations derived through random mating from Syn1, Syn2 etc. generations respectively.

SYNAPSIS. Conjugation at pachytene and zygotene of homologous chromosomes.

SYNO. The parental lines of a synthetic variety.

SYNTHETIC VARIETY. In cross-pollinated species a variety obtained by mating in all possible combinations a number of lines that combine well with each other.

SYNTHETIC VARIETY. A variety produced by crossing inter se a number of genotypes selected for good combining ability in all possible hybrid combinations., with subsequent maintenance of the variety by open pollination.

TELOPHASE. The last stage in cell division before the nucleus returns to a resting condition.

TEST CROSS. A cross of a double or multiple heterozygote to the corresponding multiple recessive to test for homozygosity or linkage.

TETRAPLOID. An organism with four basic (x) sets of chromosomes.

TETRASOMIC. An individual having one pair of chromosomes in addition to the normal somatic chromosome complement of the species (2n+2).

THERMOTHERAPY. Exposure of plants/explants to a relatively high temperature (around 40°C) for variable periods with veiw to eliminate/inactivate virus particles present in them without killing these plants. Generally, combined with meristem culture to produce virus-free plants.

THREE-WAY CROSS. A Cross between a single cross and an inbred.

TI PLASMID. The plasmid present in the tumor producing bacterium Agrobacterium tumefaciens; it contains a DNA segment, the T-DNA, which becomes integrated in the host plant genome and contains the tumor producing genes.

TISSUE CULTURE. Cultivation of plant cells and tissues in vitro on artificial media.

TOLERANCE. Ability of a host to avoid of minimize loss in productivity although it has been infected by a pathogen.

TOP CROSS. A cross between a selection, line, clone, etc., and a common pollen parent which may be a variety, inbred line, single cross etc. The common pollen parent is called the top cross or tester parent. In corn, a top cross is commonly an inbred-variety cross.

TRAIT. See, Character.

TRANSGENIC PLANTS. Plants in which a gene (or genes) from another organism has been introduced and become integrated in its genome through genetic engineering.

TRANSGRESSIVE BREEDING. Breeding methods designed to utilize transgressive segregation.

TRANSGRESSIVE SEGREGATION. Appearance in segregating generations of individuals falling outside the parental range in respect, to some character.

TRANSLOCATION. Change in position of a segment of a chromosome to another location in the same or a different chromosome.

TRIALLEL ANALYSIS. A genetic analysis based on all possible three-way crosses among n parents.

TRIPLE TEST CROSS ANALYSIS. A genetic analysis based on the progenies derived by crossing a sample of randomly selected F_2 plants with the parental F_1 and the two homozygous parents of the cross.

TRIPLEX. (See nulliplex)

TRIPLOID. An organism with three basic (x) sets of chromosomes.

TRISOMIC. An organism diploid except for one kind of chromosome which is present in triplicate, hence having 2n +1 chromosomes.

TRIVALENT. Structure formed by pairing among three homologous chromosomes during meiosis.

UNIFORM REGIONAL TRIALS. See, Coordinated Varietal Trials.

UNIVALENT. An unpaired chromosome in meiosis. Bivalents, trivalents, quadrivalents, etc. are associations of 2, 3 4... homologous chromosomes held together by chiasmata.

VARIABILITY. The amount of variation present among the members of a population or species; may have reference to one or more characters, and at genotypic or phenotypic levels.

VARIANCE. Mean squared deviation of a population of variates form their mean. The square of the standard deviation. The corresponding statistic is the mean square.

VARIATE. A single observation or measurement.

VARIATION. The occurrence of differences among individuals due to differences in their genetic composition and /or the environment in which they were raised.

VARIETAL CROSS. See, Population Cross.

VARIETAL MIXTURE. A population of a self-pollinated species constituted by mixing the seeds of two or more varieties in equal or some other proportion.

VARIETY. A subdivision of a species. A group of individuals within a species which are distinct in form or function from other similar arrays of individuals.

VEGETATIVE REPRODUCTION. Production of new individuals from vegetative parts.

VERTICAL RESISTANCE. Resistance usually immunity, towards a virulent race/pathotype; the specific virulent race/pathotype produces susceptible response. Generally, controlled by oligogenes.

VIGOROUS BIOTYPE. A biotype of an insect pest which attacks both resistant and susceptible varieties of the host more successfully than the other biotypes (cf, resistance breaking biotype).

VIRULENCE. Capacity of a pathogen to incite a disease.

VIRUS CROSS-PROTECTION. The phenomenon of reduced susceptibility of a host, plant which was earlier infected by a mild strain of a virus, to a highly virulent strain of the same virus or even of a related virus.

VIRUS INDEXING. Detection of the presence of viruses in plants using various techniques ranging from the development of symptoms in suitably inoculated susceptible genotypes of the host to serological techniques and DNA probes; it is essential while producing virus-free plants through tissue culture or whilc introducing germplasm.

WIDE CROSS. Cross between two species of the same genus or of different genera (Syn, distant cross).

WORLD COLLECTION. See, Gene Bank.

X. See, Basic Number.

X. Basic number of chromosomes in a polyploid series.

X_1, X_2, X_3. Symbols denoting first, second, third.... generations from an irradiated ancestral plant (X_o).

XENIA. Effect of pollen on the embryo and endosperm. (See metaxenia)

ZYGOTE. Cell formed by the union of two gametes and the individuals developing from this cell.

ZYGOTENE. A stage in meiotic prophase when the threadlike chromosomes pair.

Subject Index

A

B

C

D

E

F

G

H

I

L

M

N

O

P

Q

R

S

T

U

V

W

Y

Z

Species Index